T0297818

CAMBRIDGE LIBRARY COLLECTION

Books of enduring scholarly value

Botany and Horticulture

Until the nineteenth century, the investigation of natural phenomena, plants and animals was considered either the preserve of elite scholars or a pastime for the leisured upper classes. As increasing academic rigour and systematisation was brought to the study of 'natural history', its subdisciplines were adopted into university curricula, and learned societies (such as the Royal Horticultural Society, founded in 1804) were established to support research in these areas. A related development was strong enthusiasm for exotic garden plants, which resulted in plant collecting expeditions to every corner of the globe, sometimes with tragic consequences. This series includes accounts of some of those expeditions, detailed reference works on the flora of different regions, and practical advice for amateur and professional gardeners.

The Parks, Promenades and Gardens of Paris

The innovative gardener and writer William Robinson (1838–1935), many of whose other works are reissued in this series, was sent by *The Times* as its horticultural correspondent to the Paris International Exposition of 1867. As a result of his visit, he produced two books, one on gardening trends in France, and this work of 1869 on the parks and gardens of Paris and its environs (including Versailles), and on the fruit and vegetable farming which fed the famous Parisian food markets such as Les Halles. Robinson admired especially the small planted open spaces, squares and courtyards in Paris, which had no equivalent in London, and which he claimed were 'saving [its inhabitants] from pestilential overcrowding, and making their city something besides a place for all to live out of who can afford it'. This highly illustrated work will interest not only historians of horticulture but also lovers of Paris.

Cambridge University Press has long been a pioneer in the reissuing of out-of-print titles from its own backlist, producing digital reprints of books that are still sought after by scholars and students but could not be reprinted economically using traditional technology. The Cambridge Library Collection extends this activity to a wider range of books which are still of importance to researchers and professionals, either for the source material they contain, or as landmarks in the history of their academic discipline.

Drawing from the world-renowned collections in the Cambridge University Library and other partner libraries, and guided by the advice of experts in each subject area, Cambridge University Press is using state-of-the-art scanning machines in its own Printing House to capture the content of each book selected for inclusion. The files are processed to give a consistently clear, crisp image, and the books finished to the high quality standard for which the Press is recognised around the world. The latest print-on-demand technology ensures that the books will remain available indefinitely, and that orders for single or multiple copies can quickly be supplied.

The Cambridge Library Collection brings back to life books of enduring scholarly value (including out-of-copyright works originally issued by other publishers) across a wide range of disciplines in the humanities and social sciences and in science and technology.

The Parks, Promenades and Gardens of Paris

*Described and Considered in Relation
to the Wants of Our Own Cities*

WILLIAM ROBINSON

CAMBRIDGE
UNIVERSITY PRESS

University Printing House, Cambridge, CB2 8BS, United Kingdom

Cambridge University Press is part of the University of Cambridge.
It furthers the University's mission by disseminating knowledge in the pursuit of
education, learning and research at the highest international levels of excellence.

www.cambridge.org
Information on this title: www.cambridge.org/9781108075961

© in this compilation Cambridge University Press 2015

This edition first published 1869
This digitally printed version 2015

ISBN 978-1-108-07596-1 Paperback

The original edition of this book contains a number of oversize plates
which it has not been possible to reproduce to scale in this edition.
They can be found online at www.cambridge.org/9781108075961

Selected books of related interest, also reissued in the
CAMBRIDGE LIBRARY COLLECTION

Amherst, Alicia: *A History of Gardening in England* (1895) [ISBN 9781108062084]

Anonymous: *The Book of Garden Management* (1871) [ISBN 9781108049399]

Blaikie, Thomas: *Diary of a Scotch Gardener at the French Court at the End of the Eighteenth Century* (1931) [ISBN 9781108055611]

Candolle, Alphonse de: *The Origin of Cultivated Plants* (1886) [ISBN 9781108038904]

Drewitt, Frederic Dawtrey: *The Romance of the Apothecaries' Garden at Chelsea* (1928) [ISBN 9781108015875]

Evelyn, John: *Sylva, Or, a Discourse of Forest Trees* (2 vols., fourth edition, 1908) [ISBN 9781108055284]

Farrer, Reginald John: *In a Yorkshire Garden* (1909) [ISBN 9781108037228]

Field, Henry: *Memoirs of the Botanic Garden at Chelsea* (1878) [ISBN 9781108037488]

Forsyth, William: *A Treatise on the Culture and Management of Fruit-Trees* (1802) [ISBN 9781108037471]

Haggard, H. Rider: *A Gardener's Year* (1905) [ISBN 9781108044455]

Hibberd, Shirley: *Rustic Adornments for Homes of Taste* (1856) [ISBN 9781108037174]

Hibberd, Shirley: *The Amateur's Flower Garden* (1871) [ISBN 9781108055345]

Hibberd, Shirley: *The Fern Garden* (1869) [ISBN 9781108037181]

Hibberd, Shirley: *The Rose Book* (1864) [ISBN 9781108045384]

Hogg, Robert: *The British Pomology* (1851) [ISBN 9781108039444]

Hogg, Robert: *The Fruit Manual* (1860) [ISBN 9781108039451]

Hooker, Joseph Dalton: *Kew Gardens* (1858) [ISBN 9781108065450]

Jackson, Benjamin Daydon: *Catalogue of Plants Cultivated in the Garden of John Gerard, in the Years 1596–1599* (1876) [ISBN 9781108037150]

Jekyll, Gertrude: *Home and Garden* (1900) [ISBN 9781108037204]

Jekyll, Gertrude: *Wood and Garden* (1899) [ISBN 9781108037198]

Johnson, George William: *A History of English Gardening, Chronological, Biographical, Literary, and Critical* (1829) [ISBN 9781108037136]

Knight, Thomas Andrew: *A Selection from the Physiological and Horticultural Papers Published in the Transactions of the Royal and Horticultural Societies* (1841) [ISBN 9781108037297]

Lindley, John: *The Theory of Horticulture* (1840) [ISBN 9781108037242]

Loudon, Jane: *Instructions in Gardening for Ladies* (1840) [ISBN 9781108055659]

Mollison, John: *The New Practical Window Gardener* (1877) [ISBN 9781108061704]

Paris, John Ayrton: *A Biographical Sketch of the Late William George Maton M.D.* (1838) [ISBN 9781108038157]

Paxton, Joseph, and Lindley, John: *Paxton's Flower Garden* (3 vols., 1850–3) [ISBN 9781108037280]

Repton, Humphry and Loudon, John Claudius: *The Landscape Gardening and Landscape Architecture of the Late Humphry Repton, Esq.* (1840) [ISBN 9781108066174]

Robinson, William: *The English Flower Garden* (1883) [ISBN 9781108037129]

Robinson, William: *The Subtropical Garden* (1871) [ISBN 9781108037112]

Robinson, William: *The Wild Garden* (1870) [ISBN 9781108037105]

Sedding, John D.: *Garden-Craft Old and New* (1891) [ISBN 9781108037143]

Veitch, James Herbert: *Hortus Veitchii* (1906) [ISBN 9781108037365]

Ward, Nathaniel: *On the Growth of Plants in Closely Glazed Cases* (1842) [ISBN 9781108061131]

For a complete list of titles in the Cambridge Library Collection please visit:
www.cambridge.org/features/CambridgeLibraryCollection/books.htm

FOUNTAIN IN THE GARDEN OF THE LUXEMBOURG.

THE APPLE TRAINED AS A CORDON.

THE

PARKS, PROMENADES & GARDENS

OF

PARIS

DESCRIBED AND CONSIDERED IN RELATION TO THE
WANTS OF OUR OWN CITIES

AND OF

PUBLIC AND PRIVATE GARDENS.

By W. ROBINSON, F.L.S.

CORRESPONDENT OF THE "TIMES" FOR THE HORTICULTURAL DEPARTMENT OF THE GREAT
PARIS EXHIBITION.

MUSHROOM CULTURE IN SUBTERRANEAN QUARRIES.

WITH UPWARDS OF FOUR HUNDRED ILLUSTRATIONS.

LONDON:

JOHN MURRAY, ALBEMARLE STREET.

1869.

CONTENTS.

CHAPTER I.

CHAPTER VIII.

CHAPTER IX.

CHAPTER X.

CHAPTER XI.

CHAPTER XII.

CHAPTER XIII.

CHAPTER XIV.

CHAPTER XV.

CHAPTER XVI.

CHAPTER XVII.

CHAPTER XVIII.

CHAPTER XIX.

CHAPTER XX.

CHAPTER XXI.

CHAPTER XXII.

CHAPTER XXIII.

CHAPTER XXIV.

CHAPTER XXV.

CHAPTER XXVI.

CHAPTER XXVII.

LIST OF ILLUSTRATIONS.

LIST OF PLATES.

XVI LIST OF PLATES.

BY THE SAME AUTHOR.

In Preparation, and will be Published during the present Year,

ALPINE PLANTS.

THIS book will contain an explanation of the principles on which the exquisite flora of Alpine countries may be grown in all parts of the British Islands; illustrations of properly formed rockwork for these plants, with sections showing the wrong as well as the right modes of constructing it; views of the natural homes of the plants, illustrating the Author's tour in the Alps, and a description, devoid of all technicalities, of a choice selection of the more beautiful and interesting kinds, and their successful cultivation.

INTRODUCTION.

"And let it appeare that he doth not change his Country Manners for those of Forraigne Parts: But only prick in some Flowers of that he hath Learned abroad into the Customes of his own country."

BACON'S *Essay on Travel*.

THE success met with by my little book on French horticulture led me to hope that a work describing the progress of our neighbours in city improvements, and giving a detailed account of the production of the more important fruits and vegetables for the Paris market, might prove useful. Hence the present volume. In my "Gleanings from French Gardens," the question of public gardening was scarcely alluded to; in this book nearly one half is devoted to parks, wide tree-planted roads, public gardens, squares, and similar means of rendering great, ugly, gloomy, filthy human hives fitter dwelling-places for vast hosts of men. A belief that London may, without great sacrifice on our part, be made the noblest city in the world—as fair and clean as wide-spreading and wealthy—and the knowledge that the system of public gardening now pursued by us is not the one calculated to lead to this end, have induced me to give the stay-at-home public, and especially that section of it interested in city improvement, an idea of the efforts that are being made in the capital of France to ameliorate the conditions of life.

There is no need to expatiate on the necessity of a thoroughly good system of public gardening in the great cities of a wealthy and civilized race; nor to describe the want of it in our own case—this is painted but too plainly on the faces of thousands in our densely-packed cities, in

c

which the active brain and heart of the country are continually being concentrated. That London is no longer a city, but a nation gathered together in one spot, is a truism : our other great cities are almost keeping pace with it in growth; but in none of them can we see a trace of any attempt to open up their closely peopled quarters in a way that is calculated to produce a really beneficial effect on the lives and health of their workers. Parks we have, it is true ; yet they but partially supply the necessities of large cities. They would serve all our wants if the population breathed only as often as they put on holiday attire or have time to walk, it may be several miles, to a park; but, as we are constituted, room for locomotion, room for the ever-cleansing breeze to search out impurities, room for a few trees to steal away the dark and unlovely aspect of our streets—in a word, room for breathing—is a more pressing necessity than parks. The French have their parks and public gardens, and very extensive and well-managed ones, though, like some of our own, embellished in a wasteful and unnecessary manner with costly and tender plants ; but their noble tree-planted roads, small public squares and *places*, are doing more for them than parks and pelargoniums — saving them from pestilential over-crowding, and making their city something besides a place for all to live out of who can afford it.

A great many of us Britons are apt to connect real city improvement with autocratic government. One has only to speak of our backwardness, when he is instantly reminded that it is all in consequence of not being blessed with a Napoleon, and that there is for us no chance of amelioration except we can secure a ruler who, after purifying and putting our cities into decent nineteenth century order, will good-humouredly take a month's notice to quit. If the logic of such reasoners were at all in proportion to their abundance, we should move onward but little more progressively than the man-like apes. There is no natural human want or wrong that cannot be remedied by human wisdom and energy ; and the most crying evil of this period of change, when the mass of workers are steadily deserting the country for the city, is that our towns are still built upon a plan worthy of the dark ages, and barely justifiable where the breath of the meadow sweeps through the high street. Another notion is that the expense of such improvements must always prevent them from being carried out. "No labour," says Emerson—"no labour, pains, temperance, poverty, nor exercise, that can gain health must be grudged; for sickness is a cannibal which eats up all the life and youth it can lay hold of, and absorbs its own sons and daughters." And shall we spare even less in the attempt to provide for the bodily health and happiness of three millions of men closely packed in a city growing faster than the giant bamboo?

The real want is a want of plan; and that it is to be hoped Parliament will soon give us power to obtain. At present this want is glaringly apparent not only in the central and more crowded parts, but all round London, where admirable preparations may be seen for the formation of a mighty cordon of suburban St. Gileses twenty years hence. Next comes the question of expense, and from that neither autocrats nor parliaments can so readily relieve us. Is it too much to hope that a portion of our vast expenditure for arsenals, armies, fleets, and fortifications may some day be diverted to making such alterations in our cities as will render possible in them the rearing of worthy representatives of the English race? Let us hope not; but supposing that we should never see even the dawn of so desirable an era, and that money should still be profusely spent in every way but that of rendering our cities worthy of our time, our knowledge, our civilization, and our race, there yet remains a course by which we may effect some good without increasing the expenditure we bestow on parks and public gardens generally, and that is by a complete alteration in the direction of the outlay.

Our public gardening differs chiefly from that of Paris and other continental cities by keeping itself away from the very parts where its presence is most wanted. We have parks almost prairie-like in their roominess, yet locomotion is scarcely possible in those parts of the city where the chief commerce of this great empire is carried on, and square miles of densely packed regions are no more benefited by them than if they never existed. I believe that, by the diversion of all needless expenditure from the parks, and by converting this and all the future money that can be spared, to the improvement of the densely crowded parts, we may effect an admirable change for the better. The parks are now managed on a scale which is quite unjustifiable, if we take into consideration the many miserable quarters of London which are utterly neglected. It must be understood, however, that no imputation is here made against their practical management; but the system of richly embellishing them whilst paying no attention to improvements better calculated to humanize our existence in towns, is unwise in every way.

Everybody conversant with the London parks must have noticed the great display of tender flowers and costly gardening which has been presented in them for some years back. This decoration is of such a nature that it has to be renewed every year; and in every case a set of glass-houses, with all their consequent expense for fuel and labour, must be maintained for each park. On this principle a spot of ground not larger than a table may annually cost several pounds for its embellishment. There is nothing about the system more noticeable and objectionable than its growth. Each park is approaching more and more the character of a costly garden, while for the want of a few hardy trees, a patch of green sward, and a spread of gravel to act as a playground for children instead of the gutter, many close districts of London are so foul and cheerless as to be a byeword all over the world. It is perfectly natural that the superintendents of our parks should each wish to make the one under his charge as attractive as the others, from a mere gardening point of view; and it is even more natural that the authorities should accept the opinions of those officials as the most trustworthy on such matters; but it should be the duty of both to consult the public interest above all things, and that interest points to a complete alteration. It is always unpleasant to reduce an establishment, and doubtless it would be hard for the gar-

deners to part with their hundreds of thousands of tender
flowers or to endure a check in their career of converting
our parks into sumptuous gardens; but if they saw that this
reduction of expenditure would lead to a more wholesome
outlay elsewhere, they would willingly help out its adoption.
No objection could be urged against the costly system
alluded to were it not for its expense, which, as anybody
may see, is growing under our eyes every day. It is a
very good and worthy thing to display much of the beauty
of exotic vegetation in our parks and public gardens, provided we can afford it without doing injustice to those who
cannot snatch as much time from toil as suffices for an
airing in the parks. Span a piece of ornamental park water
with a crystal palace, if you will; convert it into a home
for the Great Amazonian Water Lily, and fringe it with
Palms and the richest tropical vegetation; but first be assured
that you are able to afford it, and ask yourself whether the
amount required would not do twenty times the good if
expended in green grass, and trees, and flowers that endure
the open air of Britain. Make, if you will, another ridiculous parterre of stone and water squirts like that at the
head of the Serpentine; but first consider whether it would
not be wiser to establish a little verdure and freshness in some
of the more tumid parts of what Cobbett used to call the
"great wen." The new avenue gardens in the Regent's
Park, with their griffins and artificial stonework, have certainly cost as much as would have created an oasis in some
pestilential part of the East-end. Even the annual expense
of keeping up one of these park gardens is equivalent to
what would suffice to form and plant a little square like
those so freely dotted about Paris during the past dozen
years; while the mere conversion of a strip of breezy park
into an elaborate garden effects no good whatever from a
sanitary point of view.

Let us illustrate the matter in a less general way. Last
year a number of Bay-trees in tubs were placed in Trafalgar
Square; and it need hardly be added that these require frequent attention both in summer and winter—a storehouse
during the latter season—while the wooden tubs in which

they are placed insure by rotting a perpetual, if trifling,
expense. These proved that any kind of tree may be
placed in the streets of London as safely as in any other
city; but they also showed the very short-sighted, disheartening nature of the whole scheme of our public gardening. Not one single thing could these costly green
toys do for our streets or open spaces that could not be
effected infinitely better by hardy trees, requiring no attention after planting; and when one thinks of the vast areas
of this world of London, that are almost impenetrable, miserable is the only term that can be applied to such remedies
as this! It is simply doctoring a wart while a horrid
abscess is sapping away the life of the patient. And ascending from contemptible things of this tree-in-tub sort, the
same reasoning holds good with much of our higher public
gardening.

Who would not forego the trifling gratification of seeing
large portions of our parks so elaborately decorated as to
require almost as much attention as a drawing-room, if
the small sacrifice were accompanied by the knowledge
that tenfold greater good was being carried out where the
want of it was the blackest spot on our social condition?
Are not the materials of nature in our own latitudes
good enough for us? See what is done by a few materials
in her own gardens; reflect what privileges we have in
being able to cull her varied riches from the plains and
mountains all over the temperate and cold and alpine
regions of both hemispheres; and then consider whether
it is wise to spend the public money for glass-houses and the
annual propagation and preservation of multitudes of costly
exotics. A better and a nobler system than that which is
at present the rule in our parks I have endeavoured to point
out at pages 22 to 29.

The purposes to which the greater portion of our future
expenditure in city gardening ought chiefly to be devoted are
the making of wide tree-bordered roads and small simple
squares, open to the public at all reasonable hours. The
squares should not be embellished in a costly way; but if
the persons to whose care their design may be entrusted

could not make them beautiful and grateful to the eye of taste by the use of hardy materials which require no costly annual attention after planting, they should be considered unworthy of their posts. Where space could not be afforded for a little expanse of the ever-welcome turf, even a spot of gravelled earth with trees overhead, and a few seats around, would be a real improvement. The Parisian system of managing squares, described in Chapter VI., is infinitely superior to ours, and must sooner or later be adopted with us. Of course its adoption need not necessarily interfere with the private squares, but it should be tried on a small scale at the earliest opportunity.

In connexion with small squares, we may consider the city graveyards; and nothing can be more ill-considered than the mutilations that have in several cases been considered necessary before making gardens of them. Every churchyard can be embellished, without uprooting bones, removing headstones, or anything of the kind.

In the creation of tree-planted streets in the more crowded parts both of London proper and the suburbs, they should not as a rule be formed on the site of old and much frequented streets, but, so far as possible, pierced between them, leaving the largest and most populous thoroughfares of the present day to become the secondary ones of the future. As is pointed out in the chapter on trees suitable for cities, properly selected kinds grow perfectly well in all parts of London. Indeed I know of no city where I could find finer examples of old trees, chiefly in ancient private gardens and half-hidden squares, where they never received any attention after planting. The excellent system of planting trees on every available spot practised to such a great extent in Paris, should be commenced and carried out as far as possible in our cities. It must be long before we can attempt anything like the magnificent boulevards of our neighbours, but let us insert the thin end of the wedge here and there, and perhaps some day we shall have streets to be proud of. In beginning, it is of the highest importance that we avoid as far as possible the meanness and narrowness characteristic of our style of making street, road, and footway, even in places where want of room is not a drawback. If I am not misinformed, the footway on the land side of the road that is to run alongside the Thames Embankment, near the Houses of Parliament, is to be sixteen feet wide, and probably some of that will be taken up with the proposed line of trees. In this magnificent position, to which any in Paris is insignificant, we are to have a footway that would be considered half a dozen feet too narrow for a second-class boulevard or avenue in Paris !

Whether our general scheme of city gardening be changed or not, we may carry it on with greater economy and much improvement by the adoption of a system resembling that of the public nurseries of Paris—as pointed out in the chapter on these. It is impossible to have greater need for economy than exists in this matter of public gardening; yet the public, in supplying its great London parks, does what hundreds of landed proprietors would be foolish to do, in buying its own evergreens and common nursery stuff ! Our parks are already so vast that the sums required for planting must alone form a heavy item, nearly all of which could be saved by a judicious system of public nurseries. At present, too, there is growing up in each park a nursery of glass, an expensive affair—certain to annually increase in cost if a check be not applied. All this is really unnecessary. With a sensible reduction of our expensive system of bedding out, or even as matters are at present arranged, great saving might be effected by having all the tender plants for the park gardens raised in one establishment. If the true and great principle of variety—the advantages of which as applicable to public gardening are treated of at p. 28—were adopted in earnest, this concentration of the expensive glass-house work would be all the more convenient and advantageous.

Another great improvement might be effected by a rigid exclusion from the plantings of every subject that is not likely to thrive healthfully under the influences of London smut. Many specimens of fine evergreen trees and shrubs have been planted in our parks during the last few years, though the only fate that awaits them therein is a lingering death. When it is stated that each of these costs many times more than would suffice for the purchase of a score of deciduous trees which succeed perfectly in London, the necessity for watchfulness in this respect will be apparent. I am satisfied that by adopting these reforms we could annually save as much as would suffice for the creation of a small suburban park or fresh and charming public square or garden in some overpacked region of London, into which the children could venture without rendering themselves guilty of trespassing, or making a hazardous climb over a sharp-spiked railing, as they frequently do in our amusing if unlovely Leicester Square.

We now come to practical matters relating to fruit culture, market gardening, etc., in Paris and its environs. On these matters there have recently been prolonged discussions, but many readers and disputants have been misled by confounding the comparative state of horticulture in France and England with the real point at issue—*i.e.*, the superiority of the French in certain special and most important branches of garden culture. I have never asserted, as has been assumed, that the French are our superiors in general horticulture, for I know right well that we are as far before them in horticulture, agriculture, and rural affairs generally, as we are in journalistic and magazine literature; but I do assert that in certain points of fruit and vegetable culture they are equally as far in advance of us. I am convinced, too, that more than one of their modes of culture will prove of far greater value to ourselves than ever they have been to the French. To avoid these points, and utter commonplaces about our general superiority, is completely to beg the question. Are we to ignore their good practices because we happen to be more luxurious in our gardening establishments than they are? If I were to find in use in the backwoods of America some handy tool or implement effective in saving human labour, should I be wise in refusing to adopt it because the rude inventor had not attained to the simplest luxuries of existence? If we affirm that the honey of the bee is sweet, the statement that bees are not so beautiful as butterflies is no reply. I do not write to praise the French, but to point out in what way we may learn from them. That they, too, may learn from us will be apparent when I state that intelligent Frenchmen have pointed doubtfully at plants of Rhubarb and Seakale—two of our most excellent vegetable products— and asked if it were true that we eat them in England! The general introduction into France of these two vegetables, with constitutions as vigorous as the most rampant weeds, and never failing to furnish abundant yields, would not merely be a gain to the gardens and markets of a great vegetable and fruit-eating people like the French, but a material addition to the true riches and food supplies of the country.

Of the practices which we may with advantage, and which indeed we must adopt from the French—for the fittest win the day, no matter how long the struggle—those of fruit culture command our first attention, because good fruit culture combines the beautiful and the useful in a very high degree.

There are at least six important ways in which we may highly improve and enrich our fruit gardens and fruit stores.

First, by planting against walls, with a warm southern exposure and a white surface, the very finest kinds of winter Pears—the Pears that keep, the Pears that bring a return, the Pears that cost the consumer a shilling or more each in the London markets after Christmas—the Pears of which the French now send us thousands of pounds worth annually. By doing this we shall in less than ten years have a magnificent stock of these noble fruits all over the country, and be able to export the fruit we now import so largely. Varieties of winter Pears are frequently planted in the open, in all parts of these islands, that an experienced fruit grower in the neighbourhood of Paris or even further south would never plant away from a warm sunny wall, knowing well that it would be wasteful ignorance to do so.

Secondly, by the general adoption of the cordon system of apple growing in gardens. This will enable us to produce a finer class of fruit than that grown in orchards. It may be carried out in spots hitherto useless or unemployed, and will enable us to do away with the ugly Apple trees that now shade and occupy the surface of our gardens. The system will be found the greatest improvement our garden Apple culture has ever witnessed. It should be thoroughly understood, however, that I do not recommend this system for orchard culture, or for the production of the kinds and qualities of fruits that may be gathered profusely from naturally developed standard trees.

Thirdly, by the general introduction of the true French Paradise stock into the gardens of the British Isles. Its merits are that it is dwarfer in growth than any other, and that in wet, cold soils it keeps its roots in a wig-like tuft near the surface—a most valuable quality on many of our cold, heavy soils. When well known it will be found an inestimable boon in every class of garden except those on very dry and poor soils, being wonderfully efficacious in inducing early fertility, and affording a better result without root pruning than either the Crab or English Paradise do with that attention. The knowledge that the Doucin of the French is an admirable stock for all forms of tree between the standard of the orchard and the very dwarf cordon or bush, will also be very useful. The Apple should not be worked on the Crab unless where it is desired to form large standard trees in orchards—by far the best method, if properly carried out, for market and general supplies.

Fourthly, by the practice of the French method of close pruning and training the Peach tree, as described in Chapter XIX. The system adopted in this country is an entirely different one—a loose, irregular style, the shoots not being sufficiently cut back. The Peach tree is quite as amenable to exact training as any other; and when the regular system of the French is understood among us, it will be adopted as the best for wall culture. Preference should also be given to some of the smaller forms of tree adopted by the French, as they will enable us to cover our walls with fruitful handsome trees in a few seasons instead of waiting many years, as has hitherto been the case, and then perhaps never seeing them well covered. These forms are particularly

desirable where the soil is too light and poor for the health
and full development of large wide-spreading trees. In the
last edition of the book of our most popular English teacher of
fruit culture are these words :—" A wall covered with healthy
Peach or Nectarine trees of a good ripe age is rarely to be
seen ; failing crops and blighted trees are the rule, healthy
and fertile trees the exception !" We can alter this by the
adoption of the compact cordon, U or double U forms
figured in this book, by a better system of pruning, and by
thoroughly protecting the trees in spring.

Fifthly, by adopting for every kind of fruit tree grown
against walls a more efficient and simple mode of protection
than we now use. In speaking of fruit culture, nothing is
more common than to hear our climate spoken of as the
cause of all our deficiencies—the fine climate of northern
France being supposed to do everything for the cultivator.
The value of this view of the case is well illustrated by the
fact that all good practical fruit growers about Paris take
care to protect their fruit walls in spring by means of wide
temporary copings. In this country I have never anywhere
seen a really efficient temporary coping, though endless time is
wasted in placing on boughs, nets, &c., none of which are in
the least effective in protecting the trees from the cold
sleety rains, which, if they do not destroy or enfeeble the
fertilizing power of the blossoms, prepare them to become
an easy prey to the frost.

Sixthly, by the acquirement and diffusion among every
class of gardeners and even garden-labourers of a knowledge of budding, grafting, pruning, and training equal
to that now possessed by the French. Many of the illustrations in this book show the mastery they possess over
each detail of training—the branches of every kind of
tree being conducted in any way the trainer may desire,
and with the greatest ease. This knowledge is quite common amongst small amateurs and workmen whose fellows
in this country would not know where to put a knife in
a tree. There are numerous professors who teach it in
France ; it is not taught at all or in the most imperfect
manner in this country, where it is really of far greater

importance. We require walls for our fruit trees more than the French do, and there is no way in which we need improvement more than in the matter of the proper covering and development of wall trees. With standard trees, pruning may be dispensed with to a great extent; but so long as we are obliged to devote walls to the production of our finer fruits, such knowledge as is now possessed by good French fruit growers must prove a great aid. With this knowledge, and the adoption of one of the two economical modes of wall-making described, aided by the general introduction of the mechanical aids to successful garden fruit culture now becoming so general in France, and which I have described and figured at length, we might look forward to a vast improvement in our fruit gardens both as regards their beauty and utility.

In the vegetable department we have also several important things to learn from the French, and not the least among these is the winter and spring culture of Salads—inasmuch as enormous quantities of these are sent from Paris to our markets during the spring months. During the last days of April, 1868, I saw fine specimens of the green Cos Lettuce of the Paris market gardeners selling at a high price in Nottingham, and doubtless it is the same in many of our great cities and towns far removed from London. As I write this (April 19th) the market gardens near London are faintly traced with light green lines of weak young Lettuce plants, that have been for weeks barely existing under the influences of our harsh spring. Around Paris at the same season, in consequence of the adoption of the cloche and a careful system of culture, it is a pleasure to see the size and perfect health of the crops of Lettuces—the difference in culture, and not the imaginary difference in climate, solely producing the result. Some have remarked that we are not a Salad-eating race; but the fact that large quantities of Parisian Lettuces are imported every week and every day for many weeks in spring, proves that we are so in so far as we can afford it. If the restaurants and houses of all classes in Paris had to be supplied from another country, and at about four times the price

they now pay, the Parisians would use even less than
we do.

For many years the London market gardeners, who have
long seen these beautiful Lettuces selling at high prices in the
markets—at as much as 9s. per dozen wholesale—have quietly
concluded that they came from some Eden-like spot in the
south of France, and have apparently never taken the
trouble to see how they are produced. The truth is, that
by the adoption of the French system they may be grown
to fully as great perfection near London and in the home
counties as near Paris. The fact that we have to be supplied by our neighbours with articles that could be so easily
produced in this country is almost ridiculous. It is impossible to exaggerate the importance of this culture for a
nation of gardeners like the British; and if it were the
only hint that we could take from the French cultivators
with advantage, it would be well worth consideration.
"Enormous" was the term which was made use of by a
Paris market gardener in describing to me the quantities
of Lettuces sent from his garden, and the numbers of
the traders who came in search of them. The French system
will have the first difficulty to get over—that of people
becoming used to it, and slightly changing their habits of
culture to accommodate it; but it must ere long be universally adopted with us, and nothing can prevent a great
benefit being reaped from it by the horticulturists of the
United Kingdom.

The French are also far before us in the culture and
appreciation of Asparagus, pursuing a system quite opposed to ours, and growing it so abundantly that for
many weeks in spring it is an article of popular use
with all classes. Some among us affect to ridicule French
Asparagus in consequence of its being blanched nearly to
the top of the shoot; but they forget, or ignore the fact,
that to remove this imperfection, if it be one, the grower
has merely to save himself the trouble of causing it, and that
he may adopt the superior mode of culture and root-
treatment pursued by the French without blanching the
stem if he desires it in a green state. Apart from this, their

experience of French Asparagus is frequently limited to samples that may have been cut in France a fortnight before they reach the table in England, having passed the intermediate time in travelling and losing quality in market or shop.

Having treated of Parisian market gardening generally in a special chapter, little need be said of it here except that the ground is often more than twice as dear as round London; that in consequence of close rotation and deep and rich culture a great deal more is got off the ground in the small market gardens of Paris than is ever the case in our larger ones; and that by reason of the general practice of a thorough system of watering the markets are as well supplied during the hottest summer and autumn as if the climate were a perpetual moist and genial June, whereas when we have an exceptionally warm summer supplies become scarce and dear almost immediately, as was the case during the past year. The whole system of culture of the Paris market gardens is interesting and suggestive in a high degree—especially to a people who take so much pleasure and spend so much money in their gardens as we do. There can be no doubt that the introduction of the same system of very close cropping and good culture would be a great public advantage near all our large cities, where ground is always scarce and dear. It would enable us to get at least double the quantity of vegetables off the same space of ground, and better still, tend to furnish dwellers in cities with something like the proportion of fresh vegetables that is necessary for health. Our working people do not at present use in a sufficient degree any vegetable except the universal Potato. I think I am well within the mark in stating that the poorer classes in Paris use three times as much of fresh vegetable food as the same classes in London. But improvements of our vegetable and fruit markets must precede all amelioration in this direction.

Parisian Mushroom culture is interesting and curious in a degree of which till lately we have had no conception, as will be seen by a perusal of the chapter devoted to it.

The sketches and plan that illustrate it—obtained with some difficulty—are the first that have been published on the subject, so far as I am aware, and will help the reader to obtain a fair idea of places that have been seen by very few English people, and of which most Frenchmen have only a mysterious notion. The perusal of this chapter will doubtless suggest trials of the culture to owners of mines and cavernous burrowings of any kind ; and perhaps in time to come Mushrooms may be a readily obtainable commodity in our markets, even in winter and spring, when they are usually very high priced and dear with us.

In conclusion, I may allude to a subject that is familiar to those of my readers who peruse the horticultural publications of the day—viz., the fierce attacks that have been made upon me for my advocacy of some of the practices herein described. These attacks have chiefly come from certain horticulturists who boast of having traversed France many times during the past thirty years, and who, naturally perhaps, hold that a " tyro," a " young traveller," &c. &c., who first visited France in 1867, cannot possibly have seen anything good or instructive that has escaped their experienced and sagacious eyes. The only reply I shall now or in future make to these gentlemen is in the form of a request to the horticultural public. Test such matters as interest you ; surrender not your judgment either to young or old— to the self-sufficient sage or the presumptuous student—but ascertain for yourselves who is right.

THE

PARKS, PROMENADES,

AND

GARDENS OF PARIS.

———◆———

The city swims in verdure, beautiful
As Venice on the waters, the sea-swan.
What bosky gardens dropped in close-walled courts
Like plums in ladies' laps, who start and laugh!
What miles of streets that run on after trees,
Still carrying all the necessary shops,
Those open caskets with the jewels seen!
And trade is art, and art's philosophy,
In Paris.   AURORA LEIGH.

———◆———

CHAPTER I.

THE CHAMPS ELYSÉES AND THE GARDENS OF THE LOUVRE AND THE TUILERIES.

IF not already the brightest, airiest, and most beautiful of all cities, Paris is in a fair way to become so; and the greatest part of her beauty is due to her gardens and her trees. A city of palaces indeed; but which is the most attractive—the view up that splendid avenue and garden stretching from the heart of the city to the Arc de Triomphe, or that of the finest architectural features of Paris? What would the new boulevards of white stone be without the softening and refreshing aid of those long lines of well-cared-for trees that everywhere rise around the buildings, helping them somewhat as the grass does the buttercups? The makers of new Paris—who deserve the thanks of the inhabitants of all the filthy cities of the world for setting such an example—answer these questions for us by pulling down close and filthy quarters, where the influences of sweet air and green trees were never felt, and the sun could

B

scarcely penetrate, and turning them into gems of bosky
verdure and sweetness; by piercing them with long wide
streets, flanked with lines of green trees; and, in a word, by
relieving in every possible direction man's work in stone
with the changeful and therefore everpleasing beauty of
vegetable life.

In Paris, public gardening assumes an importance which
it does not possess with us; it is not confined to parks in one
end of the town, and absent from the places where it is most
wanted. It follows the street builders with trees, turns the
little squares into gardens unsurpassed for good taste and
beauty, drops down graceful fountains here and there, and
margins them with flowers; it presents to the eye of the
poorest workman every charm of vegetation; it brings him
pure air, and aims directly and effectively at the recreation and benefit of the people. The result is so good, that
it is well worthy our attention. To understand and discuss
it with advantage we cannot do better than commence in
the Place de la Concorde, and afterwards walk up the Avenue
des Champs Elysées, and into the gardens of the Louvre and
the Tuileries—the chain of gardens about here forming a
vast open space in the very heart of Paris.

The Place de la Concorde is not a garden, but a noble
open space, admirable from its breadth and boldness, a
worthy centre to the fine streets and avenue that diverge
from it, embellished by fine fountains and some statues,
and with a terrible history. By looking to the east the
Palace of the Tuileries may be seen through the opening
made in the wood of chestnuts by the central walk, and
to the west is the Avenue des Champs Elysées. If the reader
who has not visited Paris will suppose a wide pleasure
ground flanking the lower part of Regent-street, and
having a grand tree-bordered avenue passing through its
centre straight away to the highest point of the broad walk
in the Regent's Park, and there crested by an immense
triumphal arch—the largest in the world, 161 feet high and
145 wide—he may be able to form some idea of what the
scene is, immediately after passing from the Place de la
Concorde.

PLATE 1.

THE PLACE DE LA CONCORDE.

The Avenue des Champs Elysées leads from it straight
to the Arc; and what it is and how it is laid out we have
next to see. First there is the road, well macadamized,
slightly convex, so level and easy for horses that those of
London could never again find courage to grind down
angular lumps of broken rock if they passed a few weeks
in rolling over it, and nearly 100 feet wide. There is a line
of horse-chestnuts and other trees immediately within the
footway that borders this on each side, and then more
than fifty feet clear—for the greater part a gravelled
walk, but with a well-laid footway of asphalte about
seven feet wide in the centre, which is most agreeable
to walk upon at all times, and particularly in wet
weather. Then come four rows of elm and chestnut
trees, under them about fifty feet more of gravel walks—
the other side of the central avenue being laid out in a
similar manner.

Then commences the garden, which is truly worthy of its
position. Walking up the avenue on the left side we are
in a wide and noble pleasure-ground, of which the farther-
most parts that can be seen are backed by belts of shrubs
and specimen trees. But what are these little structures one
sees quite in front? Well, simply neat little sheds for
gingerbread, cigars, and such commodities. To the British
eye this kind of thing does not seem in what is called
"keeping;" but if people will have their cigars and gingerbread they may as well be sold to them where they are
strolling or playing. Besides, you have in this case got
the gingerbread-keepers under control, and they look as
thoroughly subdued and dutiful as the sergent de ville,
who is a model of gravity and dutifulness. Talk about the
gaiety of the French! Why, you never see one of these
men smile, and yet they look thoroughly French. I once
saw a London policeman, in sheer overflow of spirits, and
probably slightly influenced by beer, throw his hat across
the street after a cat, on a bright moonlight night, and
then laugh at the fun of it; but who ever saw so much
hilarity or want of dignity as that in a Parisian policeman?
They, however, are a thoroughly efficient set of men—

earnest and alert in duty, and apparently with many shades more of self-respect than their London brethren. They keep the strictest order in these public gardens, the whole of which are as open and unprotected by fencing as the beds on the lawn of a country seat. There are no railings higher than six inches; and yet no flowers at Kew or the Crystal Palace are more valuable than these suffice to protect day and night. No doubt this results to some extent from the prompt measures of the grave policemen when occasions for their interference do occur. It is instructive and amusing to reflect that some years ago, when it was first proposed to green the heart of Paris with such beautiful open gardens as this, most wise French people considered it a foolish idea, saying : " Squares, &c., are possible in London, but not in the midst of our rough excitable people !"

Most of the stems of the trees are covered with ivy ; the wide belts of varied shrubs are encircled with the choicest flowers ; the grass, ever-welcomest of carpets, spreads out widely here and there ; great clumps of Rhododendrons and trees shroud buildings, not completely to hide them, but to prevent them from staring forth nakedly in the midst of the quiet sweetness of the garden. These buildings are chiefly for concerts, cafés, &c., and presently we come to a restaurant very agreeably situated. The plan of having restaurants in like places might be extended to London with great advantage—in such places as Kew or any of our great parks or gardens. Some captious individuals may object to such places being turned into tea-gardens ; but tea-gardens must exist somewhere, and why not have them respectably conducted under control, and well arranged to meet the public wants ? By so doing you might prevent the people from resorting to musty, and perhaps not very elevating, eating and drinking-places, and perhaps take from the charms of the lower type of music-hall entertainments now not considered so edifying as popular. On first consideration, the introduction of comfortable restaurants in a place like Kew might seem to interfere with the quietness, which is one of the best features

PLATE II.

THE AVENUE DES CHAMPS ELYSEES FROM THE PLACE DE LA CONCORDE.

of like places; but it need not be so. There is no need for placing them in competition with the glass-houses, or along a main walk, or in any position where they may in the least interfere with the beauty and peace of the scene. They might be placed in isolated yet easily accessible spots, shrouded with trees and shrubs from the garden or park, yet commanding peeps of it here and there; they might have naturally disposed groups of low spreading trees near them, under which people could sit to dine, or take tea in the summer months; they might have open-sided bowers with zinc roofs, the pillars supporting them being draped with Virginian creepers, flowering roses, and the like, and the roofs also densely covered with them. They would have all the attractiveness of open trellis-work creeper-clad bowers, and be at the same time quite impervious to showers.

As we proceed, fountains, weeping willows, and not less beautiful weeping Sophoras are seen, and so many isolated specimens of the noblest trees and plants, such as Wellingtonia, pampas grass, fine-foliaged plants, &c., that we must not mention them all; but arriving at the Palace of Industry, we make a considerable détour to the left to see a garden devoted to music—the Concert of the Champs Elysées, conducted by Musard. I draw attention to this to show that it is possible to introduce amusements into our public gardens without originating anything like the Jardin Bullier or Cremorne. I know of no place more creditably conducted than this, and any of the many English who have spent a summer evening in it will be of the same opinion. It is as quiet and free from objectionable features as a flower-show in the Regent's Park, and very tastefully arranged. In the centre a band-stand, around it a bed of flowers, then about ninety feet of gravel planted with circles of trees. Between each two of the outer line of horsechestnuts there is a lamp-post with seven lights, standing in a mass of flowers. Between this and the enclosing fence there are belts of grass, trees, and of the choicest shrubs; in one part a little lawn with its cedars and maiden-hair trees, bamboos, Irish yews, ivy-clad stems, and flower-beds;

in another spot a noble group of Indian-shot plants, with
bronzy, finely-formed leaves; an equally telling one of the
great edible Caladium springing from among mignonette;
here a pampas grass, there a broad-leaved Acanthus, with a
mass of the handsome Chinese rice-paper plant in the distance.
In its design and management it is as different from the
Cremorne type as could be desired. To compare it with
the places where the stupid and ugly cancan is performed,
and of which there are specimens near at hand, is quite out
of the question. How the young men of France, so ready

FIG. 1.

Evening Concert in the Champs Elysées.

to detect the bête in others, can go night after night to see
this performed, is beyond comprehension. I see no reason
why we should not have places managed as is this evening
concert-garden, even if it were only to counteract the evil
influences of the numerous places which cater simply
for the lowest tastes. In any case this garden will repay
a visit to those who take interest in these matters.

It was only in 1860 that the garden of the Champs
Elysées was laid out, and yet it looks an ancient affair, has
many respectable specimens of conifers, Magnolias, &c.,

PLATE III.

SCENE IN THE CHAMPS ELYSEES.

numerous large and well-made banks and beds of Rhododendrons, Azaleas, hollies, and the best shrubs and trees
generally, with abundant room for planting summer flowers,
chiefly, however, as margins to the clumps of shrubs. The
gardens end at the Rond Point, a circular open space,
in which there are large beds for flowers, fountains,
&c., disfigured, however, by the undulations which some
poor little bits of grass are made to assume. Useless and
unnatural diversification of the ground in some small spaces,
and the lumping together of too many things in one mass,
are the weak points in the gardening of Paris. Above this
Rond Point, a very wide footway of about sixty feet, shaded
by two rows of trees, divides the avenue from the houses
which here approach its sides. Instead of following the
avenue up towards the Arc side we stop at the Rond Point,
glance at the masses of Hibiscus, Caladium, and Papyrus of
the Nile which embellish it, and then descend the garden
by the side of the Rue du Faubourg St. Honoré.

Here we presently meet with a circus, a neat little

FIG. 2.

Circus in the Gardens of the Champs Elysées.

theatre, concert halls, &c., all dropped down in the quietest
way amidst the choicest trees and flowers, and many veri-

table permanently established Punch and Judy shows! I hope this will shock no well regulated mind. They are not like fugitive exhibitions tolerated at the end of obscure streets branching off from the Strand or Oxford-street, but have rights as well-established as those of the Opera. If we consider what a perennial source of amusement this Punch and Judy fun has been for children, perhaps it deserves a place as well as other more fashionable amusements. And then we have revolving circuses, on which the children of the period take their choice to ride on elephant or steed, various kinds of juvenile amusements, cafés, summer music halls, dahlia beds, fountains, Abyssinian musas, and too many similar objects to enumerate. On fine days the wide tree-shaded walks are crowded with pedestrians; all the little games are in full swing, and though it may seem a queer jumble to many, the whole thing is as orderly as could be wished.

At the top of the long avenue, the great arch is surrounded by an immense circular Place, from which straight boulevards and avenues radiate in all directions. The guide-books advise the visitor to Paris to see the lamps lit at night in the Champs Elysées, but if he should want to see the finest effect of that kind, he must go to this arch on a dark night, and standing in the centre look at their effect in the long wide avenues, which fall from where he stands, and afterwards walk around its base to see them better still. The whole scene here is magnificent, and if Paris had nothing worth seeing but what may be seen from hence, it would well repay a visit to all persons interested in the improvements of towns and cities.

Fig. 3.

Avenues and boulevards radiating from the Place de l'Etoile.

The Place de l'Etoile, with its surroundings, is precisely the reverse of our own efforts in like positions—its breadth, dignity, and airiness contrasting strikingly with the narrowness, meanness, and closeness of the

PLATE IV.

THE CHAMPS ELYSEES NEAR THE PALAIS DE L'INDUSTRIE.

best attempts in our so very much larger and busier London.

The Gardens of the Louvre and the Tuileries.

The Place du Carrousel, stretching between the Palaces of the Louvre and the Tuileries, is a large open paved square by no means attractive, but at its eastern end it merges into the narrower Place Napoléon III., to which I wish more particularly to direct attention. The Place is inclosed on three sides by the splendid buildings of the new Louvre, and is embellished with two little gardens surrounded by railings with gilt spears. The Place du Carrousel, surrounded by Palaces, is perfectly bare and without ornament, except the triumphal arch that stands at the main entrance of the court of the Tuileries, but looking towards the Louvre the eye is instantly refreshed by these little gardens, veritable oases in a wilderness of paving stone. I know of no spot more capable of teaching some of the most valuable lessons in city-gardening than this. Viewed externally from their immediate surroundings, or from the more distant Tuileries square, the gardens have a very pretty effect, and show at once the

Fig. 4.

L'Arc de Triomphe du Carrousel.

utility of such, not only for their own sakes, but also as an aid to architecture. On the one hand you have a space as devoid of vegetation as the desert—on the other, by the creation of the simplest types of garden, you relieve the sculptor's work in stone and the changeless lines of the great buildings by the living grace of vegetation, so as to make the scene of the most refreshing kind, and all by merely encroaching a little on the space that would otherwise be monopolized by paving stones. The gardens are very small and most simple in plan, a circle of grass, a walk, and a belt of hardy trees and shrubs around the whole, and an edging of ivy. No gaudy colouring of the ground—no expensive temporary decoration with tender costly flowers, but everything as green and quiet as could be desired. There are four outlets always open, so that visitors can go in and view the little gardens and the rich pavilions rising behind their small but sufficient foregrounds of verdure.

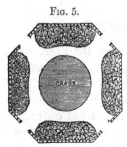

Fig. 5.

Ground plan of the small gardens in the Place Napoléon III.

It is quite common amongst landscape gardeners and others to lay down as a sort of law, that when we make a garden very near to any kind of ornamental building it is above all things necessary to make it " associate" with them —to carry the lines of the building as much as possible into the garden, to make it as angular, and it may be, as brick-dusty as possible, like some recent examples with us; but these gardens prove the fallacy of this reasoning as regards city gardens and open spaces. There are numbers of men professing taste in designing gardens who would never think of putting anything in this position, surrounded as it is, but some miserable prettinesses, expensive gewgaws in the way of trees in tubs, squirting water, vases, coloured broken gravels, &c. &c., things which in their opinion would harmonize with the work of the architect. But from the simplest materials the most satisfactory results may be obtained, as we see here ; and economical reasons also demand simplicity and permanence in all similar attempts. Ten times the amount might be spent on the space occupied, and perhaps with a far less satisfactory result, while there would of course be so much less force to expend on the ventilation and improvement of the many close and sunless quarters that still remain. The small patches of grass in these gardens are like that everywhere in Paris, deep and vividly green, and fresh at all seasons. They usually give it a top dressing of fine and thoroughly decomposed manure in April, but the secret is, dense and repeated waterings at all seasons when the natural rainfall does not serve to keep it as fresh as June leaves.

PLATE V.

THE PLACE DU CARROUSEL AND GARDENS IN THE PLACE NAPOLEON III.

Passing through the great court of the Louvre, and out on the eastern side, we see the garden of the Louvre, which is simply a rail-surrounded space, laid out with the usual very green and well-kept grass, round-headed bushes of lilac, ivy edgings, evergreen shrubs here and there, flowers at all seasons, and the best, cheapest, prettiest, and most lasting edgings in use in any garden, made of cast-iron in imitation of bent sticks. Much of this garden was once covered with old buildings and streets—even the great square just spoken of was once packed with alleys; but the recent improvements of Paris have swept all those things away, and on every side the buildings stand as free as could be desired—unlike our London ones, some of which can hardly be discovered, and which when they have an enclosed space around them, it is merely a receptacle for dead cats, &c. Against the walls of the palace numerous seats are placed, and the gardens, though not large, offer a very agreeable retreat at all seasons; for even during the colder months the old men and invalids improve the shining hours by gathering on the seats close under the great walls when the sun is out.

The main feature of the flower gardening here is a modification of the mixed border system, pretty, and also capable of infinite change. It is a combination of circle, and mixture, and ribbon, quite unpractised with us. Along the middle of the borders we have a line of permanent and rather large-growing things—roses, dahlias, neat bushes of Althæa frutex, and small Persian lilacs. The lilacs might be thought to grow too gross for such a position, but by cutting them in to the heart as soon as they have done flowering the bedding plants start with them on equal terms, and the lilacs do not hurt them by pushing out again, and make neat round heads prepared to bloom well again the following spring. Thus they have along the centre of each border a line of green and pointed subjects, which always save it from over-colouring, and then underneath they lay on the tones as thick as need be. Around each bush or tall plant in these borders are placed rings of bedding plants—Fuchsia, Veronica, Heliotropum, Chrysanthemum

grandiflorum, and fœniculaceum, the outer spaces between the rings being filled with plants of other sorts.
Then follows a straight line of Pelargoniums—scarlet, white,
and rose mixed plant for plant, and forming a very pretty
line. Outside of that a band of Irish ivy, pegged close to
the earth, and pinched two or three times a year ; and
finally, on the walk side, an edging of the rustic irons elsewhere described.

As soon as they get beyond the very primitive idea, that
because one border is of a certain pattern the others ought
to follow it, this will be found a really good plan, and it is
worth attention with us ; by its means we may enjoy great
variety in a border without any of the raggedness of the
old mixed border system. Around most of the rose trees
they place a small ring of gladioli—a good plan where the
plant grows well. Any person with a knowledge of bedding
plants may vary this plan ad infinitum, and produce a most
happy result with it wherever borders have to be dealt with.

Let us next go to the west end of the palaces to see the
gardens of the Tuileries, which stretch from the western
face of that palace to the
Place de la Concorde, bounded
on one side by the Rue de
Rivoli, on the other by the
river. Being nearly in the
centre of Paris these gardens
are as frequented as any. The
garden is very large, and laid
out in the plain geometrical
style by Le Notre, with wide
straight walks, borders round
grass plots dotted with little
lilac bushes, and flowers below
them. About one-fourth of it
near the palace is cut off for the
Emperor's private use, but this
part is merely divided from the
public one by a sunk fence and
low railing, so that the view

FIG. 6.

Statue of Winter in the Tuileries
Gardens.

of the private garden is enjoyed by all. In it they simply plant good evergreens and plenty of deciduous flowering shrubs, while the grass plots are belted by borders, one of which runs right along under the palace windows with the usual round bushes of lilac; but these borders are kept pretty gay all the year round. The private garden of the Emperor is quite open to the public when he is not at the Tuileries. It is well worth visiting should an opportunity occur, if only to see the way the ivy edgings are used. There are no beds, only borders—these touching the gravel walk, and being edged with box. Then on the bright gravel itself, or apparently so, they lay down a beautiful dark green band of ivy, of course allowing in the laying down of the walk for the space thus occupied. The effect of the rich green band adds much to the beauty of the borders. The mode of making them is elsewhere described. The flowers are kept a good deal subdued, and some trouble is taken to develope the shrubs and stronger vegetation distinctly and well. The effect is very good from the windows and the interior. Cannas are used to produce a very charming effect in mixed borders, and altogether this portion is tastefully and inexpensively planted. It is noticeable that hardy shrubs and trees predominate—I believe, by the Emperor's wish—and that, instead of the usual crowding, care is taken to give even the commonest kinds room to grow and become respectable specimens.

A very wide walk crosses the garden just outside the private division; at about its centre are a large basin and fountain, from which another wide walk goes straight towards the Place de la Concorde, and by looking in that direction we see the whole length of the magnificent Avenue des Champs Elysées, terminated on the crest of the hill by the Arc de Triomphe. This walk cuts the garden into two portions chiefly planted with chestnuts and other forest trees, which have not been sufficiently thinned, but are allowed to run up very tall, and thus afford a high arched shade in summer, the ground being gravelled underneath, so that it is comfortable to walk or play upon. There is a slight narrow terrace on both sides, an orangery, the contents of which are placed out in summer, an alley arched over with lime trees by the side of the Rue de Rivoli, and at the western end there are terraces which afford a capital view of the bright and busy scene around and the noble avenue towards the west. There is a great deal of sculpture, both copies of celebrated works and original ones, but as for fresh horticultural interest there is little or none to be seen; and a passing glance is all the visitor need bestow on the public part of the garden of the Tuileries, though it is only fair to add that its general effect is very

FIG. 7.

The Rhone and the Saone, by G. Coustou, in the Tuileries Gardens.

good, and that it in all respects answers its purpose as a play and promenading ground and a " lung " to the city.

A few words must be devoted to those long lines of large orange trees in tubs—they are so very conspicuous that they force themselves upon our attention. There are many ignorant and hopeless ways of spending money in gardens, but few more so than this,—indeed it is one of the most familiar instances of unworthy outlay that is known. Consider for a moment the enormous expense incurred by those lines of finely-grown old orange trees in the gardens of the Tuileries, at Versailles, the Luxembourg, and in other gar-

PLATE VI.

THE GARDENS OF THE TUILERIES.

dens, public and private! Every one of them has cost
more to rear to a condition that is presentable than the
education of a surgeon or barrister, and all in order to produce a deep round tuft of not very healthy green leaves at
the end of a black stem seven feet high or thereabouts.
Costly tubs that rot periodically; costly storing in large
conservatories in winter; costly
carriage from the house to
open garden, and from open garden to house, and all to no good
purpose whatever. The foliage
differs not at all, or in but a
trifling degree, from evergreens
common in our shrubberies;
the clipped head of green is far
inferior to that afforded by the
hardy and elegant spineless Robinia, the flowers are few or
none, the whole thing is a relic
of barbarism, and as such should
be excluded from the tasteful
and well-arranged garden. The
kind of effect they produce is
afforded in a far higher degree
by perfectly hardy subjects.

Fig. 8.

Group in the Tuileries Gardens.

But an orange is an orange : and suppose we wish to
have a little grove of them ? Then make the grove at once,
and, by planting them in an elegant conservatory, grow
them ten times as well and ten times as cheaply as you can
by this absurd process of carrying in and out, and never
withal seeing them in good condition. What a potato is
without tubers, an orange is minus flowers and fruit. By
planting them in a conservatory you may enjoy all the
beauty of leaves, flowers, and fruit—by carrying out of
doors, hoping thereby to embellish what you only disfigure,
you enjoy nothing but imperfectly healthy leaves. The
conservatory must exist to hold them in any case, and one
only big enough to contain, say half those in this garden,
would, if planted with orange trees, afford the Parisians

more gratification by showing them what orange trees really
are, than all they have ever enjoyed through the vast sums
that have been spent upon orange trees for several hundred
years past. They were all very well in an age when exotics,
and above all such attractive exotics as the orange, were
rare, and when good glass-houses were unknown, and bad
ones impossibly dear; but now, when we have thousands of
choice exotics grown in perfection everywhere around us, the
present condition of these fine old trees should not be tolerated. They should be planted out in a conservatory
worthy of the city, or be done away with.

There are, however, some circumstances in which the
culture of plants in tubs for placing in the open air in
summer may not only be tolerable, but desirable. At
Geneva I once saw, opposite a restaurant, the finest specimen
of the fragrant Pittosporum Tobira that I ever met with,
and was informed it had been in a cellar all the winter.
Such as the orange trees are, however, they have admirers,
most of whom believe that they cannot be grown to such perfection by the same method in England. This is not the
case : the method pursued in northern France (which is
described in another chapter) will succeed almost equally
well in the south of England and Ireland.

Let us wait a moment to look at these people feeding the
birds, so much to their own amusement and also that of the
lookers-on. It is a pretty sight, and seems to afford great
pleasure to many people, and doubtless much more to the
successful feeders. It is quite a little scene in the gardens
every day, and on fine days it attracts numbers of people,
though it is an every-day occurrence there. The Jardin des
Tuileries is inhabited by a great number of the common
ringdove, or "quest"—those wild pigeons which in
Britain and elsewhere, when in a wild state, flash away from
man like an arrow from the bow. In these and other
gardens in Paris they seem perfectly at home, and perch at
ease in the trees over the heads of the multitudes of children
who play, and of people who walk on fine days. Their in-
timacy does not extend further, except with their friends
who come to feed them now and then. Here is an instance.

A man, evidently a respectable mechanic, comes to a certain spot, near the private garden of the Emperor. Presently some of the pigeons fly to their friend. He is an old acquaintance, and a bird alighting on his left arm gets a morsel of bread to begin with ; others follow. He has previously put a few crumbs of bread into his mouth, of which the birds are well aware, and, arching their exquisitely graceful necks, they put their bills between his lips and take out a bit turn about. Perhaps one alights on his head, and he may accommodate two or three on his right arm. There are others perched on the railings near at hand, and they come in for their turn by-and-by. A dense ring of people stand a few yards off, looking on, especially if it be a fine day, but they must not frighten the birds, and this persistent feeder looks daggers at a small boy who allows an audible yell of delight to escape. Presently the sparrows gather round the feeder's feet, and pick up any crumbs that may fall while he is transferring the bread from his pocket to his mouth. The sparrows, sagacious creatures, do not as a rule light upon the arm, and never even think of putting their heads in the mouth of the man, but flutter gently so as to poise themselves in one spot about fifteen inches or so from the hand of the feeder. He throws up bits among them, and they invariably catch them with slight deviation from their fluttering position, or at most with a little curl. Sometimes the sparrows pluckily alight on the hand, and root out crumbs held between the finger and thumb, but this only in the case of very old friends.

CHAPTER II.

THE BOIS DE BOULOGNE AND THE BOIS DE VINCENNES.

The Bois de Boulogne.

THIS park illustrates how we improve by friction, so to speak. Till 1852 the Bois was a forest; but Napoleon III., in his admiration for English parks, determined to add their charms to Paris, or rather to improve upon them, and the Bois is one result. In concert with the municipality, the Emperor dug out the lakes, and made the waterfalls. As a combination of wild wood and noble pleasure garden, it is magnificent. The deer are placed in an enclosed space. The Bois is splendid too as regards size—containing more than 2000 acres, of which nearly half is wood, a quarter grass, one-eighth roads, and more than seventy acres water. Though with large expectations in other directions, the reader will hardly be prepared for the statement that the French beat us in parks. When first entered this may not be much liked, the numerous Scotch pines around one part of the water giving it a somewhat barren look, but a few miles' walk through it soon dispels this idea. It has more than the beauty and finish of any London park in some spots, but, on the other hand, vast spreads of it are covered with a thick, small, and somewhat scrub-like wood, in which wild flowers grow abundantly, unlike the prim London parks. There are plenty of wild cowslips dotted over even the best kept parts of it in spring, while the planting on and near the islands is far superior to anything to be witnessed in our own parks. To see what the Bois de Boulogne really is, the visitor should keep to the left when he enters from Passy or the Arc de Triomphe, and go right to the end of the two pieces of ornamental water. Then, standing with his back to the water, he will notice an

PLATE IX.

THE LAKE AND ISLANDS IN THE BOIS DE BOULOGNE.

elevated spot, and by going to that spot he will enjoy one of the finest views he has ever seen in a public park—the water in one direction looking like an interminable inlet, beautifully fringed with green and trees, while in the other several charming views are opened up, showing the hilly suburban country towards Boulogne, St. Cloud, and that neighbourhood.   Then, by turning to the right and returning to Paris by the west side of the water, he will have a pretty

Fig. 9.

One of the small lakes in the Bois de Boulogne.

good idea of what a noble promenade, drive, and garden this is.

It is in all respects worthy of its grand approaches, of the width and boldness of which those who have not seen Paris can have no conception.   There is some bold rock-work attempted and well done about the artificial water; and very creditable pains are taken to make the vegetation along it diversified in character, so that at one place you meet conifers, at another rock shrubs, in another Magnolias, and so on ; without the eternal repetition of common things which one too often sees at home.   At Longchamps, near the racecourse, which attracts half Paris to this part of the wood on fine Sundays, there is a large and ambitious cas-

c 2

cade. Above the spring or shoot of the cascade is an
arch of rustic rocks, over which fall ivy and rock shrubs,
the whole being backed with a healthy rising plantation.
Although made at great expense, this cascade cannot be
pronounced a happy one; to me it is less pleasing than the
less pretentious ones at the head of the large lakes.

The fault of the most frequented part of the Bois de
Boulogne is that the banks which fall to the water are in
some parts a little too suggestive of a railway embankment,
and display but little of that indefiniteness of gradation and

Fig. 10.

Grand cascade in the Bois de Boulogne.

outline which we find in the true examples of the real
"English style" of laying out grounds. But you do not
notice this from the position above described, from whence
indeed the scene is charming. The fault just hinted at is
common to almost every example of this style to be seen
about Paris; and in most of their walks, mounds, and the
turnings of their streams, you can detect a family likeness and
a style of curvature which is certainly never exhibited by
nature, so far as we are acquainted with her in these latitudes.
But it is only justice to say that, taking the park as a whole,
it is far before our London ones in point of design.

Apart from the perfect keeping of the whole, the chief
lesson to be learnt here by the English planter is the
value of paying far greater attention than we at present do
to artistic planting of choice hardy trees and shrubs.   The
islands seen from the margin of the lakes are at all times
beautiful, in consequence of the presence of a varied collection of the finest shrubs and trees tastefully disposed.   They
show at a glance the immense superiority of permanent
embellishment over fleeting annual display.   The planting of
these islands was expensive at first, and required a good

Fig. 11.

Winter scene on the lake in the Bois de Boulogne.

knowledge of trees and shrubs, besides a large amount of
taste in the designer ; but it is so done that were the hand
of man to be removed from them for half a century they
would not suffer in the least.   Nothing could be easier
than to find examples of gardens quite as costly in the first
instance, which, while involving a yearly expenditure, would
be ruined by a year's neglect.   It is summer, and along the
margins of these islands you see the fresh pyramids of
the deciduous cypress starting from graceful surroundings of
hardy bamboos and pampas grass, and far beyond a group
of bright silvery Negundo in the midst of dark-green vegeta-

tion, with scores of tints and types of tree-form around. It is spring, and the whole scene is animated by the cheerful flush of bloom of the many shrubs that burst into blossom with the strengthening sun, and while the oaks are yet leafless the large swollen flower-buds of the splendid deciduous Magnolias may be seen conspicuous at long distances through the other trees. In autumn the variety and richness of the tints of the foliage offer a varied picture from week to week; and in winter the many picturesque and graceful forms of the deciduous trees among the evergreen shrubs and pines offer the observant eye as much interest as at any other season.

Looking deeper than the immediate results, we may see how the adoption of the system of careful permanent planting enables us to secure what I consider the most important point in the whole art of gardening—variety, and that of the noblest kind. Mr. Ruskin tells us that "change or variety is as much a necessity to the human heart in buildings as in books; that there is no merit, though there is some occasional use, in monotony; and that we must no more expect to derive either pleasure or profit from an architecture whose ornaments are of one pattern, and whose pillars are of one proportion, than we should out of a universe in which the clouds were all of one shape and the trees all of one size." These words apply to public gardens with even greater force. In them we need not be tied by the formalism which comfort, convenience, and economy require the architect to bear in mind, no matter how widely he diverges from the commonplace in general design. In garden or in park there is practically no noticeable tie; in buildings there are many. Vegetation varies every day in the year. In buildings more than on any other things unchangeableness is stamped. In the tree and plant world we deal with things by no means remotely allied to ourselves—their lives, from the unfolding bud to the tottering trunk, are as the lives of men. In the building we deal with things much less mutable, which have a beginning and ending like all others, but their changes are much less apparent to our narrow vision. Therefore the opportunity

PLATE VIII.

CASCADES IN THE BOIS DE BOULOGNE.

for variety is beyond comparison greater in public or private gardening than in the building art, or indeed in any other art whatever.

Without the garden, Lord Bacon tells us, " Buildings and pallaces are but grosse handy works : and a man shall ever see that when ages grow to civility and elegancie, men come to build stately sooner than to garden finely : as if gardening were the greater perfection." As yet we are far from perfection as builders, and the garden holds still the relationship to the building art which is described by Bacon. Indeed, it is more backward ; for in a day when building has eloquent champions to put in some such pleas as that quoted, and, moreover, give us practical illustrations of their meaning, we can find no proof that any knowledge of the all-important necessity for variety exists in the minds of those who arrange or manage our gardens, public or private. And yet this unrecognised variety is the life and soul of high gardening. If people generally could see this clearly, it would lead to the greatest improvement our gardening has ever witnessed. Considering the variety of vegetation, soil, climate, and position which we can command, it is impossible to doubt that our power to produce variety is unlimited.

The necessity for it is great. What is the broadly marked bane of the public as well as private gardening of the present day? The want of variety. What is it that causes us to take little more interest in the ordinary display of " bedding out," fostered with so much care, than we do in the bricks that go to make up the face of a house ? Simply the want of that variety of beauty which a walk along a flowery lane or over a wild heath shows us may be afforded by even the indigenous vegetation of one spot in a northern and unfavourable clime. But in our parks we can, if we will, have an endless variety of form, from the fern to the grisly oak and Gothic pine—inexhaustible charms of colour and fragrance, from that of the little Alpine plant near the snows on the great chains of mountains, to the lilies of Japan and Siberia. And yet out of all these riches the fashion for a long time has been to select a few kinds which have the property of producing dense masses of their particular colours on the ground, to the almost entire neglect of the nobler and hardier vegetation. The expense of the present system is great, and must be renewed annually, while the gratification is of the poorest kind. To a person with no perception of the higher charms of vegetation the thing may prove interesting, and to the professional gardener it is often so; but to anybody of taste and intelligence, busy in this world of beauty and interest, the result attained by the above method is almost a blank. There can be little doubt that numbers are, unknown to themselves, deterred from taking any interest in the garden; in fact, it is a blank to them. They in consequence may talk or boast of having a "good display," &c., but the satisfaction from that is very poor indeed, compared with the real enjoyment of a garden.

The one thing we want to do to alter this is to break the chains of monotony with which we are at present bound, and show the world that the "purest of humane pleasures" is for humanity, and not for a class, and a narrow one. Eyes everywhere among us are hungering after novelty and beauty; but in our public gardens they look for it in vain as a rule, for the presence of a few things that they are already as familiar with as with the texture of a gravel walk, must tend to impress them with an opinion that our art is the most inane of all. In books they everywhere find variety, and some interest, if high merit is rare; the same is the case in painting, in sculpture, in music, and indeed in all the arts; but in that which should possess it more than any other, and is more capable of it than any other, there is as a rule none to be found. This is not merely the case with the flower-garden and its adjuncts; it prevails in wood, grove, shrubbery, and in everything connected with the garden. What attempt is made in our parks and pleasure grounds to give an idea of the rich beauty of which our hardy trees are capable, although these places afford the fullest opportunity to do so? How rare it is to see one-tenth of the floral beauty afforded by deciduous shrubs even suggested! Hitherto our gardening has been marked by two schools—

PLATE X.

STREAM AND ROCKS NEAR LONGCHAMPS IN THE BOIS DE BOULOGNE.

one in which a few, or comparatively few, " good things" are grown; the other, the botanic garden school, in which every obtainable thing is grown, be it ugly or handsome. What we want for the ornamental public garden is the mean between these two; we want the variety of the botanic garden without its scientific but very unnatural and ugly arrangement; we want its interest without its weediness and monotony.

There is no way in which the deadening formalism of our gardens may be more effectually destroyed than by the system of naturally grouping hardy plants. It may afford the most pleasing results, and impress on others the amount of variety and loveliness to be obtained from many families now almost unused. To suggest in how many directions we may produce the most satisfactory effects, I have merely to give a few instances. Suppose that in a case where the chief labour and expense now go for an annual display, or what some might call an annual muddle, the system is given up for one in which all the taste and skill and expense go to the making of features that do not perish with the first frosts. Let us begin, then, with a carefully selected collection of trees and shrubs distinguished for their fine foliage— by noble leaf beauty, selecting a quiet glade in which to develope it. I should by no means confine the scene to this type alone, as it would be desirable to show what the leaves were by contrast, and to vary it in other ways—with bright beds of flowers if you like. It would make a feature in itself attractive, and show many that it is not quite necessary to resort to things that require the climate of Rio before you find marked leaf beauty and character. It would teach, too, how valuable such things would prove for the mixed collection. Many kinds of leaf might be therein developed, from the great simple-leaved species of the rhubarb type to the divided ones of Lindley's spiræa, and the taller Ailantus, Kolreuteria, Gymnocladus, &c. The fringes of such a group might well be lit up with beds of lilies, irises, or any showy flowers; or better still, by hardy flowering shrubs. An irregularly but artistically planted group of this kind would prove an everlasting source of interest; it might be improved and added to from time to time, but the original expense would be nearly all.

Pass by this rather sheltered nook, and come to a gentle knoll in an open spot. Here we will make a group from that wonderful rosaceous family which does so much to beautify all northern and temperate climes. And what a glorious bouquet it might be made, with American and European hawthorns, double cherries, plums, almonds, pears, double peaches, &c., need hardly be suggested. You would here have a marked family likeness prevailing in the groups, quite unlike the monotony resulting from planting, say, five or six thousand plants of Rhododendron in one spot, as is the fashion with some; for each tree would differ considerably from its neighbour in flower and fruit. Then, having arranged the groups in a picturesque and natural way, we might finish off with a new feature. It is the custom to margin our shrubberies and ornamental plantings with a rather well-marked line. Strong-growing things come near the edge as a rule, and many of the dwarfest and prettiest spring-flowering shrubs are lost in the shade or crowding of more robust subjects. They are often overshadowed, often deprived of food, often injured by the rough digging which people usually think wholesome for the shrubbery. Now I should take the very best of these, and extend them as neat low groups, or isolated well-grown specimens, not far from, and quite clear of the shade of, the medium-sized or low trees of the central groupings. The result would be that choice dwarf shrubs like Ononis fruticosus, Prunus triloba, the dwarf peach and almond, Spiræa prunifolia fl. pl., the double Chinese plum, and any others of the numerous fine dwarf shrubs that taste might select, would display a perfection to which they are usually strangers. It would be putting them as far in advance of their ordinary appearance, as the stove and greenhouse plants at our great flower shows are to the ordinary stock in a nursery or neglected private garden. It would teach people that there are many unnoticed little hardy plants which merely want growing in some open spot to appear as beautiful as any admired New Holland plant. The system might be varied as much as

PLATE VII.

ISLAND AND RESTAURANT IN THE BOIS DE BOULOGNE.

the plants themselves, while one garden or pleasure ground need no more resemble another than the clouds of to-morrow do those of to-day.

In the rich alluvial soil in level spots, near water or in some open break in a wood, we might have numbers of the fine herbaceous families of Northern Asia, America, and Europe. These, if well selected, would furnish a type of vegetation now very rarely seen in this country, and flourish without the slightest attention after once being planted. In rocky mounds quite free from shade we might well display true Alpine vegetation, selecting dwarf shrubs and the many free-growing, hardy Alpines which flourish everywhere. To turn from the somewhat natural arrangements, as the years rolled on, occasional plantings might be made to show in greatest abundance the subjects of greatest novelty or interest at the time of planting. In one select spot, for example, we might enjoy our plantation of Japanese evergreens, many of them valuable in the ornamental garden; in another the Californian pines; in another a picturesque group of wild roses; and so on without end. Were this the place to do any more than suggest what may be done in this way in the splendid positions offered by our public gardens and parks, I could mention scores of arrangements of equal interest and value to the above. If the principle of annually planting a portion of a great park or garden of this kind were adopted instead of giving all the same routine attention after the first laying out, I am certain it would prove the greatest improvement ever introduced into our system of gardening. The embellishment of the islands in the Bois de Boulogne is very successful, but it is merely one of many fine results that artistic planting would secure. Plantations as full of interest and beauty might be made in other portions, and the fact is the vegetable kingdom is so wide that, although the combination of plant knowledge and taste necessary to success might not often be found in the designer, the materials for any number of varied pictures in vegetation could never fail.

The principle here advocated should not only be applied to the details of one garden, but on a greater scale, and

with even more satisfactory results, to all the gardens of any
great city.

Take a city with half a dozen parks, a score of squares,
and perhaps numerous avenues and open places where
trees or flowers might be grown—take, in fact, the public
gardening of Paris or London at the present day. Now, in
the ordinary course of things, several kinds of trees and
plants, or several dozen kinds, will be found to do best in
all these places, and under the usual management the same
subjects will predominate in each. To the people who
live in the neighbourhood of each the effect will be perhaps
agreeable ; but it must become monotonous. To prevent
people endeavouring to see any life or interest in vegetation, the true way is to make a few things predominate
everywhere. It is also a simple and easy way for the
superintendents ; there is no " bother with it," but there is
also little pleasure, and little of that enthusiastic effort which
is the highest of pleasures, and one only enjoyed by those
who work at things for their own sakes. Innumerable beds
of Cannas and Pelargoniums are better than nothing, no
doubt, but are bad where the opportunity for a higher kind of
embellishment exists. For the credit and encouragement of
our city gardening, it is necessary that we confine ourselves
to the better kind of trees, as many good kinds do not grow
well in streets ; but when it comes to the parks and open
gardens, it is a very different matter. If each park and
square in a city were arranged entirely different from every
other, the enjoyment of those in the immediate neighbourhood of each would be none the less, while the gardening
treasures of the town would be greater in proportion to the
number of parks or squares. A walk in any direction
would reveal new charms to those having the slightest
sympathy with nature, and help to sow the seed of love for
it, were the ground ever so barren. A walk to distant
parks or squares would furnish an object to the many, who
might be expected to take an interest in gardens under
such management ; and objects for walks in towns and
cities cannot be too numerous.

One park might display minute floral interest in all its

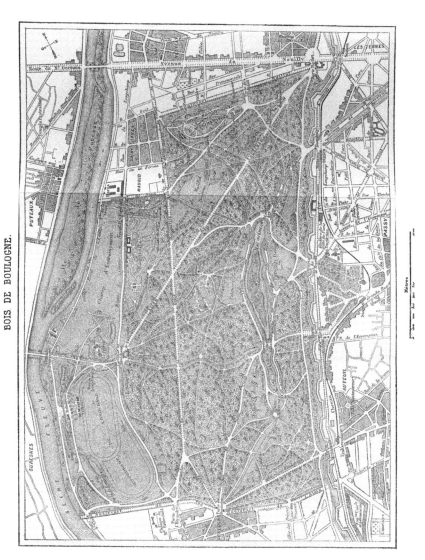

BOIS DE BOULOGNE.

The material originally positioned here is too large for reproduction in this reissue. A PDF can be downloaded from the web address given on page iv of this book, by clicking on 'Resources Available'.

variations, with the larger subjects only used as the necessary setting, shelter, and greenery. Another, with a good soil and favourable exposition, might be made to show the dignity and variety of the forest trees of northern and temperate Europe, Asia, and America. One square might, like Berkeley-square in London, or the little squares in the Place Napoléon III. in Paris, be made very tasteful and effective from simple inexpensive materials—such as green grass, hardy shrubs, and trees. Another might display leaf-beauty so as to remind one of the vegetation of the South Sea Islands; another, chiefly the dwarf prairie and hill flora of cold and temperate countries; and so on—each class of vegetation to be considerately adapted to soil, conditions, and surroundings of the place as regards shelter, liability to foul vapours, position in relation to other gardens and avenues, and so on. In fact, this great principle of variety is capable of doing so much for public gardens, that it should be made compulsory on the heads of these establishments to make each as different from its brother as it possibly could be made. Carried out, then, as I have slightly indicated, both in the private and public place, gardening would be nearer to proving the "greatest refreshment to the spirits of man" than it has ever been in any age.

There is one feature in the Bois de Boulogne which cannot be too strongly condemned—the practice of laying down here and there on some of its freshest sweeps of sloping grass enormous beds containing one kind of flower only. In several instances, near the very creditable plantations on the islands and margins of the lake, may be seen hundreds of one kind of tender plant in a great unmeaning mass, just in the positions where the turf ought to have been left free for a little repose between the very successful permanent plantations. This is done to secure a paltry unnatural and sensational effect, which spoils some of the prettiest spots. Let us hope that some winter's day, when the great beds are empty, they may be neatly covered with green turf.

The Bois being rather level, heavy rains used to lie a

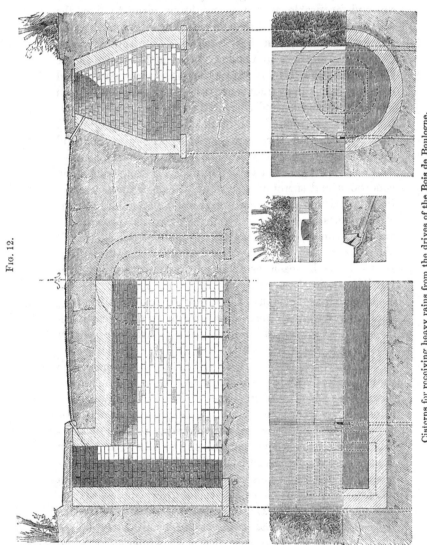

Fig. 12.

Cisterns for receiving heavy rains from the drives of the Bois de Boulogne.

long time on the surface of the roads, &c., before being absorbed; to have remedied this by means of sewers would have cost about 160,000*l.*, so the plan was adopted of constructing a number of tanks at intervals, on an average, of 200 metres, and capable of containing from ten to twenty cubic metres of water each. These tanks are generally circular in form and crowned by a truncated cone—a form which of course requires less mason's work than the rectangular, the latter being adopted only when large trees interfere with the plan. These tanks are shown in fig. 12. The rectangular cisterns measure from four to six metres in length, one to two metres in width, and two to three metres in depth; they are arched at the top, and, like the circular ones, provided with a trapped hole, which serves, first, to withdraw the centrings, and afterwards to clean out the cisterns if they become choked with refuse carried down by the water; the floor is uncovered, and barbicans are left in the footwalls to aid the escape of the water. These cisterns are placed either under the footpaths or in side alleys, so as not to interfere with the grass or the flower beds. The water is conveyed to the cisterns by means of drain pipes 4in. exterior diameter, the first joint being embedded in a mouthpiece of Portland cement, shown in the engraving. These mouthpieces are nearly 20in. in length; they are cast in wooden moulds, and cost 2f. 90c. per metre.

Not far from the lower lake, and at about the centre of the Bois, occurs the Pré Catalan—an enclosed space, occasionally the scene of fêtes, having several refreshment rooms, an open-air theatre, and a peculiar feature in the form of a cow-house, containing about eighty milch cows. The milk is sold to those who frequent the place, especially to horsemen who ride out from Paris for exercise in the early morning, and call here on their way to have a draught of new milk. These features, however, are kept well in the background, and the place generally bears the appearance of an ornamental garden, well worthy of a few minutes' inspection from any horticultural visitor who is traversing the Bois or on the fashionable drive, which is near at hand.

Gardeners may be interested to learn that every year,

on the 30th day of August, the fête of their order is held here, the patron saints being St. Fiacre and St. Rose. Here the gardeners of Paris and their friends assemble to the number of three or four thousand, and amuse themselves with dancing, games, and the usual accompaniments of a Parisian fête, including fireworks, of course. As a garden, the Pré Catalan is distinguished by good specimens of standard Magnolias, both the evergreen grandiflora and the deciduous kinds, and large masses of flowers and fine-leaved plants.

Apart from these, which are well known and extensively employed elsewhere about Paris, I noticed that fine aquatic, Thalia dealbata—usually grown in stoves in England—in robust condition in the midst of a shallow running stream, the canna-like leaves large, handsome, and 22 inches long by 12 broad, and the flower stems 7 and 8 feet high (17th September). It is one of the handsomest and most distinct of all aquatic plants, quite different from the normal type, and should be much used with us. Erianthus Ravennæ, an ornamental grass, was in flower at the same date, and 10 or 11 feet high. Lantana delicatissima was used as margining carpeting to some beds here. Simple and inconspicuous thing as it is, it is multiplied to the extent of from 12,000 to 20,000 every year, which may serve to give another idea of the way in which ornamental gardening is carried on by the municipality of Paris.

FIG. 13.

Ostriches in the Jardin d'Acclimatation.

Garden of Acclimatization in the Bois de Boulogne. — This is a pretty garden and a most interesting place. In it you may study many things, from the culture of the oyster to the numerous breeds of domestic fowls, from ostriches to the different plants used for bee feeding. There is here an interesting hybrid ass—a neat cross between the domestic and wild varieties, which proved useless for the carriage, and kicked it and the harness into " smithereens" when yoked, in consequence of the virus, or what an Irishman would call the " divilment" of the exotic parent predominating. I was not insensible to the claims of a Russian dog, with a coat like a superannuated door-mat; I laughed at a duck which had a velvet-looking head remarkably like a hunting-cap, and nearly as big, but with a body no larger than a debilitated blackbird; and was amazed to see a Chinese dog having no hair except on the top of his head; but we must let all such curiosities pass, and confine ourselves exclusively to vegetable life, now as always of great importance, since Man first regaled himself upon fruits and green-meat.

Doubtless one of the first things that sagacious creature pitched upon was the grape—at least, the best varieties of grapes and the best varieties of men are supposed to have originated in much the same place. To-day the vine is more important than ever, and the garden here has a magnificent collection of 2000 varieties! This collection is the famous one formed in the gardens of the Luxembourg, and fortunately saved from destruction by M. Drouyn de Lhuys, acting upon the urgent request of a friend of horticulture. The vines were actually about to be thrown away when the recent mutilation of the Luxembourg garden took place. So by authority they were ordered to the gardens in the Bois de

Fig. 14.

Streamlet in the Jardin d'Acclimatation.

Boulogne, where, let us hope, they will be well looked after, as it would be a great pity if a collection embracing, as far as could be gathered, nearly all the varieties cultivated in

D

the world, should be lost to horticulture and to science. I
saw a man carrying manure on his back to the vines, and
sat down and contemplated him going through the interesting task ; the basket (panier) was placed on a slightly elevated board supported by three sticks, from which he could
readily hook on to it when it was filled.   I looked at him with
respect and some sympathy, just as we should at a living
specimen of the Dodo or any other animal supposed to be
extinct.   It occurred to me at the time that the acclimatization of a handy useful species of wheelbarrow would not

Fig. 15.

Conservatory in the Jardin d'Acclimatation.

be unworthy of the Society.   However, it is only fair to add
that this kind of basket would prove useful in town gardening, where soil has often to be taken through the house,
also for carrying vegetables, and for conveyance of manure
between close rows of vines, and like uses.

Although the glass-houses in the garden afford but little
interest, rockwork and the planting out of fine foliage plants
tend to make the conservatory very pleasant and refreshing.

PLATE XI.

VIEW IN THE BOIS DE VINCENNES.

The Lycopodium is used with charming effect to form a turf in the conservatory, and nothing can look better than the New Zealand flax, and several palms and tree ferns, planted near the margin of a winding piece of water in that structure. Musa Ensete too looked nobly in the same position. Those who visit it during the winter, cannot fail to be much struck with the effect produced by beds cut in the rich green of Lycopodium denticulatum, and filled with Primulas, Cinerarias, and spring flowers generally. The whole floor of the house, walks excepted, was effectually covered by the Lycopodium.

Fig. 16.

Restaurant in the Bois de Vincennes.

The Bois de Vincennes.

The west end of Paris has its Bois de Boulogne for drives, promenades, quiet walks, fêtes, races, &c.; it has, in fact, its Kensington-gardens, Hyde Park, Green Park, and St. James's Park, and more than all these in one; but the east end is equally well off in having the extensive and noble Bois de Vincennes, which in some respects is quite equal to the Bois de Boulogne, and in one or two even superior.

It contains well designed sheets of water about forty acres in extent; a wide, open plain, about 755 acres, and of which about 284 are devoted to a drill-ground; between 700 and 800 acres of forest; 110 of shrubbery and select plantation; 110 of roads;—in all nearly 1800 acres. The same care in keeping, the fine roads and walks, and the breadth of design, which are seen in the Bois de Boulogne, are also seen here, though this is entirely distinct from that as regards plan. But as there is no feature in it that we have not discussed or shall not discuss with more profit elsewhere, a detailed description of it is not given. Opening up the city by means of airy, open roads, little squares, &c., is of far greater importance than the creation of vast domains outside a city, where people may enjoy a little fresh air once a week or so.

It has quite a novel feature, in the fruit-garden of the city of Paris, recently formed. This is described at length in another chapter.

The lake nearest to the fruit-garden and the Avenue Daumesnil entrance is beautifully disposed, and its margins and islands are well planted. To walk completely round it, starting from the neighbourhood of the fruit-garden and returning to the same position, will well repay the visitor; few public parks offering anything so refreshing and agreeable of a warm summer evening.

A restaurant near one of the lakes illustrates admirably how like conveniences may be introduced into public parks without in the least rendering them objectionable. It commands excellent views of the park and water from the groups of trees by which it is hidden, and which perfectly prevent it from obtruding upon the quietness of the park. It would be well if like care were always taken to veil such structures. The restaurant figured on the preceding page is not quite so happily placed, but nevertheless forms a not objectionable feature in the park.

The Bois de Vincennes contains also the city nursery for herbaceous plants, &c., alluded to in the chapter on the Public Nurseries.

Cercis australis and Planera acuminata have been tried

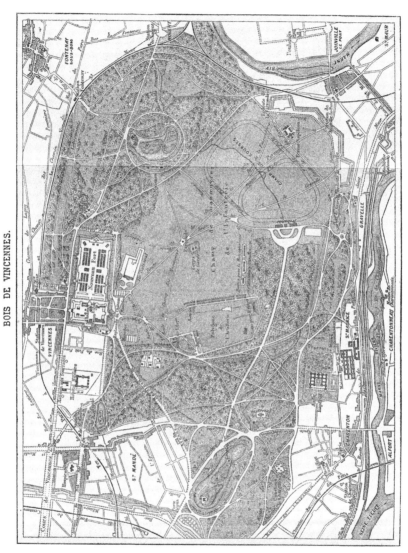

BOIS DE VINCENNES.

The material originally positioned here is too large for reproduction in this reissue. A PDF can be downloaded from the web address given on page iv of this book, by clicking on 'Resources Available'.

as boulevard trees in and near the Bois de Vincennes, and promise well. The Planera, it is hoped, will replace the elm in places where that is destroyed by the Scolytus; and the Cercis looked very fresh and well about the middle of September, and at the end of the very trying season of 1868. A plantation of about seven acres of Wellingtonias was made here about three years ago, and the plants are strong and good. Were it not for the ver blanc this would even now be a fine feature; but unfortunately very few specimens remain uninjured by this most terrible of pests. Some of the trees had formed good specimens, and showed what a noble wood of Wellingtonias would have been seen here were it not for this grub. Hares are rather plentiful here, and may be seen scampering over the open parts—quite an uncommon occurrence in a public park.

To connect the Bois with the promenades in the neighbourhood, the plains of Bercy and St. Mandé, lying between the old boundaries of the wood and the walls of the fortifications of Paris, were bought up, so that the new promenade, like the Bois de Boulogne, now begins at the very gates of the city. The pieces of water in the Bois de Vincennes, as well as the pipes by which the gardens are watered, are supplied from the river Marne. Here, as in other parks and gardens, the hottest and most arid weather merely makes the grass and plants greener and healthier, in consequence of the admirable arrangements for watering both turf, trees, and flowers.

Watering the Parks.

The climate of Paris being dryer than that of London, and the soil less conducive to the growth of grasses, the verdure maintained in the more ornamental parts of the Paris parks is naturally a source of some surprise to visitors. It is difficult to give the reader, who has not seen it himself, an idea of how perfectly the watering is done. The contrast between the parks and gardens of London and Paris is in this way by no means flattering to our way of managing them. It will be better to quote one of our journals to represent our own side of the question. " We have re-

peatedly called the attention of the authorities during the
summer to the melancholy state into which the parks were
falling. The mischief we desired to guard against is now
done. The grass is of the colour of hay, and the little of
it that remains is being so rapidly trodden down that in
many parts what used to be greensward is now nothing
better than hard road." So wrote the *Pall Mall Gazette*, one
day last summer; and really, about the end of July and
the beginning of August, nothing could look more unat-
tractive than the London parks. These parks are supported
at heavy public cost; and it is a great mistake to let
them be rendered as brown and uninviting as the desert
by an exceptional drought, which of course will happen
at the very season when the grounds ought to be in perfect beauty and attractiveness. The French system of
watering gardens, &c., is excellent, or at least the generally
adopted system; for at the Jardin des Plantes there are
yet watering-pots made of thick copper, which are worthy
of the days of Tubal Cain, but a disgrace to any more
recent manufacturer, and a curse to the poor men who
have to water with them. Generally Parisian lawns and
gardens are watered every evening with the hose, and most
effectively. It is so perfectly and thoroughly done, that
they move trees in the middle of summer with impunity;
keep the grass in the driest and dustiest parts of Paris as
green as an emerald, the softest and thirstiest of bedding
plants in the healthiest state; and as for the roads, the
way they are watered cannot be surpassed. They are kept
agreeably moist without being muddy, while firm and crisp
as could be desired. Of course all this is effected in the
first instance by having abundance of water laid on; but
that is not all. With us, even where we have the water
laid on, we too often spend an immense amount of labour
in distributing it. In Paris generally it is applied with various modifications of the hose, which pours a vigorous stream,
divided and made coarse or fine either by turning a cock,
by the finger, or even by the force of the water.

This is the way they apply it to roads, the smaller bits of
grass about the Louvre, and other places; but when water-

ing large spreads of grass in the parks the system is different. One day in passing by the racecourse at Longchamps I saw it carried out in perfection. The space had become very much cut up by reviews and races; but in any case it is watered to keep it as green as possible in summer. At first sight it would appear a difficult thing to water a racecourse, but two men were employed in doing it effectually. Right across the whole open space from east to west stretched an enormous hose of metal, but in joints of say about six feet each. The whole was rendered flexible by these portions being joined to each other by short strong bits of leathern hose, each metal joint or pipe being supported upon two pairs of little wheels. Fig. 17 shows a section of the apparatus at work. By means of these the whole may be readily moved about without the slightest injury to the hose in any part. At about a yard or so

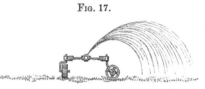

FIG. 17.

Section of perforated self-acting hose on wheels.

apart along this pipe jets of water came forth all in one direction, and at an angle of about 45 deg., and spread out so as to fully sprinkle the ground on one side; and thus four feet or so of the breadth of the whole plain of Longchamps was being watered from one hose. There were two of these hoses at work, one man attending to each of them; the only attention required being to pass from one end of the line to the other, and push forward the hose as each portion became sufficiently watered. The

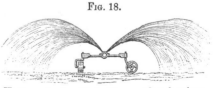

FIG. 18.

Hose on wheels with double row of perforations.

simplest thing of all is the way they make the perforations for the jets along the pipe. They are simply little longitudinal holes driven in the pipe with a bit of steel. They must be made across the pipe, or the water will not spread in the desired direction. The wind causes the water to fall in the most divided form possible. With an apparatus thirty metres long a man can easily water 1500 square metres per hour, moving the hose three times. Of course the quantity of water depends on the force in the conduits and the length of the tubes. With a pressure of 22 metres and hose 320 metres long the quantity of water per metre and per minute is nearly two litres. The hydrants in the grass are placed about fifty metres apart, and the wheels of the trucks are of wood, in order not to cut the grass. There are many modes of spreading water in use about Paris, but none of them half as good as this simple method. More than a mile of this kind of hose may be seen at work at one time and with hundreds of jets playing.

The hose for watering the roads is arranged on wheels also, but, as it must be at all times under command when carriages pass by, it has only one rose or jet, which is directed by a man who moves about among the carriages with the greatest ease, and keeps his portion of the road in capital condition. Of course it is a much cheaper way than carrying the water about as we do, as then we must have horse and cart, wear and tear, and man also ; whereas, by having the water laid on, all the men have to do in watering is to attach the hose and commence immediately. In the same way as much work can be done in a garden in a day as with ten men by the ordinary mode ; so that in the end it is much cheaper to have the water laid on. There can be no doubt that to the efficient watering much of the success of the fine foliaged plants in Paris gardens is to be attributed.

As a good system of watering is of the highest importance to cities and towns in every region of the earth a more detailed and technical account of the watering of Paris gardens may prove useful to some. The article first appeared in the *Engineer,* and refers chiefly to the arrangements for the Bois de Boulogne, but the system is the same for all other places.

The watering is performed chiefly by means of long hose with a copper branch, the latter being provided with a stopcock, so that the delivery of the water may be arrested instantly, without having to turn off at the plug. The hose

PLATE XII.

MODE OF WATERING THE GRASS IN THE PARKS, WITH PERFORATED HOSE ON WHEELS.

is generally twelve metres long and 2in. in diameter; it is constructed either of leather, vulcanized india-rubber or canvas; the first and second costing from 6s. to 6s. 8d. per yard, and the last only 10d. or 11d. The screw connecting pieces, which are made of gun metal, cost about 6s. The leather hose, losing the oily matters from its pores, through the pressure of the water, soon becomes brittle, but it lasts on an average two years; the rubber is light and has no other fault but that of wearing out in twelve months, while the canvas hose soon cuts to pieces on the gravel. A system of mounting such tubes on small trucks so as to keep them from trailing on the ground, and consequently making them lighter to handle and more durable, was tried for a long time, but this has been superseded by a very simple and inexpensive invention, that of tubes made of sheet iron, lined with lead and bitumen, and connected together by means of leather joint pieces, the whole being mounted on small wooden trucks. The cost of this apparatus complete, with the single exception of the branch, is only 70f., or 5f. 20c. per metre, and it will last on the average four years, while the old hose on trucks costs 127f., or nearly double.

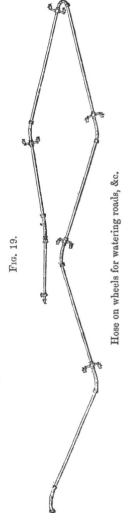

Fig. 19.

Hose on wheels for watering roads, &c.

The cost of that now in use is made up as follows :—Eleven metres of iron tubes, 19f. 25c.; leather junction pieces, 25f. 60c.; ten trucks 20f.; ligatures, 5f. 15c.; total, 70f. The apparatus in use at the present moment in Paris consists of five tubes, each about 6ft. long, and a shorter one to which the

branch is attached, so that only five trucks are required;
the trucks also in practice consist of a piece of plain wood,
a little more than a foot in length, the tube being bolted
on to the upper
side and the run-
ners fixed to the
lower. As regards
the connexion of
the joints, this is
made sometimes
with brass flanges,
but a joint which
answers equally
well, and is much
cheaper and lighter,
is that made with
copper wire; for
the branch joint,
however, brass
flanges are always
used, as the branch
itself is removed
and carried away
when not in use,
while the tubes are
simply folded toge-
ther, fastened with
a piece of cord, and
left in any conve-
nient corner.

It is found in
practice that a man
cannot manage an
apparatus of this
kind, which is more
than about 40ft.

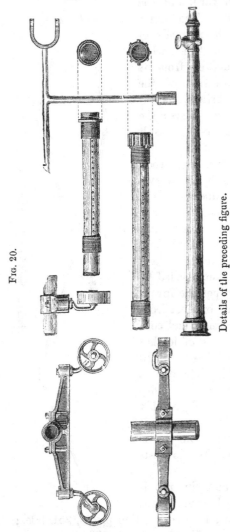

Fig. 20.

Details of the preceding figure.

long; but for watering grass, in which case the hose is left
stationary in one place for some time and then moved to
another, several apparatus are, if necessary, screwed on to

each other. The effects of these tubes or hose have been
carefully studied. The following is a table of results with
a twelve metre apparatus, the inner diameter of the nozzle
of the branch being 0·012 metres, or rather less than half
an inch, and the branch itself being held at an angle of
45 deg.:—

Pressure at the surface.	Quantity of water given per second.	Extent of the jet.	Quantity of water given when the branch is not on.
Metres.	Litres.	Metres.	Litres.
8	0·90	10	1·80
12	1·25	12	2·40
15	1·40	14	2·75
20	1·60	15	3·10
25	1·80	15	3·40
30	1·90	15	3·60
35	2·00	16	3·80
46	2·10	16	4·00

These results, it is stated, are averages, for some apparatus give superior or different results, although all the
conditions appear the same. Experience shows that with
the same amount of pressure in the pipes the extent of the
jet is enormously reduced by the lengthening of the hose.
Of course the diameter of the nozzle of the branch depends
on the pressure within the tubes, but it was thought necessary to have a uniform model, and 0·012 metres was adopted
as distributing the water most advantageously with a pressure of eight to fifteen metres. An apparatus twelve metres
long, with a branch one metre in length, and giving an
average jet of twelve metres, is effective over a radius of
twenty-five metres. The plugs or hydrants are placed at
intervals of thirty metres on roads twenty metres wide, and
forty metres apart in narrower roads, when they are all on
one side of the road.

Formerly all the roads in and about Paris were watered by
means of carts which held one ton of water. It required
twenty-four tons to water the Avenue de l'Impératrice properly,
the road round the lakes, and some few others. The whole of
the roads in the Bois de Boulogne, as they now stand, would
require ninety tons of water, which would cost, men, horses,
and carts included, 13f. per ton, or 200,000f. (8000l.) for the

six summer months. The new system of watering by hose
costs for the whole of the Bois but 55,000f., or little more
than a quarter of the expense under the old system. In
this estimate, however, no account is taken either of the cost
of the water itself or of the capital expended for its conveyance. Finally, it is remarked, as regards the Bois de
Boulogne, that the cost is, in fact, little more than that of
the maintenance of the apparatus in repair, or about 250*l.*
a year, the work being done by the body of men called
cantonniers, who have little else to do during the summer
months.

A water cart drawn by one horse, in cases where the
hydrants are 400 metres apart, will water 1300 metres an
hour over a width of four and a half metres—that is to say,
a cart will water about 6000 square metres, using in the
operation three tons of water. But in the parks it was
found that the cart should pass over every spot once in the
hour, and this gives, with an average of seven hours' effective work, an expenditure of three and a half litres, or more
than seven pints per day per square metre. The cost of
labour, cart, and horse is given at about 10f. per day, so that
the actual expense per ton and per square metre stands thus,
$\frac{1}{6000} = 0\cdot00165$f. In calculating the cost of watering by
means of hose and branch, the hydrants or plugs must
necessarily be much more numerous, the intervals between
them being in the case of watering by cart 400 metres,
while in the case of the hose the intervals are on an
average only thirty-five metres. The total length of the
roads to be watered in the Bois de Boulogne is 53,000
metres, and the number of hydrants 1500, whereas under
the old system 132 would have sufficed, a difference of
1380 hydrants, costing 4*l.* each, or 4s. a year for interest,
and, in addition, 4s. for repairs, &c. The latter is contracted for at the following rate—namely, eight centimes
per metre, or about three farthings a yard run of conduit,
and 4s. per hydrant.

A hundred and twenty men are required for watering
the 540,000 square metres of road in the Bois; in five
hours a man waters 4500 metres of road three times over,

PLATE XIII.

MODE OF WATERING ROADS, DRIVES, FOOTWAYS, AND THEIR MARGINS.

besides watering the side paths once, which the carts of course did not touch. The cost is given as follows :—

	Francs.
Interest and Maintenance of hydrants . . .	13,800
Cost and repair of hose, &c.	6,200
Wages of 120 men at half a day for six months . .	35,000
Total	55,000

The surface watered being, in round numbers, 600,000 square metres, and the average number of days 180, the cost per square metre and per day is

$$\frac{550,000}{180 \times 600,000} = 0 \cdot 00051,$$

showing a great economy as compared with the expense of watering by cart. The hose and branch dispense (making allowance for interruptions caused by traffic and by moving the apparatus) a litre of water per second, or 18,000 litres in five hours; the quantity is therefore about the same as that dispensed by cart, only it is effected in five instead of seven hours. Previous to the general adoption of the hose and branch, experiments were tried with small handcarts containing a quarter of a ton, and drawn by two men, but these were found to cost more than the old carts.

Another method of keeping roads and pathways in order, namely, by the application of deliquescent salts, is interesting from its novelty. The salts used are chloride of magnesium or of calcium. The former salt does not exist in commerce, but large quantities have been obtained from the residue of the manufacture of carbonate of soda, at a cost of 15f. the 100 kilogrammes; it may, however, be produced for less than a third of that rate. The salt is well calcined (in order to make it lose as much of its water as possible), and then coarsely pulverized; it is sprinkled over the road by hand. The effects of this deliquescent salt, as compared with those of water, are not uniform; in the case of roads with much traffic the salt is twice as dear as water, because of the necessity of constant renewal, but in side paths and roads with little traffic the salt was found far more economical. The use of deliquescent salts has this great advantage, namely, that it does not interfere in any way with the circulation, and maintains the pathways clear of dust or mud, while of course in places where there is no grass to be watered the whole of the cost of water-pipes and hydrants would be saved.

The surface of grass which has to be watered with Seine water in the Bois de Boulogne is about 250 acres, and the quantity of water required to keep it in good condition averages ten litres, or more than two gallons, per square metre, every third day. To water this surface in the same manner as the roads would require more than a hundred

F ig. 21.

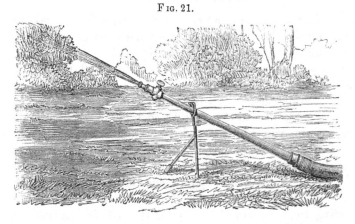

Hose allowed to play on the grass and shifted from time to time.

hose working ten hours a day, and this would entail a very heavy cost. But as the grass does not require to be treated with the same regularity as the roads one system adopted is to place a branch on a stand at an angle of 45 deg., and allow it to play over the grass for a certain time, when it is removed to another spot: in this way one man can manage ten apparatus.

The total amount of water taken from the Seine for the purposes of the Bois never exceeds 240 litres, or about fifty-four gallons, per second. The natural meadows by the side of the Seine form about 400 acres, but the soil here is alluvial, and therefore irrigation is only necessary in very hot weather, whereas the soil upon which the artificial grass is planted is nearly all sand, and the greatest care is required to keep the turf in order. The total cost of the arrangements of conduits and pipes for the supply of water to the Bois and the avenues leading to it is given at 1,520,000f., or 60,800*l*.; the number of stop-cocks is 385, and of hydrants 1600; and the length of the conduit is 66,200 metres. It results from these figures that the cost of the whole has amounted to 22f. 97c., or about 18s. 5d. per metre.

CHAPTER III.

THE PARC MONCEAU.

THIS is on the whole the most beautiful garden in Paris, and well shows the characteristics of the system of horticultural decoration so energetically adopted in that city. It is not large, but exceedingly well stored, and usually displays a vast wealth of handsome exotic plants in summer. In spring it is radiant with the sweet bloom of early-flowering shrubs and trees, every bed and bank being covered with pansies, Alyssum, Aubrietia, and all the best known of the spring flowers, while thrushes and blackbirds are whistling in the adjacent bushes, as if they were miles in the country, instead of only a few minutes' walk from the Rue du Faubourg St. Honoré. This park was laid out so long ago as 1778 for Philip Egalité as an "English garden," and passed through various changes, till it at last fell into the hands of the Municipality of Paris, a very astute corporation, who have converted it into a charming garden, and are not likely to part with it in a hurry.

The system of planting adopted here as well as in the other gardens of the city is often striking, often beautiful, and not unfrequently bad. It is striking when you see a number of that fine showy tree, Acer Negundo variegata, arranged in one great oval mass, silvery and bright; it is beautiful when you see some spots with single specimens and tasteful beds, every one differing from its neighbour; and bad when you meet with about a thousand plants of one variety stretched around a collection of shrubs, or flopped down in one large mass, or when a number of plants too tender for the climate are put out for the summer months amidst those that grow with the greatest luxuriance. "The subtropical system will never do for England!" say some practical men. The truth is, that it requires to be done very carefully in Paris, and there is a great mistake made by putting out a host of tender plants merely because they are exotics, unless indeed you wish to contrast healthy beauty with ragged ugliness. In the Parc Monceau there is usually a group of Musa Ensete worth making a journey to see, and masses of Wigandia, Canna, and such Solanums as Warcewiczii, that are worthy of association with it; but I have also seen there beds of Begonias without a good leaf or a particle of beauty—scraggy stove plants, with long crooked legs, and a few tattered leaves at the top, and poor standard plants of the sweet-verbena at the same time. If it were an experimental ground, one would not mind, of course; but this, in a garden where its omission would leave almost nothing to be desired, is too bad. In some respects this park is really unequalled, and therefore one regrets the more to see these blemishes, which let us hope will not be repeated.

What first excites the admiration of the visitor used to the monotonous and highly-toned type of garden now seen so much with us is the variety, beauty of form, and refreshing verdure which characterize this garden—good qualities that are so often absent in too many of our own. The true garden is a scene which should be so delightfully varied in all its parts—so bright, so green, so freely adorned with the majesty of the tree, the beauty of the shrub, the noble lines of the fine-leaved plant, the minute beauty of the dwarfer plants of this world; so perpetually interesting, with vegetation that changes with the days and seasons, rather than puts the stamp of monotony on the scene for months; and so stored with new or rare, neglected or forgotten, curious or interesting plants—that the simplest observer may feel that indefinable joy which lovers of nature derive from her charms amidst such scenes, but which few, except those of a high degree of sensitiveness and power of expression, like Shelley, can give utterance to. It would be teaching him to use the words of Goethe—

"To recognise and love
His brothers in still grove,
Or air or stream."

E

If any good at all is to be done by means of flowers and gardens, you must give men a living interest, a lasting curiosity in them, and some other objects than those which can be taken in by the eye in a moment.   Numbers are occupied and delighted with gardening as it stands at present, but it can hardly be doubted that a system with something like an aim at true art would be sure to attract many more; and it is patent that there are numbers even among the educated classes who take no interest whatever in the garden, simply because they can in few places find any real beauty or interest in it.   To confine ourselves to a single phase of the subject, it is certain that if all interested in flower gardening had an opportunity of seeing the charming effects produced by judiciously intermingling fine-leaved plants with brilliant flowers, and of which there are such handsome examples in this park, there would be an immediate revolution in our flower-gardening, and verdant grace and beauty of form would be introduced, and all the brilliancy of colour that could be desired might be seen at the same time.   The beauty and finish of many of the finer beds here, are of the highest order, in consequence of the adoption of the principle of variety.   Here is a bed of Erythrinas not yet in flower : but what affords that brilliant and singular mass of colour beneath them, a display which makes the visitor pause when he comes near the bed ?   Simply a mixture of the lighter varieties of Lobelia speciosa with variously coloured and brilliant Portulaccas.   The beautiful surfacings that may thus be made with annual, biennial, or ordinary bedding plants, from mignonette to Alternanthera, are infinite.   At the risk of driving off the general reader we must now begin to use hard names, and go deeper into purely technical and horticultural matters, for we shall not elsewhere meet an opportunity of doing so with so much advantage.   It is only fair to warn the reader that this is a purely horticultural chapter.

The following are a few examples of these graceful mixtures seen in this garden during the past year :—A bed of Arundo Donax versicolor, springing from Lobelia speciosa; a bed of Ficus elastica, the ground beneath perfectly hidden by

PLATE XIV.

THE PARC MONCEAU: VIEW FROM THE DRIVE.

luxuriant mignonette; Wigandia, springing from the little silvery sea produced by the mixture of the blue and white varieties of Brachycome iberidifolia; Caladium esculentum, from a rich surface of flowering Petunias; glowing Hibiscus, from Gnaphalium; graceful dwarf Dracænas, from very dwarf Alternantheras; Aralias, from Cuphea; taller Dracænas, from a deep and richly-toned mass of Coleus Verschaffeltii; Erythrina, from a sweet low carpet of soft purple Lantana; tall Solanums, on mats of that most finished little plant Nierembergia; sea-green Bocconias, from the dwarf dark-toned Oxalis corniculata var., and so on. Reflect for a moment how consistent is all this with the best gardening, and the purest taste. Your bare earth is covered quickly with these free-growing dwarfs; there is an immediate and a charming contrast between the dwarf-flowering and the fine-foliaged plants; and should the last at any time put their heads too high for the more valuable things above, they can be cut in for a second bloom, as was the case with some Petunias here which had got a little too high for their slow-growing superiors. In the case of using foliage plants that are eventually to cover the bed completely, annual plants may be sown, and they in many cases will pass out of bloom and may be cleared away just as the large leaves begin to cover the ground. Where this is not the case, but the larger plants are placed thin enough to always allow of the lower ones being seen, two or even more kinds of dwarf plants may be employed, so that the one may succeed the other, and that there may be a mingling of bloom.

It may be thought that this kind of mixture would interfere with what is called the unity of effect that we attempt to attain in our flower-gardens. This need not be so by any means; the system could be grandly used in the most formal of gardens laid out on the massing system pure and simple; besides, are there not positions in every place where such arrangements could be made without interfering with what is sometimes called the "flower garden proper"? Some may say we cannot grow the fine-leaved plants in England. But this is not so. The most beautiful

E 2

bed of those above enumerated was that composed of variegated Arundo and Lobelia—the former a plant that may be
readily grown on good soils in Britain, and merely requiring
the protection of a little ashes, refuse, or an old mat over
the crown in winter, even in soils that are not particularly
favourable, while the Lobelia is one of the many fragile and
delicately pretty little plants that do perhaps best of all in
England.   The fact is, we can find numbers of plants
among the hardy and free-growing kinds, which will enable
us to enjoy all the desired variety and diversity, even if we
cannot wisely venture to plant out Wigandias and coloured
Dracænas except in the more favoured districts of southern
England and Ireland.

One of the most useful and natural ways of diversifying
and dignifying a garden, and one that we rarely or never
take advantage of, is abundantly illustrated here, and as it
is perhaps the most important lesson to be learnt in the
garden, we will discuss it at some length.   It simply consists in placing really distinct and handsome plants alone
upon the grass, to break the monotony of clump margins
and of everything else.   They may be placed singly or in
open groups, near the margins of a bold clump of shrubs
or in the open grass ; and the system is applicable to all
kinds of hardy, ornamental subjects, from trees downwards,
though in our case the want is for the fine-leaved plants
and the more distinct hardy subjects.   Nothing, for instance, can look better than a well-developed tuft of the
broad-leaved Acanthus latifolius, springing from the turf
not far from the margin of the walk through a pleasure
ground ; and the same is true of the Yuccas, Tritomas, and

FIG. 22.

Groups and single specimens
of plants isolated on the
grass.

other things of like character and
hardiness.   We may make attractive
groups of one family, as the hardiest
Yuccas ; or splendid groups of one
species like the Pampas grass—not
by any means repeating the individual, for there are about twenty varieties of this plant known on the
Continent, and from these half a

dozen really distinct and charming kinds might be selected to form a group. The same applies to the Tritomas, which we usually manage to drill into straight lines : in an isolated group in a verdant glade, they are seen for the first time to best advantage ; and what might not be done with these and their like by making mixed groups, or letting each plant stand distinct upon the grass, perfectly isolated in its beauty !

Let us again try to simply illustrate the idea. Take an important spot in a pleasure ground—a sweep of grass in face of a shrubbery, and see what can be done with it by means of these isolated plants. If, instead of leaving it in the bald state in which it is often found, we try to place distinct things in an isolated way upon the grass, the margin of shrubbery will be quite softened, and a new and charming feature added to the garden.

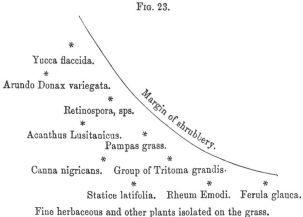

Fig. 23.

Fine herbaceous and other plants isolated on the grass.

If one who knew many plants were arranging them on the ground, and had a large stock to select from, he might make no end of striking effects. In the case of the smaller things, as the Yucca and variegated Arundo, groups of four or five good plants should be used to form one mass, and everything should be perfectly distinct and isolated, so that a person could freely move about amongst the plants without touching them. In addition to such arrangements, two or three individuals of a species might be placed here

and there upon the grass with the best effect. For example,
there is at present in our nurseries (I once saw quantities
of it preparing for game covert at
Mr. Standish's, of Bagshot) a great
Japanese Polygonum, which has never
as yet been used with much effect in
the garden. If anybody will select
some open grassy spot in a pleasure
ground, or grassy glade near a wood
—some spot considered unworthy of
attention as regards ornamenting it—
and plant a group of three plants of
it, leaving fifteen feet or so between
the stools, a distinct aspect of vegetation will be the result. The plant is
herbaceous, and will spring up every
year to a height of from six feet to
eight feet if planted well; it has a
graceful arching habit in the upper branches, and is covered
with a profusion of small pale bunches of flowers in
autumn. It is needless to multiply examples—the plan is
capable of infinite variation, and on that account alone
should be welcome to all true gardeners. The diagram with
the names is far too formal, and merely given to more
fully explain the system. The little plans show better the
irregular way in which the plants ought to be disposed.

The preceding part of this chapter was written in 1867;
but as this park is so full of interest and instruction for all
practically interested in the decoration of the flower-garden,
the following description, written on the spot during the
early part of last September, may be of some interest to the
horticultural reader :—

Entering the park from the Boulevard Malesherbes we
pass along an avenue of plane trees that leads from the high
and ornamental gates. The walk on each side is bordered
with roses in lines of different colours—the front row well
pegged down. They form long borders on each side, and
are very ornamental in early summer. A carriage road
leads through the park, so that it may be seen by those

Fig. 24.

Portion of plan showing
Yuccas, graceful dwarf
pines, &c., irregularly
isolated on the grass.

who drive through—but imperfectly, as the more interesting objects are along the shady side and boundary walks. On each side of the central drives glimpses are caught of very diversified and graceful foliage and flowers, but conspicuous on the margin is a great mass of Caladium, with leaves three feet long and two and a half feet wide, springing from a groundwork of blue Lobelia.

You can have no real beauty in an ornamental garden without the aid of full grown trees, their majesty producing an effect which cannot be dispensed with. Here they approach the drive in groups, sometimes overshading plantations of dense shrubs, at others springing clean from the grass. In some places they are so crowded as to make one wish for a little breath, in others they disappear, and spreads of grass and dwarfer plants permit the eye to range. On one side of the route may be noticed a hardy bamboo with black polished stems, and rods ten, twelve, and fourteen feet high; on the other, one with yellow stems of about the same height. An old specimen of the Abyssinian Musa is vigorously pushing up a massive flower shoot scarcely yet seen through the leaves, and in consequence they are by no means so ornamental as those of younger plants which devote all their energy to foliage. Tree ferns, and the curious and graceful Beaucarnea with the great swollen base, are seen here and there, the Beaucarnea apparently not a first-rate subject for placing in the open air. Next to the great Musa Ensete, the best Banana is the well-known edible Musa Cavendishii : it is in perfect health, emerging from a mass of Tradescantia zebrina; the leaves twenty-four to thirty inches long, and not often lacerated. A great mass of the variegated Acer—several hundred trees—is margined with rose-coloured geraniums, and all the space between filled with Dahlias, Salvias, and the like : a good plan, inasmuch as it prevents a naked base. Groups of palms, single specimens of birch (as graceful as any exotic), and fine out-arching specimens of the hardy Polygonum Sieboldi form the most notable features of the central drive. Palms from regions comparatively temperate, like the dwarf fan palm of the south of Europe, the Palmetto of the Southern

United States, the Seaforthia, and some others, bear the
open air of summer without injury, and add a very
striking and valuable aid to the scene. From the cross-
drive groups of Yuccas, rather thinly placed in masses of
dwarf flowers and plants, a large specimen of the Angelica
tree in flower, a mass of the Papyrus of the Nile, and
tall specimens of Colocasia odorata, are the most conspicuous
of the objects that approach the margin.

Again, commencing at the Boulevard Malesherbes entrance, and this time turning to the left, we meet with
masses of Musa rosacea, Blechnum, Lomaria magellanica,
the older specimens with stems two feet high; Nicotiana
wigandioides; a telling, dark bronzy mass of Canna atro-
nigricans, with some of the larger leaves two feet long, and
the stems nearly seven feet high; groups of Latania plunged
in the grass; and large leaved Begonias dotted amongst dense
masses of Tradescantia zebrina. These Begonias do not grow
well enough to warrant their being put out in our latitudes
except under the most favourable conditions. Next come
masses of Hibiscus, rather sparing of their great red flowers;
numerous specimens of handsome plants isolated on the
grass, from double scarlet Pomegranates to Thuja aurea
and Clianthus Dampieri; masses of india-rubber plants
with groundwork of mignonette, of Wigandia macrophylla
with groundwork of Coleus, of silvery Solanum marginatum
with groundwork of dwarf herbaceous Aster, of Tupidanthus
in carpet of Cuphea, and of variegated Arundo in one of
German Aster. A mass of Caladium bataviense, with
leaves three and a half feet long and dark stems, is very
imposing. As a foliage plant, it is second to no other
employed in Parisian gardens, though hitherto C. esculentum has generally been considered to be the best. Here
there are large masses of both it and bataviense. Usually
C. bataviense makes leaves larger than C. esculentum, and
as a rule its leaves are the largest this year, but the
biggest specimens of the year were of esculentum, of which
the largest measured four feet seven inches long, bataviense
reaching four feet one inch. C. esculentum best withstands
the winds, the leaves of C. bataviense often getting broken

PLATE **XV.**

FINE-LEAVED PLANTS IN THE PARC MONCEAU.

Caladium.   Aralia.   Phœnix.   Musa Ensete.   Yucca.

by them, so that many of the finer leaves made during the
season were lost before September, their great stumps
showing how vigorous they had been. It is usually and
from the same cause denuded of leaves about the base; C.
esculentum retaining them. The leaf-stalks of bataviense
are of a dark hue, by which it is easily distinguished from
esculentum with its pale green leaf-stalks. The stems of
bataviense are also much larger than those of the esculentum, a few of those growing here being ten inches in
diameter.

Of the Ficuses grown here, the best is yet the old
F. elastica; but Chauvieri is also good, and Porteana has
done well this season, though the Parisian summers are
usually too cold for it; its leaves were fifteen inches long.
Yucca aloifolia is hardy here. A fine old plant of it, ten
feet high, and with a considerable portion of the stem
naked, was in perfect health. Every winter the stem is
protected as far as the leaves, and the snow prevented
from remaining on these. Melia Azederach is also hardy
here—at least, it has stood out during the past winter;
and as its large compound leaves would prove so useful in
the flower-garden, it should be tried out in favourable
parts of England. Andropogon formosum does well here,
and a group of Dasylirions are plunged in the grass. The
Erythrinas are a fine feature, the old E. crista-galli being
considered the best on the whole; but E. ruberrima is
very fine from its hue of scarlet and crimson. Bocconia
frutescens is five and a half feet high, with leaves two and
a half feet long; and an Encephalartos is fine as an isolated
specimen. Agave americana is left in the garden during
winter and protected, but with more trouble and cost than
would be incurred by taking it indoors. A mode of training various flowering climbers up the stems of trees is
worthy of special notice. Clematises, honeysuckles, various
kinds of ivy, everlasting peas, and many other kinds of
climbing plants may be used in this way with good effect. There is one plant grown here in quantity, which is
rarely seen in England, but which should be in every
English garden—Funkia subcordata, a dwarf, hardy

plant with snowy white flowers sweeter than orange-blossom.

Two large carriage drives, laid out so as to interfere as little as possible with the old plantations, run through the park from one end to the other, and form a continuation of the boulevards leading to it. These drives are closed by iron gates of a highly ornamental character. The area of the park is about twenty-two English acres, of which thirteen are in turf, and five planted with flowers, shrubs, and trees, the remainder being devoted to walks and the small and unhappy piece of water. The total cost of alteration was over 48,000*l.* The work was begun in the month of January, 1861, and finished in August of the same year.

CHAPTER IV.

THE PARC DES BUTTES CHAUMONT.

THIS is the boldest attempt at what is called the picturesque style that has been attempted either in Paris or London. It is hardly wise to attempt expensive and extraordinary works in places of this sort, at least till all the densely populous parts of a city are provided with open, well-planted spaces. Thus in London it is a mistake to devote great expense to a few parks, and leave so many square miles of population without a green spot. But in this instance an unusual attempt was to some extent invited by the peculiar nature of the ground. The whole park may be described as a sort of diversified Primrose Hill with two or three "peaks and valleys," and an immense pile of rock seen here and there. At its hollow or lower end there was a quarry, and this has been taken advantage of to produce a grand feature. They have cut all round three sides of this quarry, smoothed it down, leaving intact the great side of stone, and adding to it here and there masses of artificial rock.

This forms a very wide and imposing cliff, 164 feet high, or thereabouts, in its highest parts, and from these you may gradually descend to its base by a rough stair, exceedingly well constructed, and winding in and out of the huge rocky face. At the base of the cliff, and widely spreading round it, there is a lake. This ponderous cliff has several wings, so to speak, and in one bay has been constructed a large stalactite cave, about sixty feet high from its floor to the ceiling, and wide and imposing in proportion. At its back part the light is let in through a wide opening, showing a gorge reminding one of some of those in the very tops of the Cumberland mountains, and down this trickles the water into the cave, ivy and suitable shrubs being planted along its course above the roof of the cave.

The effect is remarkably striking, though it is hardly the kind of thing to be recommended for a public park. By all means let us leave the luxuries of gardening out of the question, till we have provided the necessaries for the population of great towns, and these are green lawns, trees, and wide open streets and ways, with their necessary consequence, pure air. On one of the buttes, or great mounds here, they have planted 500 or 600 deodars—forming it a hill of deodars in fact. This is a mistake, for though Paris is not as foggy as Spitalfields, it is a great city, as may be seen from this park, and with many a vomiting chimney too, so that the better plan would be to pay double attention to deciduous trees, using only such evergreens as are certain to grow. In one wide nook, perfectly sheltered on the three coldest sides, M. André planted a collection of subjects mostly tender in the neighbourhood of Paris.

From this park, the surroundings of which are by no means attractive, you can look over nearly all Paris. The approach to it from the central parts is shabby for Paris, and on the way some idea of what the city was before the splendid improvements of the past ten years may be caught; but this approach, like most objectionable things there, is simply tolerated till more important ones are finished. Of the quick way in which they proceed with them, the reader can scarcely have a notion. I have seen acres of land removed to a depth of several yards without any fuss, and in a few weeks; miles of trees planted in the course of a single week; old suburbs blown up by hundreds of mines a day, and levelled into commanding terraces fit for princely mansions. One June day, bright, dry, and very warm, they were planting trees in this park, and large ones too—trees that required great machines to lift them—while they were marking the ground for fresh plantings. Do you plant after this date? I asked. Every day in the year! Of the larger trees some seem not to take well, and doubtless in consequence of summer-planting, for which there seems little excuse.

PLATE XVI.

LAKE AND CLIFFS IN THE PARC DES BUTTES CHAUMONT.

The entrance is not promising—a hard-looking porter's lodge, and a mass of badly-made rockwork face a mound, and from the rockwork springs an apparently quite unnecessary bridge. The rockwork is bad because, although superior in general design to the masses of burnt bricks that sometimes pass for it with us, it shows radical faults —presumption and unnaturalness. Instead of a true rockwork, something like a very puny attempt at reproducing the more insignificant ribs of Monte Campione is the result of plastering over a heap of stones. A hole is left here and there in this mass from which may spring a small pine or an ivy, but the whole thing is incapable of being divested of its bald artificial character. One-fourth the quantity of natural blocks of stone, visible through the breaks in a mass of evergreens, would have been far better. By this means one could get the necessary elevation, concealing the basis of the stones with evergreens and trailing plants, and not sealing up the thing with cement in any part. The plastering of the joints merely makes the "rocks" look truly artificial, especially when it begins to drop out.

Bold high green mounds meet us immediately after passing under the ugly bridge at the entrance—here and there patched with very presentable shrubs—as is not rarely the case in Paris gardens. One girdle seems to bind both French and English, however, as regards the compact and formal outlines of these shrubberies and plantings. We know very well that in nature nothing of the kind ever occurs; that away from the wood strays the clump of low shrubs which do not seem to be gregarious like their pillared fellows of the forest; that indeed anything like straitlacing is unseen. Why then should we draw a cordon of regularity and sameness round our shrubberies in the shape of a line of some showy flower, making the whole thing changeless as possible? What calls for this definiteness? I know not unless it be that the mowing machine may have the less trouble in cutting the grass around. Imagine the British Museum or the Louvre arranged chiefly for the convenience of the dusters! The sooner everybody having the

interests of gardening in mind proclaims that variety and
not formality should be the aim of all high gardening, the
better for the progress of the art. In their clumps the
French seem as straitlaced as ourselves, but in the newer
gardens they have adopted a system of dotting about single
specimens of individual beauty, which is very successful in
breaking up formalism, and is well worthy of imitation.

The chief feature of the place, as previously indicated, is
the great cliff, and unhappily the chief feature of the rock
is plaster. You can hardly approach it in any place without perceiving the seams of plaster giving out, and where
this is not the case it is all palpably plastered. And why?
Perhaps the plasterer who made it could supply a reason ;
but, whether he can or not, the sooner plasterers are dispensed with as imitators of nature in her grandest workings
the better. There never was in a garden such a chance
of presenting walls of rock-plants almost as striking and interesting as those one meets with in the pass over the Simplon ;
yet it is entirely lost. By leaving the chinks and filling
them here and there with turf, by chopping back or leaving
the face of the high rocks sloping in some places so that
they would be well exposed to the rainfall, by trickling a
little streamlet over the face of the cliffs here and there,
and by scattering a few packets of seeds over the face of the
cliffs in spring, they would have given rise to an alpine
vegetation of great beauty. The great long-leaved Saxi-
frage of the Pyrenees might have spread forth its silvery
rosettes here, so might its smaller relatives, its big brother
of the Piedmontese valleys, and little Campanulas, Thymes,
Erinuses, Brooms, Stonecrops, Houseleeks of many kinds,
with hundreds of the prettiest plants of northern and temperate climes might have been grown here. Now all is
daubed over and plantless, save a bit of ivy and wiry
grass in some few spots ; and the face of the high rocks is
suggestive of little but suicide.

One of the few attempts to cultivate alpine plants out
of pots that I have ever seen made in France is here, but
it has been done on a mistaken principle. A tasteful and
desirable practice in some of the newer gardens and parks

PLATE XVII.

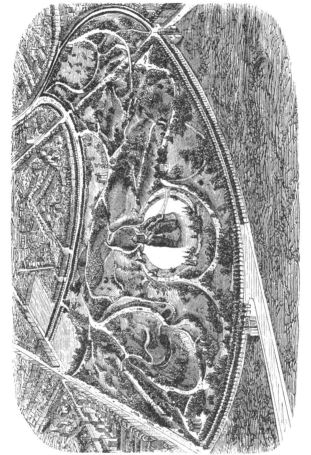

BIRD'S EYE VIEW O   THE PARC DES BUTTES CHAUMONT.

of Paris, is that of conducting a tiny streamlet irregularly through the grass, and bordering it with water and marsh plants; here there are a few examples of it for the most part creditable.   In one case, however, the streamlet instead of coming from any probable source of higher rock or brushwood, starts out of a plastered hole in the grass, in a way one cannot admire.   By the side of this and a neighbouring streamlet alpine plants are placed, to grow here and there in little beds along the stream, and indeed now and then on a plastered spot in the middle.   They are associated with such lowland marsh plants as the loosestrife; and in one instance a willow had started up and shaded some choice dwarf Saxifrages and Rhododendrons. It is creditable to attempt the cultivation of these plants here, but alpine plants can never be grown thus.   If they could, it would be difficult to enjoy their native beauty or their tiny character alternated with such things as the bullrush and the flag!   With the supply of water that these parks command, nothing could be easier than the creation of a rocky mound healthfully covered with true alpine plants. However, as no English landscape gardener has yet presented us with a rockwork well covered with its proper ornaments, instead of merely ivy, Virginian creeper, &c., it would be captious to find fault with the French for failing in a branch which requires so much taste and knowledge of plants.   Not a few of the minor masses of rock—and there are many of them—are in better taste; and being less ponderous, they will some day no doubt display the plant life without which a rockwork is a poor affair.   A piece of very bad taste is shown in bringing a café right to the edge of the walk commanding one of the best views of the rocks and water.   Restaurants and refreshment places are wanted, but they should not be thrust in face of the most important spots.   People should never go to such places for the sake of the café, however interesting it might be as an accessory.   There are unobtrusive and readily accessible positions where they may be situated.

One feature deserves denunciation—the glaring way in which the walks are exposed.   There can hardly be two

opinions about the desirability of concealing the walks of a
naturally disposed garden as much as may be convenient.
A marked feature in many new French gardens is the way
they are exposed. In the plans of the best French landscape-

Fig. 25.

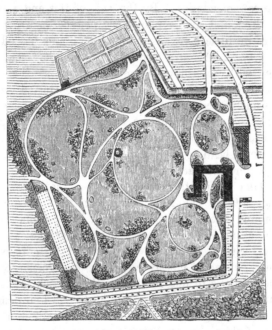

Plan of garden, showing how fond the French landscape-gardeners are of
describing sections of eggs while laying out their walks.

gardeners it is quite ridiculous to see the way the walks
wind about in symmetrical twirlings, and, when they have
entwined themselves through every sweep of turf in the
place, seem to long for more spaces to writhe about in.
Most glaring instances of this are seen here, and particularly on the top, the highest rock, where a small temple
is seated.

Near one of the entrances, here is a mixture of Indigofera Dosua and the holly-leaved Mahonia, the first predominating and full of flower in summer, having the
delicate beauty and profusion of flowers characteristic of

New Holland, and greenhouse plants: it is worthy of being extensively used with us, and Indigofera floribunda should be everywhere used as a flower-garden wall plant. There is not much in the summer decoration of the place that is worthy of note. Some kinds of Cannas in flower look almost as showy as beds of Gladioli, but their real value will always be greatest as fine-leaved ornaments. The common artichoke was very effective in one spot as an isolated specimen of a "foliage plant," nothing being finer than the nobly formed silvery leaves of this plant. Indeed, there is nothing to surpass it among subjects suited for single specimens on the green grass. A well-developed example would be sufficient in a private garden; and if nobody else plants it, schools of art would find it to their advantage to have a specimen of it somewhere near at hand.

The Parc des Buttes Chaumont was made on the site of old and abandoned plaster quarries. It forms a curvilinear triangle, having an area of nearly forty-five acres included between the Rue de Crimée and two boulevards running between Belleville and Puebla. Before the park was made, the ground, which was divided by the Chemin de Fer de Ceinture and the Rue Fessard, was an arid wilderness of clay mounds and of excavations left by the quarrymen, many of which were so deep as to form miniature precipices. It was proposed to turn this waste into a public promenade by taking advantage of the natural irregularities of the ground, by forming paths, laying turf, and making a piece of water. To obtain this result, the natural hollows of the ground in the part nearest to Paris were deepened, paths leading to the top of the hills and mounds were laid down, the general surface was made more regular and covered with garden earth and flower-beds, and plantations were formed where necessary. The improvements made were of an important character only as far as it was necessary to bring the boundary of the park into harmony with the Boulevard de Ceinture, which runs through a trench nearly sixty feet deep. The other portion of the park, in which are situated the cutting through which the Chemin de Fer de

Ceinture passes, and the old plaster quarries, which now forms the most picturesque part, necessitated works of a much more considerable cost.

The line of rocks, which in some places are much over 100 feet in perpendicular height, was luckily terminated by a craggy promontory looking down into the old excavations. This promontory was separated from the general mass in such a way as to form an isolated rock rising out of the lake which surrounded it on all sides. The lake is supplied by two rivulets which run through the two valleys of the park. One of them flows out of the lower wall of the upper boulevard, and falls down into a large cavern forming a cascade over 100 feet in height. The wall and grotto were formed to support the neighbouring land towards Belleville which was gradually falling into the excavations left in the quarries. The marly soil which lies above the gypsum in a layer of forty-eight feet thick, the slightly sloping surface of which was gradually crumbling away under the action of the air, has been dug out so as to allow the slopes to sustain the mould forming the plantations. At the highest point of the promontory, however, where it was necessary to have a bold mass of rock hanging over the water, an embankment of masonry built in imitation of the rocks at the base has been found necessary to support the crumbling soil. A suspension bridge more than 200 feet long thrown over the lake and the path surrounding it joins this portion of the park to the other, and obviates the necessity of a long walk round. A large number of carriage roads twenty-two feet wide, the inclines rarely reaching 6 in 100, allow carriages to drive all over the park in spite of the great difference of level existing in various parts.

The paths, whose inclination seldom exceeds 10 in 100, but which are sometimes cut into steps, afford foot-passengers the means of making short cuts between the carriage-drives in order to reach the heights of the park more expeditiously. Four bridges have been built over some of the deeper hollows, also a wire bridge has been thrown across the railway, a stone bridge, forty feet in span and sixty feet high, above a road and a small arm of the lake, the suspension bridge already mentioned, and a skew bridge fifty-six feet in span, made of iron resting on stone piers.

The park being surrounded by large roads is enclosed with an open iron railing, so that the view is never obstructed. Besides this, wherever it has been possible, the garden has been so arranged as to be looked down upon from the boulevards above. The boulevard itself is supported by a wall forming a terrace over one part of the park, upon which it looks down almost perpendicularly over an escarpment 120 feet high. The water which supplies the cascades and the pipes by which the garden is watered is pumped by a special engine belonging to the Canal de l'Ourcq into a reservoir situated at the side of the upper boulevard which surrounds the park. As for the end of the park nearest to Paris, it is, on the contrary, much higher than the boulevards. It has therefore been laid out in such a way as not to interfere with the panorama of Paris seen above the tops of the houses which will be built in the intervening thoroughfares. The works, which were commenced early in 1864, are now finished. The cost of the bridges, roads, and gardens amounted to something near 120,000l. The architectural work, including a first-class and two second-class restaurants, one double and eight single park-keeper's lodges, a rotunda, and the surrounding railing, will amount to nearly 20,000l., making the entire cost close upon 140,000l.

CHAPTER V.

WE have nothing in the British Isles like the Jardin des Plantes. It is half zoological, half botanical, and nearly surrounded by museums containing vast zoological, botanical, and mineralogical collections. The portion entirely devoted to botany is laid out in the straight, regular style, while the part in which are the numerous buildings for the wild animals, has winding walks, and some trifling diversity here and there. The place is really an important school of science, and as such it is great and useful. In

FIG. 26.

Conservatories and Museums in the Jardin des Plantes.

addition to able lecturers on botany, culture, and allied matters, there are, I believe, a dozen on various other scientific subjects, some of these gentlemen being among the ablest and most famous naturalists in Europe. Here Buffon, Cuvier, Jussieu, and other great men have worked; and here at the present day, even in minor departments, are many men of well known ability.

Although the Jardin des Plantes is quite inferior in point

of beauty to any of our large British botanic gardens, it contains some features which might be introduced to them with the greatest advantage. Its chief merits are that its plants are better named than in any British garden; it possesses several arrangements which enable the student to see conveniently, and most correctly, all obtainable useful plants infinitely better than in any British botanic garden; and it displays very fully the vegetation of temperate and northern climes, and consequently, that in which we are the most interested, and which is the most important for us. Its chief faults are that it has a

Fig. 27.

Aquatic birds in the Jardin des Plantes.

bad position in an out-of-the-way part of the town; the greater part of its surface is covered with plants scientifically disposed; the houses are poor and badly arranged compared to those in our own good botanic gardens; and there is no green turf to be seen in its open and important parts. It has, in addition, a very bad atmosphere for pines and evergreens, and there is a ridiculous kind of maze on the top of an otherwise not objectionable mound. Half way up this elevation stands a tolerably good Cedar of Lebanon, the first ever planted in France. It was planted by Jussieu, to whom it was given by the English botanist Collinson. Beyond this there is not much tree-beauty in the Jardin des

Fig. 28.

Animals in the Jardin des Plantes.

Plantes. There are fine collections of palms and other subjects of much importance for a botanic garden, and the

Fig. 29.

house collections are on the whole good, but the plants in a great many cases are very diminutive and poorly developed, therefore we will pass them by.

There is one admirable feature which must not be forgotten, and that is the fine collection of pear trees. M. Cappe has had charge of this section for about thirty-five years, and is now a very old man, but still he attends to his trees, and has them in fine condition, though contending with much difficulty, because the space upon which the trees

Cedar planted by Jussieu in the Jardin des Plantes.

stand is really not enough for one-half the number, and thus he is obliged to keep lines of little trees between and under big ones, and so on. There are few things in the horticultural way about Paris better worth notice than this collection of pears.

Remarking that they have a graceful way of commemorating great naturalists by naming after them the streets in the immediate neighbourhood of the garden, I will pass on to the more important feature of the garden; that is, its very extensive and well named collection of hardy plants. The only species of Pelargonium that ventures into Europe (P. Endlicherianum) is grown here, and it is quite hardy. The first of the principal arrangements of hardy herbaceous plants, &c., is a curious and distinct one. It is simply two large and wide spaces planted with masses of ornamental species; and looks pretty well, though far from being arranged in a way to develope fully the beauty of its contents. Edgings composed of the several

varieties of Iris pumila look well in early spring, and
many plants are used for edging which we are not accustomed to see so employed in England. Thus the good
double variety of Lychnis Viscaria has been very pretty as

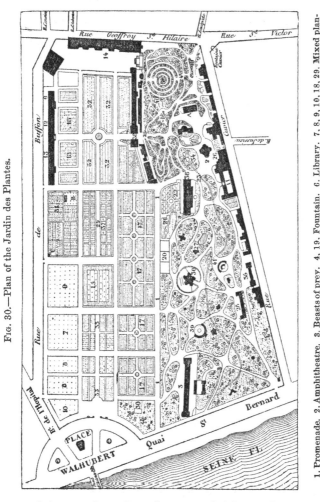

FIG. 30.—Plan of the Jardin des Plantes.

1. Promenade. 2. Amphitheatre. 3. Beasts of prey. 4, 19. Fountain. 6. Library. 7, 8, 9, 10, 18, 29. Mixed plantations and nurseries. 11, 12, 13, 14. Museums of Anatomy, Botany, Mineralogy, and Zoology. 15. Parterre. 16. Cedar. 17. School of Botany. 20. Bears. 21, 22, 23. Nurseries, &c. 24, 25. Labyrinths. 26. Cuvier's house. 27, 28. Birds. 30, 31, 32, 33, 34. Aquatic, Edible, Herbaceous, Officinal, and Tropical Plants. 35. Reptiles. 36, 37, 89. Animals. 38. Glass-houses. 40. Monument to Daubenton. A. Offices. B. Café. C. Cloak-rooms.

an edging, and so has the neat, bright, and pure white
Silene alpestris—an alpine plant not half so popular as it
ought to be, though I observe that some seedsmen, while
not offering it, sell a pretty fair proportion of the weeds

that belong to the genus. Then there is a large space devoted to plants used for the decoration of the parterre, all
or chiefly tender plants or annuals. This is not so successful or useful as some of the other arrangements, though
it displays numbers of popular ornamental subjects.

Let us pass on to a large division devoted to the culture
of plants used as food, and in commerce. It is at once
successful, useful, and complete. The chief varieties of all
garden crops, from Radishes to Kidney Beans, are to be
seen ; the various species of Rhubarb, all important varieties of Lettuce—in a word, everything that the learner
could desire to see in this way. It is not merely the plan
of the thing that is sensible and good, but its carrying out.
The annuals are regularly raised and put out; the ground
is kept perfectly clean, and it is, in fact, the best place I
have ever seen in which to become acquainted with useful
plants. Such arrangements well carried out, and cut off by
judicious planting from the general verdure and chief area
of any of our great public gardens, would be of the
greatest service. The ground is thrown into beds about
six feet wide, and each kind is allotted six feet run of the
bed. The sweet potato is grown here, as indeed are all
interesting plants that may be grown in the open air.

Below this arrangement, and near the river end of the
garden, is another very interesting division. It is chiefly
devoted to medicinal and useful plants of all kinds, arranged
in a distinct way. First we have the Sorghums, Millets,
Wheats, and Cereals generally—all plants cultivated for
their grains or seeds. Then come plants cultivated for
their stems, from Polymnia edulis to Ullucus tuberosus.
Next we have the chief species and varieties of Onion, such
plants as Urtica utilis, the Dalmatian Pyrethrum rigidum,
and in a word almost everything likely to interest in this
way, from Lactuca perennis to the esculent Hibiscus.
Here again the plants are well named and kept clear and
distinct, each having full room to develope, the general
space devoted to the subject being sufficiently large; and
the practice of giving each plant a certain portion of the
whole breadth of each bed to itself is better than the more

crowded arrangements adopted in our British botanic gardens. All these divisions we have just passed through cover an oblong expanse of ground, the effect of which is of course anything but beautiful from an ornamental point of view; but yet, in consequence of the ground being well kept, each subject grown well and vigorously, and all the squares bordered with roses and summer flowering plants, the effect is better than might be expected. This great oblong space is bordered on each side by double rows of lime trees planted by Buffon. Between these are wide walks, agreeably shady on hot days.

The second great oblong space to the north is entirely devoted to the school of botany, and this is simply a large portion of ground planted on the natural system, remarkable for the correctness of its nomenclature and the richness of its collection. Here again everything is well taken care of and kept distinct; the aquatics are furnished with cemented troughs, in which they do quite luxuriantly, one of the singular and handsome Sacred Beans (Nelumbium speciosum), and Limnocharis Humboldtii being well grown in the open air. The whole is most satisfactory, with one exception—that they place out the greenhouse and stove plants in summer to complete the natural orders. These poor plants are stored pell-mell in winter in a great orangery, from which they are taken out in early summer literally more dead than alive. They make a few leaves during the summer, and are again put into their den to sicken or die. The medicinal and other plants for special uses are indicated by variously coloured labels.

Among many handsome hardy plants which I met with here, and which are deserving of being more largely grown with us, are Hibiscus militaris, Crambe juncea, Verbascum vernale, Heracleum latisectum, Yucca lutescens, flexilis, Treculeana, angustifolia and stricta (all hardy), Spiræa decumbens, Iris nudicaulis, Antirrhinum rupestre, Merendera Bulbocodium, Colchicum montanum, Magydaris panacina, Sorghum halapense, Panicum bulbosum, altissimum, and virgatum, Epilobium sericeum, Gundelia Tournefortii, Dahlia arborea, imperialis, and Decaisneana (out only during the summer of

course), Datura fastuosa alba-duplex, Pyrethrum Tchihat-
chewii, of the south of Europe—a capital plant for covering
the dryest of banks with dark green; it is very low in
habit, produces white flowers in spring, and for banks
and other positions so dry and arid that grass or anything
else fails to grow upon them, it will probably prove highly
useful. Anemone alba, Ficaria calthœfolia, Echinophora tenuifolia, a graceful umbelliferous plant with hoary leaves; Gly-
ceria Michauxii, a pretty grass; and a collection of the genus
Asparagus, among which one, A. Broussonetii, is remarkable
for its great vigour and rapidity of growth—it quickly
runs up with dense vigour to a height of ten feet in spring,
its foliage is glossy and dense, and it might be used with
success as a covering for bowers or to make pyramids in a
highly diversified garden of hardy plants, and of course it
would be valuable in such a place as the subtropical
garden at Battersea Park. Asparagus tenuifolius is as
graceful and elegant as the one before-named is vigorous
and rampant in its climbing power.

Iris Monnieri, of Western Asia, is a really fine, bright
yellow kind. Among the larger Compositæ are some likely
to prove useful for the subtropical garden; notably Rha-
ponticum scariosum, and cynarioides. Serratula pinnatifida
is elegant in leaf; and particularly fine is a silvery-leaved
Tanacetum (T. elegans), with finely divided and elegant
frond-like leaves. Dipsacus laciniatus is fine in its line
when well grown, and it will prove really well worth raising
annually, somewhat like the Castor-oil plants, for the
garden where distinction is desired. Sideritis syriaca is
hardy here, and fairly tried might make a useful edging
plant in the way of Gnaphalium lanatum, than which it is
a shade more silvery. Phlomis herba-venti is a pretty
and distinct herbaceous plant, medium-sized, and Eremo-
stachys iberica is a yellow species, well worthy of association with laciniata. Acantholimon venustum is prettier
and more elegant than the admired A. glumaceum, the
dwarf cushion of leaves being of a glaucous tone, and the
large rose-coloured flowers being well thrown out on bold
graceful stems; it is one of the prettiest dwarf plants I

have ever seen, and for a well made and tasteful rockwork it will prove one of the best summer ornaments. Geranium platypetalum is very good here, and one of the best of the family. Erodium carvifolium is so elegantly cut that I should not hesitate to place it beside Thalictrum minus, from which it is, of course, quite distinct in character.

There is a capital collection of the very neat Semper-vivum family, planted in the open air, where they do remarkably well. Seseli gummiferum is a pretty umbelliferous plant, of a peculiarly distinct and pleasing glaucous hue. Thapsia villosa is also fine, and so is Aralia edulis. Vicia tenuifolia formosa is a very handsome climber; and Orobus rosea is one of the most elegant and pretty of its family, having arching and drooping shoots, and being well suited for a large rockwork. There are many others in various departments, but as the subject is not of interest to a very wide class, it must not be enlarged upon further.

For the information of curators of botanic gardens, and those taking a botanical interest in curious plants, I may state that Cuscuta major is luxuriantly grown here upon the nettle, C. Epithymum upon Calliopsis tinctoria, C. Engelmanii upon a Solidago, and Orobanche grows upon Hemp. I have grown O. minor upon perennial Clovers, and O. Hederæ may be readily grown upon the Ivy at the bottom of a wall (I once saw it growing freely on the top of a wall near Lucan, in Ireland); so that there ought not to be the difficulty which our botanic gardeners find in growing these curious plants. Orobanche ramosa is also grown here upon Calliopsis tinctoria. The safest way with the Orobanches is to scrape away the soil till you come near the root of the plant on which you intend it to be parasitical, and then sow the seed.

A very old and fine pair of dwarf fan palms, given to Louis XIV. by Charles III., Margrave de Bade, are usually placed in summer one at each side of the entrance of the amphitheatre. They have straight clean stems, and are more than twenty feet high. They escape the notice of many visitors, but are well worth seeing by all plant-lovers, not only from their age, but their exceptional height. Should any

visitor to the Jardin des Plantes wonder at the poor external
aspect of its houses and some other features as compared

Fig. 31.

with those at Kew, he would
do well to bear in mind that
money has a good deal to do
with such things; and that the
grant for museums, lecturers
(the lectures are free), the expensive collection of animals,
and everything else in the
Jardin des Plantes, is miserably
small.   On the other hand, the
gardens and plants of La Ville
de Paris are plentifully provided with money; the municipality of Paris often spending

The Amphitheatre in the Jardin des
Plantes.   On each side of the entrance there is a very tall and old
specimen of the "dwarf fan palm."

prodigious sums for the purchase of plants, and even for the
plant decoration of a single ball.

One ball at the Hotel de Ville
during the festivities of 1867 cost considerably over 30,000*l.*,
while the poor Jardin des Plantes gets from the State not more
than one-third of that sum to exist upon for a whole year.

The Luxembourg Garden.

The beautiful old garden attached to the Palais du Luxembourg—the favourite resort for many years of the Parisians
of the left bank of the Seine—has lately been almost entirely
remodelled, much to the indignation of the Parisian public
and journalists; but it is still a pretty garden.   Geometrical
gardens are seldom capable of affording any prolonged interest
or refreshing beauty; very rarely so much so as that of the
Luxembourg.   Before the recent alterations there was a good
botanic garden—an irregular sort of English garden, which
the French call the " never to be forgotten nursery"—and
much miscellaneous interest now passed away.   At present
matters are much more concentrated, and we shall find less
to speak of than of old, but yet enough to make the place

PLATE XVIII.

THE GARDENS AND PALACE OF THE LUXEMBOURG.

worth a short notice. The garden used to be famed for its roses, and for perhaps the largest collection of vines ever accumulated, but recent changes have altered all this. The vines were removed bodily to the Jardin d'Acclimatation, in the Bois de Boulogne, and thus it lost some of its interest. The glass-house department, however, retains most of its attractions, and to the horticultural visitor will present a good deal of interest. It contains the best collection of Orchids in any public garden about Paris, fine Camellia-houses in which the specimens attain great perfection, and miscellaneous collections. The object and limits of this book will not permit us to enter into particulars of this department, and therefore we will go in the open air and look at the broader features of the scene.

Usually in geometrical gardens the portion nearest the building is a terrace commanding the surroundings—here, on the contrary, the part nearest the palace and stretching away from its face is a basin flanked by balustraded terraces. Above these terraces are seen numerous marble statues and horse-chestnut groves. The lower portion, however, is from a gardening point of view the most interesting, and we will glance at the mode of decoration pursued therein.

The grass banks that rise from the lower garden to the balustrade—such slopes as may be seen in most places of the kind—are not left naked, but planted with two rows of dwarf rose bushes, and the effect of these is very pretty. There seems no particular reason why like spots should be left naked with us. Continuous borders, not beds, run round the squares of grass, &c., and from the dawn of spring to the end of autumn these are never without occupants—never ragged, never flowerless. The system adopted is one of bedding and herbaceous plants mixed, but all changed every year. They steal out a spring flower this week, and put in a fine herbaceous or bedding plant, or strong growing florists' flower in its stead, and with the very best success. Stocks of good bedding and herbaceous plants are always kept on hand to carry this out, and the placing of the herbaceous plants into fresh ground every year causes them to flower as freely as the bedders.

But these borders also contain permanent things—Lilac bushes, Roses, &c., which give a line of verdure throughout the centre of the border, and prevent it from being quite overdone with flowers.   Among those woody plants there were others very beautiful and very sweet for many weeks

Fig. 32.

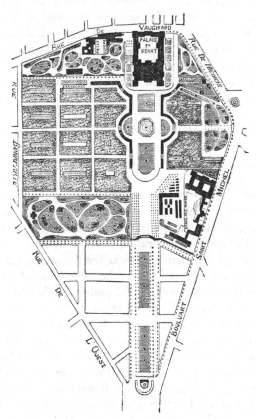

Plan of the Luxembourg Garden as recently altered.

through the better part of the season, and these were low standard bushes of the common Honeysuckle! English flower-gardeners would perhaps scarcely ever think of that for such a position ; but alternating between a Rose and a Lilac, or other bush, and throwing down a head of free-growing and flowering shoots, very few subjects look more pleasing in the flower garden. The mixture of Phloxes, Gladioli, Œnothera speciosa, Fuchsias, Pelargoniums, large yéllow Achillea, &c., to be seen here every summer and autumn, is quite attractive, and much more varied than is now often the case. They also have the subtropical system, and rather more tastefully than elsewhere. Thus in one part may be seen a graceful mixture of a variety of fine-leaved plants with an edging of Fuchsias, instead of the ponderous mass of 500 plants of one variety of Canna, which you sometimes meet with in other places about Paris. M. Rivière is fond of having mixed beds of ferns in the open air, isolated specimens of tree ferns, Woodwardias elevated on moss-covered stands, &c. and their effect is usually very good. The planting of the vases too is good. Instead of using only flat-headed subjects, as many do with us, they place in the centre of each a medium-sized plant of the New Zealand flax, with its long and boldly graceful leaves, and then set geraniums, &c., around, finishing off with the ivy-leaved geranium, the Tropæolum, &c., for drooping over the margin.

The effect of the fountain of Jacques Debrosse and its surroundings is the most satisfactory of the sort I have ever seen. The frontispiece, engraved from a photograph, almost does away with the necessity for a written description of it.

Stretching from the foot of the fountain there is a long water-basin, a walk on each side of that bordered with Plane trees, which meeting overhead make a long leafy arch, so that the effect of the fountain group at the end, representing Polyphemus discovering Acis and Galatea, is very fine. It is of course heightened by the leafy canopy of Planes, but very much more so by the way in which the Ivy and Virginian creeper are made to form graceful wreaths from tree to tree. Between the trees the Irish ivy is planted, and then trained up in rich graceful wreaths, so as to join the stems at about eight feet from the ground. At about a foot or so above the ivy another and almost straight wreath of Virginian creeper is placed, and the effect of these two simple wreaths from tree to tree is

quite refreshing at all times. The wreaths seem to fall
from the pillar-like stems of the Planes rather than to
grow from the space beneath them, the bottom of the
lower wreath resting on the earth. An adoption of this
or a similar plan would add verdure and grace to many a
formal grove, bare and naked-looking about the base.

In these gardens the Oleander is grown into large bushes
like the orange-trees, and put out with them during the
summer months. They become perfect beds of flowers.
I have seen plants or rather trees of those oleanders in
flower here, quite ten feet across, and with the flowers as
thick upon them as on a bed of Pelargoniums. They are
simply treated like the orange-trees, the culture of which
is fully described elsewhere in this book. Doubtless the
plan would succeed in England, and it is worth a trial.
Even indoors the Oleander is not often flowered well with us,
though quite worth the trouble of cultivation. Probably
the complete rest during winter that the plants get in an
orangery, and the making of all their growth out of doors
in the full light and free air, are more conducive to their
well-being than the careful culture they receive in our
glass-houses. On the Continent they are abundantly grown.
M. Rivière fils has obligingly written a short article on their
cultivation for me, which will be met with further on.

On the 5th of July, 1867, the men were busily employed in these gardens moving large chestnut and plane
trees in full leaf. They take them up with immense balls
of earth, by powerful machinery, and very successfully, but
this system should not be pursued more than is barely
necessary in private gardens or public either. It may be
very desirable for Paris to move common trees of goodly
size to complete and rearrange straight avenues here and
there, but the plan is not worth the expense in any other
case.

Numerous amateurs and others go to the Luxembourg to
hear M. Rivière, the superintendent, deliver his free lectures,
which are thoroughly practical, and illustrated by the aid
of living specimens and all the necessary material. The
lecturer goes through the theory and practice of the subject

PLATE XIX.

WOODWARDIAS IN THE CONSERVATORIES OF THE LUXEMBOURG GARDEN.

before an attentive class, consisting of several hundred persons, and elucidates the subject in a way which cannot fail to highly benefit the numerous amateurs who attend. It is interesting to see such a number of people here at nine o'clock in the morning, and the deep interest taken in the matter, speaks much for the excellence of the professor. As botanical professors lead their pupils on occasional excursions over meadow and hill, so M. Rivière takes his classes to famous horticultural establishments from time to time,— to Montreuil, famous for its peaches; Thomery, for its vines, and so on. There are many lectures delivered in England on like subjects, but none so directly useful to the horticulturist as these.

M. Rivière, being an admirer of Woodwardias, pays special attention to their cultivation, and succeeds in growing them to great size in small baskets, balls of moss, &c. The accompanying plate will show how effective they are when thus treated. Some of the specimens are placed in the open garden on rustic stands or in vases during the warmer months, and thus they grace the flower garden in summer as well as the conservatory in winter.

G

CHAPTER VI.

MOST of us are familiar enough with the aspects of the London squares, with their melancholy loneliness, and frequent filthiness—their highest efforts being in the planting of Privet, &c., so cleverly that any view of the interior is impossible. If by way of contrast we glance at the state of one of the most central and best known squares in Paris before entering on the general question, we may be able to get an idea of the different system pursued in each city, and I trust also of the great advantages and superiority of the Parisian one. The square and Tour St. Jacques illustrate judicious city improvements better than anything else that I am acquainted with. This tower—originally part of an old church, and hidden from view by tall, narrow, dirty streets which crowded around it, is now one of the most beautiful and interesting objects in Paris—striking to every one who passes by it, and with the garden a source of much pleasure and benefit to the people who live in this central neighbourhood. It was made so by clearing away narrow old streets and buildings and making a garden.

The first thing that strikes the visitor in this square is

FIG. 33.

Portion of the Plan of margin of a Parisian square.

its freshness, perfect keeping, and the numbers of people who are seated in it, reading, working, or playing. " The same reason," it is said in ' Guesses at Truth,' " which calls for the restoration of our village greens, calls no less imperatively in London for the throwing open of the gardens in all the squares. What bright refreshing spots would these be in the midst of our huge brick and stone

PLATE XX.

VIEW OF THE SQUARE AND TOUR ST. JACQUES FROM THE RUE DE RIVOLI.

labyrinths, if we saw them crowded on summer evenings with the tradespeople and mechanics from the neighbouring streets, and if the poor children who now grow up amid the filth and impurities of the alleys and courts, were allowed to run about these playgrounds, so much healthier both for the body and the mind ! We have them all ready, a word may open them. At present the gardens in our squares are painful mementoes of aristocratic exclusiveness. They who need them the least monopolize them. All the fences and walls by which this exclusiveness bars itself out from the sympathies of common humanity must be cast down." The aspect of this square with its wide walks lined with chairs, on which hundreds of people sit and enjoy the scene at all hours, but particularly in the evenings, would have well realized this writer's ideal of what a square should be. Nor have the richest potentates more beautiful or diverse objects in their gardens than are here spread out for all who will enjoy them. It is almost as attractive to the passer-by in the street as to those inside, for instead of a clump of shrubs of commonplace character, cutting it off from the view of the passer-by, there is a belt of grass of varying width, kept perfectly fresh and green, and on it here and there large beds and masses, usually distinct from each other. Now it is a fine bed of the dwarf fan-palm, edged with Carludovica, as much exposed to the street as to the square ; now a group of shade-giving hardy trees, furnished beneath with neat evergreens, and finished off with a line or two of flowers, next, a mixed bed of variegated Dahlias and other tall autumn flowers, and so on. On the carpets of fresh grass between these various clumps there are here and there isolated trees—chestnuts, and the like, to give the necessary shade and dignity, and to flower in their season. In nearly every case the stems of these are neatly clothed with climbers, generally ivy, occasionally Aristolochia and Clematis. Very pretty effects may be worked out by using the best climbers. But the grassy carpet is also ornamented by smaller, though no less noble, things than the large trees just mentioned. It is sparsely dotted with plants having fine leaves, or distinct character. On one sweep we have a tree Pæony, the tall Japanese Polygonum, and a large-leaved Solanum. Passing two clumps of shrubs, and between them an entrance, we meet with another strip of green grass, adorned with four distinct plants—the Pampas grass, the Irish Yew, Melianthus major, Hibiscus roseus, and so on. It should be distinctly understood that these plants stand singly and isolated on the grass, so that their character may be seen. In the mixed clumps and plantations near there are plenty of opportunities of seeing the effects of things when grouped or massed.

Between the walk and the beautiful old tower there is a little lawn, and in one nook of that deep green carpet, sheltered on three sides, but coming boldly into the view from the greater part of the square, is a specimen of the noblest of fine-leaved plants, the great Abyssinian Musa. It is about twelve feet high : the base appears quite two feet in diameter, the young leaves made during the season are perfectly intact, eight feet long each, a great red tapering midrib, like a huge billiard cue, running from base to point of each, and from this supply-pipe the gracefully waved venation curls away towards the margin. Backed by the foliage of the trees of our own latitudes, it forms a striking and noble object indeed. Then, in the immediate foreground, there is a mass of a scarcely less striking plant, the edible Caladium, which springs from a groundwork of fragrant mignonette, edged with the woolly Gnaphalium ; and so in like manner are sparsely scattered over the green (they wisely keep the central parts clear to secure a little breadth and repose), striking specimens or groups of specimens, some of which it would pay the city to grow, if it were only to give art students living specimens of Nature's finest leaf forms. I know some botanic gardens ten times the size of this little square, which fail to furnish anything like so good an illustration of the diversity and beauty of the vegetable kingdom, and others where huge, tasteless and formal arrangements prevent an equally agreeable impression from being obtained.

Amidst the whole stands the famous old tower, with its leaves and figures in stone, a thing of beauty and interest

PLATE XXI.

THE SQUARE AND TOUR ST. JACQUES.

of itself, but greatly enhanced by being set so sweetly in a
green and brilliant garden. At every step the tower presents a fresh face, and the square a new charm. People
who sneer at what they call Haussmannization would do well
to ponder on such facts as this : a little reflection might lead
them to discover numerous objects more worthy of satire.
About this Tour St. Jacques were tried for the first time
the Wigandias, now the admiration of so many in both
French and English gardens, the Cannas, the Musas, Palms,
Ficuses, and others of the better kinds of what may be
termed the flora of Parisian gardens. What a change from
the filth and consequent unwholesomeness of its ancient
state! How different from the small squares around our
churches and monuments with their naked slimy earth and
doleful aspect! Surely they might as well bloom with
verdure and life as be so suggestive of all that is opposite!
A visit to the Tour St. Jacques and its surroundings,
especially if accompanied by some idea of what the spot was
before the improvement was carried out, could not fail to
leave a deep impression of the great advantages to be
derived from the execution of similar improvements in our
cities. The old tower belonged to the ancient church of
St. Jacques, which was built in 1508. It is 175 feet high,
and affords a fine view of the greater part of the capital.
It was this tower that was used by Pascal in his experiments
on the variation of the barometer at different heights. The
works belonging to the garden were executed in 1856, the
total cost being nearly 6000*l.* for the alterations and planting.

Although so far in advance of our own squares in
every way, it is interesting to note that the idea was first
taken from London ; but while we still persist in keeping
the squares for a few privileged persons, and usually without
the faintest trace of any but the very poorest plant ornament, they make them as open as our parks, and decorate
them with a variety and richness of vegetation with which
it is only fair to say the choicest spots in our own great
gardens, public or private, cannot be compared. The whole
subject is treated of in such a judicious way by M. R.
Mitchell in the "Constitutionnel" that his remarks may be

appropriately quoted here, dealing as they do fairly with both sides of the question.

" It has been often remarked, and with great reason, that the English have carried their material civilization further than we have. Comparisons have frequently been made between Paris and London that were not at all to our advantage, and we are obliged to allow that the sort of accusation brought against us was not wanting in justice. It is not many years since the boundaries of Paris inclosed an old city that was a disgrace to our civilization; streets, or rather fissures, without ventilation, and unhealthy districts where an entire population of poor people were languishing and dying. Now, however—thanks to the useful and important works that have been lately carried out—the sun shines everywhere; streets have been enlarged, and every one has sufficient air to breathe. Paris contains but few unhealthy alleys, whilst in London the existence of such localities as Bermondsey, Soho, St. Giles's, Spitalfields, Whitechapel, &c., &c., is still to be deplored.

" We are far from forgetting the immense development of material civilization in England. We simply mean to say that our neighbours frequently invent for the sake of privilege, and that when their ideas are good we take advantage of them and popularize them. We will take a single example: every one knows how justly the English pride themselves on their gardens called squares, which are the admiration of every foreigner. Our unfortunate public places that the pedestrian cannot cross in summer without being grilled by the sun or blinded by the dust only serve as examples of our inferiority in this respect. The square, that is to say, a little park surrounded by a railing, is the representation at once of a question of health—a question of morality, and perhaps even of national self-respect. We certainly could boast of the Place Royale, which, however, much more closely resembled an unsuccessful attempt than the first step in a happy way. At present, however, Paris need envy London for nothing. The Emperor, who understands that for an idea to be adopted in France it is not indispensable that it should be French, was struck with the happy results that

PLATE XXII.

THE SQUARE AND FOUNTAIN DES INNOCENTS.

would accrue from the naturalization of the square amongst us. He understood the necessity of a place of refuge, rest, and freshness for those who have never carried their desires even so far as the Passy omnibus, or even the railway to the Bois de Boulogne. He has consequently bestowed on our capital the squares of St. Jacques la Bouchérie, St. Clothilde, the Temple, Louvois, des Arts et Métiers, and the Parc Monceaux. These masses of vegetation widely distributed amongst the most populous neighbourhoods cleanse the air by absorbing the miasmatic exhalations, thus enabling every one to breathe freely.

" The time has passed when a plate of copper exposed to the air in one of the streets now demolished, would become covered with oxide in a single night. This is a question of public health that it is most important to bring forward. Before the establishment of the Paris squares the existence of a great number of children was passed in confined and unwholesome districts. The fresh air for them was only the threshold of that vitiated atmosphere that we have just been speaking of. They were obliged to take a long walk before they could find a patch of verdure or a bit of country. The children went out but little ; it was useless to dress them or make them clean, because they never went out of their own neighbourhood, and in this way their early years passed away. How many times have we not noticed with painful emotion these little, ragged, pale creatures, who never apparently thought of the filth in which they were obliged to live !

" Now, thank God, this dark picture has become bright. Within a couple of steps of the poor man's house there are trees, flowers, and gravel-walks where his children can run about, and clean and comfortable seats where their parents may sit together and talk. Family ties are strengthened, and the workman soon understands that there are calmer and more moral pleasures than those he has been used to seek in the wine-shop. Again, the different degrees of the members of the working classes meet together on common ground, and parental feeling is developed by emulation. A child must not be allowed to be ragged for fear of its being remarked, and we will answer for it that a woman in whose breast maternal instinct has not been entirely smothered will never take her child into a public place without first paying attention to the cleanliness which is the ornament of the poor. Some time ago, while walking through the Square du Temple, where hundreds of children were running and jumping and filling their lungs with the country air that has thus been brought into Paris, we could not help saying to ourselves that strengthened and developed by continual exercise these youngsters would one day form a true race of men, which would give the State excellent soldiers, good labourers for our farms, and strong artisans for our factories.

" It has already been stated that the English originate privileges and that we popularize and perfect their ideas. We shall prove what we advance by comparison. The Parisian Ediles have made squares wherever a too crowded population threatened to contaminate the atmosphere, and in all the parts of the city, farthest from the Tuileries, the Luxembourg, or the Bois de Boulogne, so that those living in the neighbourhood might be able to get to them easily. In London, on the contrary, with but few exceptions, there are no squares worthy of the name, except in rich and open neighbourhoods. The largest and most beautiful gardens are found at the West-end in Belgravia, or at Brompton, that is to say, at the very gates of Hyde Park. With us trees are planted for sanitary reasons, and the squares have been established, more especially in those neighbourhoods where the atmosphere most required to be constantly purified, and to this end trees of a particular sort were chosen for their power of absorption. Fountains too were built, and small pieces of water, which spread that pleasant freshness through the air that is so grateful to the workman who has passed the whole day in the heavy atmosphere of the workshop.

" In London they appear to have been above everything anxious about the health of the trees; a healthy and warm climate was chosen for them in open neighbourhoods close to the parks, so that they should not suffer too much from

PLATE XXV.

THE GARDENS OF THE PALAIS ROYAL.

home sickness.   We do not mean to say that the city, for
instance, or the other parts of the town, are completely un-
provided with squares, but simply that they are so small
and mean that they give one the idea of having been
blown into their position by the wind.   But the headquarters of misery that we spoke of a short time ago—
those masses of crumbling houses—those networks of dark
alleys,—in a word, all that most needs pure air and daylight
has been forgotten, or rather neglected while the richer
parts have been improved.   In Paris the squares are open
to every one; in England they are locked up, surrounded
by a railing surmounted with spikes, and planted with
bushes so as to impede the view of all that is going on
inside.   By the payment of a small sum, generally a pound
a year, each inhabitant of the houses forming the four
sides of the square has the right to a key of the gate.   So
that for a poor man to walk with his family in any of these
gardens, he must first live in a square and pay a high rent
for the privilege, and then contribute a pound a year towards the expense of maintaining it.   Practically these
squares are useless, and nearly always deserted.   In
London the squares are private property with which
the State cannot meddle.   With us, on the contrary,
it is the Government that takes the initiative in these
municipal improvements.   It is to the city of Paris
that we owe their construction; they have cost a great
deal, and the Imperial idea has only as yet been partially
carried out.   We have already transformed the Bois de
Boulogne, the Bois de Vincennes, and we shall soon have
many more public promenades in different parts of the capital.   Before long Paris will be one vast garden.
   " It is only necessary to walk in the neighbourhood of any
of the squares of Paris towards the middle of the day to see
with what pleasing readiness they are patronized by the
working classes.   To give only an example, the Square des
Arts et Métiers is so crowded with people after four o'clock
that it is impossible to pass through it.   It was at one time
said that the establishment of a public garden was an idea
that was perfectly practical in London, but not in Paris,

where the inhabitants were so turbulent and revolutionary
that they would soon pull down the trees, pluck the flowers,
and pull up the plants by the roots. Experience, however,
has shown how utterly this opinion was devoid of foundation. At the inauguration of the Parc de Monceaux all the
gates were thrown open to the crowd. No surveillance was
exercised over the 50,000 persons who crowded the walks.
At the end of the day the total amount of damage done
only amounted to some forty-five francs for a few turf
borders that had been trampled upon. This fact is perfectly conclusive. Besides, the squares have now been
opened for a long time, and the numberless frequenters
of them have conducted themselves with admirable order
and decency. The people evidently understand that they
are at home; that it is for their especial behoof that the
gardens have been constructed; they know that in pulling
up a flower it is their own property they are destroying;
and, moreover, they evince a respectful gratitude for the
hands that have given them these pleasant places of resort.
The establishment of public squares in Paris is an eminently
social idea. We repeat it, it tends to regenerate the human
race by the development of the physical forces; by exercise
in the open air it improves the morals of the people, by
allowing the working man to change the dirty wine-shop
by a pleasant walk and an agreeable resting-place; and,
lastly, it proves our readiness to adopt in our own country
whatever appears good and useful to our neighbours."

It is to be hoped that we in our turn shall show an equal
readiness to profit by the excellent example shown us in city
squares. There are many private squares in London which
merely occupy space that otherwise would be devoted to the
gardens of the houses around; but, on the other hand,
there are not a few which seem to invite a trial of the
system found to work so well in Paris. I have very little
doubt that if we could set one of these sweet little Parisian
squares down in the centre of London, it would induce
many who would now oppose with all their might any
attempt to open their square to the public to ask for the
change. And eventually it would come to this, that even

persons having a claim over the smallest squares in London—those that have been substituted for the little private gardens—would see that it was to their interest and for the benefit of everybody living near the square that it should be cheerfully decorated, well kept, open to the public at all reasonable hours, and a place where a working man, too tired to walk to a distant park, could sit down to rest without the necessity of resorting to the public-house or any like place.

The Square St. Jacques, already alluded to, is so placed that every visitor to Paris must see it. The next to be noticed is rather out of the usual route of the English visitor.

The *Square des Batignolles* is one of the largest and best worth seeing in Paris. Entering it from its lower side, the general scheme is seen to be that of a little vale, down which meanders a streamlet, ending in a small round piece of water. The margins of this streamlet are variously embellished with suitable plants : the rich grassy sides slope up till they end in dense plantations of the choicest shrubs, so well planted and watered that they look as fresh as if growing twenty miles from a large city. Let us walk round—the margin of the shallow grassy vale to our right, the boundary shrubberies and the railing to our left. The walk expands from a breadth of ten or a dozen feet to forty, in the first corner of the square, so that the children find little playgrounds without going on the vividly green grass. The first attraction to the eye on the right is a group of the variegated maize springing out of a mass of dwarf Phlox Drummondi. Beyond it is a group of

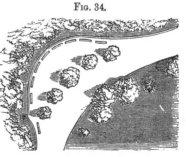

Fig. 34.

Portion of plan of Parisian square, showing the widening of the walk to form a playground, with seats and shade-giving trees.

Plane trees, Honeysuckles being trained up their stems by the aid of rings of galvanized wire.

Next on the right again comes a magnificent group of Caladium esculentum, springing out of Lobelia Paxtoni; behind it a dense mass of the Pampas Grass, in front of groups of Poplars and Cedars.   On the left a profuse variety of the very best shrubs, flowering and otherwise; all these groups of shrubs being edged with some kind of summer flower.   Indeed it is these margins that afford the floral display; and the absence of all attempt to make a species of extensive coloured cotton handkerchief of the place makes it almost as fresh and free from vulgarity and gaudiness as a ferny dell in a forest.   The keeping is perfect, and there is no fence between the public and the flowers but the very neat edging of rustic iron, which rises about five inches above the gravel, and is placed about two inches outside the grass.   The only bed without any green relieving it in the whole place was one of Centaurea ragusina, planted thinly and springing out of a ground-work of variously coloured and brilliant Portulacas.   Again we come to another angle of the ground, and the walk once more widens to forty feet, with lots of seats in its back portion.   Behind all, to the left, is the well diversified dense shrubbery; to the right Cedars and Thujopsis on the grass, of the freshness, softness, and verdure of which latter I can give no adequate idea.   Here and there, isolated on the turf, was a single plant of the red-stained variety of the common Castor-oil plant, of which the fruit, leaves, and stems were all effective, the former strikingly so.   The Bananas planted out here are in a poor state, except Musa Ensete, which is, as usual, superb.   At another corner there is again a widening of the walk to forty feet.   A few Chestnuts are planted on these wide spots for shade; on the right there is a bank of choice shrubs and low trees, margined with a belt of scarlet Pelargoniums—the only ones on the spot; but as it probably took more than 800 plants to form this belt, I do not think anybody could complain of the scarcity of them.

We will next pass up the walk by the streamlet that runs through the centre of the grass.   This is tastefully margined with tufts of water plants; but a novel and

praiseworthy feature is added. At some distance from
the margin—from four to ten feet—are planted here and
there single specimens of plants which, while not of
the water or the marsh, assimilate more or less in character
with the plants of those places—hardy Bamboos, Yuccas,

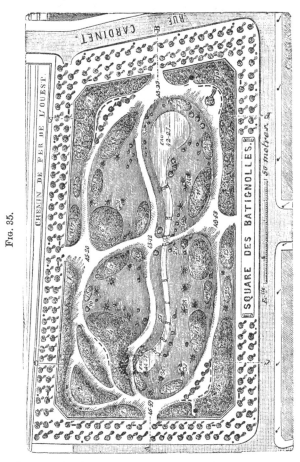

Fig. 35.

Erianthus, and other large grasses, some truly fine Acanthus
latifolius, the Pampas Grass, Tamarix, Funkia grandiflora,
&c. Finally, we arrive at a mass of ivy and creeper clad
rockwork, from which issues the source of the rivulet: this
rockwork has its rear hidden amongst trees.

The Square des Batignolles, constructed on the once
open space in front of the church belonging to the commune, is the largest of all the squares belonging to new
Paris; it contains over three acres of ground, without
taking into consideration wide promenades planted with
trees outside. It cost no less than 60,000*l.* The works were
commenced in 1862, and were finished the following year.

In a work of this kind minute details, especially of
commonplace subjects, are very rarely desirable, and for
this reason I avoid as much as possible describing the
contents of the squares and gardens in full. Nevertheless,
some may wish to know about the details of the planting,
and in the case of this square it is given. It will be noted
that each group of Section 1 is divided in three—the first
being trees; the second the shrubs that adorn the outer
sides of the plantation; the third the decorative plants
of the margin. The numbers answer to those of the plan
on page 93.

Sect. 1.—*Groups of trees, shrubs, and flowers.*—1. Æsculus rubicunda,
Æsculus hippocastanum, Tilia europæa, Padus virginiana.—Ligustrum ovalifolium, Berberis vulgaris, Ribes sanguineum, Virgilia rosea, Lonicera tartarica —
Phlox decussata, Coleus Verschaffeltii.—2. Paulownia imperialis, Catalpa
syringæfolia, Platanus occidentalis, Negundo fraxinifolium.—Forsythia viridissima, Ribes (in var.), Spirea (in var.), Sambucus nigra, Symphoricarpus
(in var.)—Pelargonium zonale inquinans, var. Prince Imperial.—3. Æsculus
hippocastanum, Sorbus aucuparia, Cytisus laburnum, Acer platanoides, Alnus
communis.—Ligustrum ovalifolium, Ligustrum spicatum, Cydonia japonica,
Buxus sempervirens angustifolius, Prunus lauro-Cerasus.—Chrysanthemum pin-
natifidum.—4. Alnus communis, Kœlreuteria paniculata, Padus virginiana,
Paulownia imperialis.—Ligustrum spicatum, Ligustrum ovalifolium, Cytisus
sessilifolius, Mahonia Aquifolium, Berberis vulgaris.—Pelargonium zonale inquinans, var. Christinus.—5. Juglans nigra, Sorbus aucuparia, Tilia europæa,
Acer platanoides, Platanus orientalis, Robinia viscosa.—Lonicera tartarica, Sam-
bucus racemosa, Mahonia Aquifolium, Euonymus japonicus, Deutzia scabra,
Kerria japonica, Weigelia rosea.—Phlox decussata.— 6. Robinia Pseud-Acacia,
Acer striatum, Cytisus laburnum, Catalpa syringæfolia, Eleagnus angustifolius.—
Hibiscus syriacus, Philadelphus coronarius, Ligustrum ovalifolium, Ligustrum
spicatum, Viburnum Lantana, Tamarix indica, Chionanthus virginica.—
Ageratum cœlestinum.—7. Catalpa syringæfolia, Alnus glandulosus, Cytisus
Laburnum, Sophora japonica, Juglans nigra, Robinia Pseud-Acacia.—Berberis
vulgaris, Viburnum Opulus, Ribes sanguineum, Euonymus japonicus, Philadelphus
inodorum, Deutzia scabra.—Veronica var. Gloire de Lyon.—8. Tilia argentea
Acer striatum, Æsculus hippocastanum, Sophora japonica, Robinia Pseud-Acacia,
Fraxinus excelsior var. aurea.—Ribes sanguineum, Forsythia viridissima, Malus
spectabilis, Prunus japonica, Cytisus sessilifolius, Kerria japonica, Deutzia scabra.
—Achyranthes Verschaffeltii.—9. Alnus fulva, Æsculus hippocastanum, Sophora
japonica, Tilia europæa, Cytisus laburnum, Sorbus aucuparia, Acer platanoides.—
Mahonia Aquifolium, Deutzia scabra, Forsythia viridissima, Philadelphus grandi-
florus, Kerria japonica, Sambucus laciniata, Chionanthus virginica.—Pelargonium
zonale inquinans, var. Eugénie Mézard.—10. Sorbus aucuparia, Acer plata-

noïdes, Juglans nigra, Paulownia imperialis, Alnus glandulosus, Catalpa syringæ-folia.—Euonymus japonicus, Forsythia viridissima, Philadelphus coronaria, Mahonia Aquifolium, Cornus alba, Robinia hispida.—Gazania splendens, Phlox decussata.—11. Negundo fraxinifolium, Populus fastigiata, Juglans nigra, Catalpa syringæfolia, Cytisus laburnum, Sorbus aucuparia.—Symphoricarpus alba, Forsythia viridissima, Ribes sanguineum, Euonymus japonicus, Deutzia scabra, Syringa (in var.).—Chrysanthemum frutescens.—12. Paulownia imperialis, Negundo fraxinifolium, Tilia europæa, Æsculus hippocastanum, Æsculus rubicunda, Catalpa syringæfolia, Acer striatum.—Ribes Gordonii, Weigelia rosea, Mahonia Aquifolium, Syringa inodorum, Kerria japonica, Hibiscus syriacus.—Phlox (in var.), Ptarmica flore pleno, Calceolaria rugosa.—13. Æsculus hippocastanum, Æsculus rubicunda, Robinia viscosa, Paulownia imperialis, Acer platanoides.—Berberis foliis purpureis, Deutzia scabra, Forsythia viridissima, Rhus Cotinus, Prunus lauro-Cerasus, Euonymus japonicus.—Phlox decussata, Lantana var. Queen Victoria.—14. Sophora japonica, Juglans regia, Acer rubrum, Ailantus glandulosus, Cytisus laburnum, Robinia viscosa.—Bupleurum fruticosum, Prunus lauro-Cerasus, Euonymus japonicus, Spirea (in var.), Hibiscus syriacus, Tamarix indica, Rhus Cotinus, Viburnum Opulus.—Phlox decussata, Coleus Verschaffeltii.—15. Acer platanoides, Paulownia imperialis, Cytisus laburnum, Sorbus aucuparia, Robinia Pseud-Acacia, Acer pseudo-Platanus.—Ligustrum ovalifolium, Prunus colchica, Sambucus racemosa, Berberis vulgaris, Rhus glabra, Kerria japonica, Ribes aureum.—Chrysanthemum frutescens.—16. Paulownia imperialis, Acer striatum, Catalpa syringæfolia, Tilia argentea, Sophora japonica, Æsculus hippocastanum.—Amorpha fruticosa, Ligustrum spicatum, Euonymus japonicus, Sambucus nigra, Prunus Mahaleb, Kerria japonica, Cornus alba.—Fuchsia (in var.).

Section 2.—Beds for foliage plants and flowers.—17. Pelargonium zonale inquinans.—18. Hibiscus rosa sinensis, Nierembergia frutescens.—19. Senecio platanifolia, Centaurea candidissima.—20. Heliotropium var. Anna Thurel, Koniga maritima var. foliis variegatis.—21. Colocasia bataviense.—Calceolaria rugosa, Gazania splendens. — 22. Ficus Cooperii, Cuphea platycentra. — 23. Colocasia esculenta, Koniga maritima.—24. Campanula pyramidalis, var. cærulea et alba.—25. Musa paradisiaca, Lobelia erinus.—26. Plumbago scandens, Dianthus var. Seneclauzii.

Section 3.—Isolated trees and plants.—27. Bambusa aurea.—28. Pinus uncinata.—29. Araucaria imbricata.—30. Salisburia adiantifolia.—31. Pinus excelsa. — 32. Thujopsis borealis.—33. Cupressus funebris.—34. Cedrus deodora.—35. Thuja occidentalis Warreana.—36 Abies Pinsapo.—37. Thujopsis borealis.

The Square de Montrouge.—Although our island is in good repute for its natural verdure, I feel pretty sure that there are few Britons who would not be persuaded of the necessity of more efficient watering in our public gardens if they had seen this square during the last days of the month of August of the past year. To say it was green would be to give the faintest idea of the glistening, deep, and refreshing verdure displayed by everything in it, from the trees to the grass. It is a very small place, not so big as Leicester-square, but quite a gem in its way. It is simply laid out with belts of low trees and shrubs; the centre of the little lawn left unadorned, while all around its edges really distinct and good things are dotted about. The Acanthuses were very fine here in consequence of the constant and thorough waterings. Previous to visiting this garden, I had no idea that they would under any treatment look so well at the end of a hot season. A plant of A. latifolius here was six feet in diameter, not three feet high, and of the deepest and freshest green. I can compare it to nothing but the young pushing leaves of a healthy Camellia in point of glistening verdure. Bocconia frutescens I never saw in such prime condition as here—the leaves were three feet long and fifteen inches wide, the plant four and a half feet high. Perhaps the handsomest grass I have ever seen, as regards its foliation, was here also. I do not except the Pampas. It was a species of Cinna, which had just shown flower for the first time; but it was the grace and position of the leaves that were the most conspicuous. The central shoots gave off a lot of leaves near the base, as grasses usually do, and continued ascending till seven or eight feet high, giving off arching leaves all the way to the summit. The falling spray of a fountain is not more graceful than were these leaves. The effect of such things isolated on the grass is by no means sufficiently appreciated by us. A handsome specimen of Bambusa aurea, planted alone on the grass, helps to show what may be expected of these tall, shrub-like grasses in the time to come; I believe they will impart to our gardens an entirely new aspect, and that of the most desirable sort. The one we suppose to be the hardiest of all is tenderer than several other species grown in Parisian gardens, and which are enumerated elsewhere in this book. Cyrtanthera carnea is used in this and other Parisian squares as a tall edging plant, and is effective when so employed.

On one of the grass plots here is a group in bronze. Though in an out-of-the-way part of the town, the keeping is quite as good and the plants quite as choice as in the most fashionable parts.

The Square du Temple.—This, although a pretty square, has scarcely the finish of those previously noticed, but it is a great advance on anything we possess in the same way, and, as usual, was, on a very hot day in the beginning of last September, as fresh as if it had not endured a scorching

summer. A great advantage of the system of watering employed is that walks and every surface may be washed and
saturated with ease; for on hot days it is desirable that the
whole garden be moist and cool. A very splendid effect was
afforded here by a great mass of Caladium esculentum,
planted in a groundwork of the deep crimson Amaranthus
tricolor, the whole edged with a wide band of silvery
Gnaphalium. There is also a small pond with water plants,

FIG. 36.

The Square du Temple.

a piece of rockwork, and two fine examples of the weeping-
willow—always among the best ornaments of a garden.
The larger specimen is said to be four centuries old. Small
ponds in city squares, however, are in very doubtful taste, as
usually arranged. In a town possessing an abundant
supply of water it is possible to make some grand features
with it occasionally; but the number of small ponds should
not be increased. They are usually dirty-surfaced, and
besides seem out of place in a square from which the
buildings around are not hidden. This square was formed

H

in 1857 on the site of the old palace of the same name. It has a surface of about 8000 square yards, and cost 6000*l.*

The Square des Arts et Métiers is like not a few others of the minor ones, more of a playground than a garden, gravel and trees being the main features. A low balustraded wall encloses it, and at intervals vases for aloes and like plants are placed upon this. The appearance of this enclosure may suggest that a kind of fence different from what we

Children at Play in the Square des Arts et Métiers.

usually think necessary in London, might be at once more elegant, and certainly not more expensive. It would not do in the case of the large enclosures, but should we open small squares to the public there is no reason why it might not be tried. Few of us could have believed that Bay trees in tubs would remain intact in that playground for the London Arab—Trafalgar-square—as they did during the past year. In the centre of the Square des Arts et Métiers there is a small but elegant Crimean monument, and there are two oblong fountain basins with statues. Around the whole runs a narrow flower border, backed by a few shrubs. It is like all Parisian squares, full of people, both during the day and till late in the evening. The square occupies a space of about 5000 square yards, and cost 12,800*l*.

Place Royale.—This out-of-the-way square is chiefly a playground for the infants, and a lounging and chatting place for the children and old men. There is an equestrian statue of Louis XIII. in the centre, and around it a group of horse-chestnut trees. Underneath, and indeed

Fɪɢ. 38.

The Place Royale.

nearly the whole square, is a gravelled surface, except a slight belt of flowers which encircles the fountains that are placed in each corner. Between the bed of flowers and the fountain basin there is a belt of grass, a mere strip four feet wide, and on this were planted at intervals single specimens of the dark-leaved Canna. They of course backed up the bedding-flowers, and came between the eye and the fountain basin. This place was opened in the reign of Henry IV. A couple of centuries ago it was the fashionable quarter of Paris—now comparatively few but those who live in the neighbourhood know of its existence. Richelieu, Marion

ʜ 2

Delorme, and Victor Hugo amongst other noted persons have inhabited the old houses around this old square.

The garden of the Palais Royal, probably the best known spot in Paris to the English, must not be confounded with the Place Royale. It existed long before many of the squares herein mentioned were thought of, is inferior to most of them in beauty, and possesses no noticeable features except it be ugly lines of clipped trees.

The Square des Innocents.—This square which was formed in 1859 and 1860 surrounds the celebrated Fontaine des Nymphes, which was built in 1550 by Pierre Lescot, and decorated by Jean Goujon. In 1860 it was completely restored. The square measures internally 6800 square yards. The expenses attending the construction of this square amounted to the sum of 8000*l.*, of which more than three quarters was spent in architectural embellishments. Like all the new squares of Paris, it is tastefully embellished with trees, shrubs, and flowers, and forms a fresh and pleasant resort for the inhabitants of the busy neighbourhood in which it is placed.

Square de la Chapelle Expiatoire de Louis XVI.—This square, which is situated between the new Boulevard Haussmann and the Rues d'Anjou, Pasquier, and Neuve des Mathurins, surrounds the chapel that was erected in the year 1825 on the site of the cemetery containing the remains of the unfortunate Louis XVI. and Marie Antoinette. The total area is about 7500 square yards, of which about 5000 are open to the public. The square cost altogether nearly 7500*l.*, and was completed in the year 1865.

Square de Belleville.—In the centre of the old *place* where the fêtes of Belleville were formerly held, which was planted with lime trees trained in the form of an arbour, there existed an open space a hundred yards long by sixty yards broad. This piece of ground has been transformed into a square by excavating the earth and planting it with flowers and shrubs, thus creating a very pretty public garden in a not by any means charming neighbourhood. The works were executed in the year 1861, at an expense of 800*l.* The

making of squares on this very inexpensive scale is much
more desirable than turning them into costly gardens.

The Square Montholon.—The Square Montholon is situated
on the Carrefour de la Rue Montholon in the Rue Lafayette,
and was constructed in 1863. It is composed of a central
grassplot, sunk below the level of the garden, and is ornamented at the further end with a rock, from a fissure in
which a stream of water is constantly falling into a little
basin. The size of the square is about 5000 square yards,
and it cost the sum of 7500*l.* The introduction of the
miniature lake is here, and in all squares of like extent, a
mistake.

Fig. 39.

Square and Fountain Louvois.

The Square Louvois is formed on the site of the old
Théâtre de l'Opéra, which stood there until 1820. After
the assassination of the Duc de Berri, which took place in
the February of that year, the theatre was pulled down,
and a Chapelle Expiatoire was built on the spot. The
building, however, was hardly completed when the Revolu-

tion of 1830 burst forth.   The chapel was pulled down and
the ground was turned into a public square and planted with
trees.   Later on a beautiful fountain, from the designs of
Visconti the architect, was built in the middle of the spot.
The square consists principally of a grassplot which surrounds the fountain, and of two rows of the old trees, and
a few simple ornaments, but notwithstanding the effect is
very pretty.   The Irish Ivy is used here in a peculiar way:
—trained so as to form low pyramids, Phloxes being planted
between.   Usually, it is embellished by a few ornamental
exotics in summer, and is at all times a graceful spot.

Fig. 40.

View in the Garden of the Palais des Thermes.

The square or garden around the Hôtel Cluny and Palais
des Thermes is quite distinct from all its fellows, and rightly
so.   Inclosing ruins and a museum of antiquities, the character of both has been imparted to it by arranging some of
the rougher and more permanent objects in it; and being
green and shady, the effect of the whole is quiet and charming, though situated alongside the busy boulevard St. Michel.
As in most cities there are old ruins and buildings bearing
some resemblance to those in this garden, it may not be

PLATE XXIII.

VIEW OF THE OLD PALAIS DES THERMES FROM THE GARDEN.

amiss to say that they are always greatly enhanced by being surrounded with the simplest kind of garden. Ivy, grass, and a few hardy trees and shrubs are sufficient to change their aspect from grimness, hardness, and decay to living interest. A few shillings' worth of the seeds of alpine plants shaken in the tufts of moss or cracks of mortar would give rise to a dwarf vegetation interesting in itself, but doubly so from so markedly illustrating the ceaseless spring of life even in the most unlikely places. For the additional embellishment of gardens round old buildings, abbeys, &c., there are usually *disjecta membra,* not of importance enough to be preserved indoors, in sufficient abundance, and if arranged somewhat as they are here, the result will prove satisfactory. The grounds of the museum at York afford an admirable example of good taste in this kind of garden.

The Square Vintimille.—This square, situated in the centre of the *place* of that name, is but of small extent, its area being only 650 square yards. The cost of construction and restoration amounted to less than 600*l.* It shows that the smallest spots in dusty cities may readily be converted into oases of verdure and sweetness.

It would be useless to enumerate all the small squares and places that, like the last, are little more than mere specks in the city. Enough has been said to show that there is life and merit in the Parisian system of keeping squares. It may not be perfect, and fault may easily be found with the best of them, but considering how short a time the municipality has had such works in hand, nobody can doubt that they are a credit to it, and well worthy of imitation by other people interested in the improvement of cities. Some may say, Look at the expense—it must be given up some day. Not so; intelligent Britons and others of means have such a keen appreciation of a well-ordered city, and go to Paris in such numbers to enjoy it, that they pay a very considerable portion of the expense. I trust, however, that the day is coming when all hindrances to making London a clean, airy, and noble city will be cleared away, and when it will be made habitable for those who must live in it at all times, as well as suited to the wants of men

of business who can live out of town, or men of pleasure
who can leave it at will.

As to the order kept in these squares, nothing can be
more perfect. Being as a rule small and compact, the eye
of the guardian is a thorough protection, if protection were
required ; but the people seem to require no looking after.
Of course there are many who will say that these open
and sweetly embellished squares would not be possible in
London—which is precisely what the Parisians used to say
before squares were tried there. At the hours fixed the
guardians of the squares are " instructed to politely invite the
promenaders to retire, and the public ought immediately to
conform to this invitation." The gates of the squares,
gardens, &c., enclosed by railings, are opened to the public
from the 1st May to 1st October from six in the morning
to ten in the evening, and from seven in the morning to
eight in the evening at all other seasons. It is, however,
added, that in case of great heat, of snow, or of bad weather,
or when the wants of the department may require it, the
hours above indicated may be altered.

Church Gardens and Cemeteries.

There is no place in which a fresh little garden can
be made in better taste than round a city church; and
in Paris, where the difficulty would be to find an open
spot that is not planted, it is not likely that the spaces
around churches are neglected.

There are several instances of very pretty little gardens
being associated with churches in Paris, and they are so
successful that doubtless the system will be extended. The
best known is that in front of the new parish church of the
Trinité, a large and attractive building. An oval space,
three times as wide as the church, is enclosed in front of it.
The gradually ascending carriage-way is cut off from the
garden by a white stone balustrade, as shown in the plate.
From the garden to the church ascent is gained by two
flights of steps, and between these steps three curvilinear
cascades fall from three groups of statues, the waters unit-

ing in one semicircular basin.   The place is very tastefully
disposed and embellished.   The effect of this fine new
church, with its sweet little garden in front, is something
quite sparkling even for Paris ; and the place is, like the
squares, freely open to the public and much frequented.

Another exceedingly pretty garden intimately associated
with a church is the Square St. Clothilde.   The view,
engraved from a small photograph picked up by chance

Fig. 41.

The Square and Church of St. Clothilde.

in Paris, will show at a glance how much the beauty of
like buildings may be enhanced by a little judicious gardening.   It is only justice to state that the tasteful planting, the neatness, and, above all, the refreshing verdure
which is sustained everywhere by profuse waterings, make
the place leave quite a different impression to anything of
the kind seen in the British Isles.

It is an unfortunate fact that when we do attempt
any sort of garden round our churches it is usually
of the poorest character.   The gardens just alluded
to have not been made on the site of cemeteries, and

therefore the designer was free to do as he liked with the ground.   In this country there are numbers of city grave-yards which, now disused, ought sooner or later to be turned into gardens, but gardens of a peculiar kind.

In some places they have commenced rooting up the graveyards, not merely where the tunnelling power of a railway company is brought to bear, but in places untouched in this way, and where the thing is done for mere love of " improvement."   Evergreen shrubs are proverbially fond of London smut.   The visitor to London who observes such matters can hardly fail to be struck with their luxuriance in front of Tattersall's, and many other spots in which they have been planted at some expense.   The verdant and luxuriant aspect of these places has had its effect upon the churchwardens and powers that be, and accordingly they have set to work to beautify our graveyards.   Ever-greens are to be substituted for headstones, and lamentable bits of cockney-gardening for the memorials of the dead. The most notable instance of this kind with which I am acquainted is around the church in Bishopsgate-street. Tombs and headstones appear to have been cleared out of the way and all obstructions removed, so that a level surface might be obtained on which to set a few hundred evergreens, which have little more chance of flourishing in Bishopsgate-street than if planted in the Salt Lake.   To have the bones or memorials of one's friends disturbed for the ill-digested schemes of a jobbing gardener is bad enough; but when it is considered that this sacrilege is performed to plant subjects that have no chance of thriving, then the wisdom of the change is fully seen. It is true the sculpture in our cemeteries is anything but Greek, and the inscriptions are not quite so simple and elegant as those in the catacombs; but the rudest and most mono-tonous of them tell of love and death " where human harvests grow," and to all but the most vulgar minds must be sacred and beautiful.   What, then, must be the feelings of those who have had the memorials of friends and ancestors disturbed for such a purpose ?   It is enough to draw an anathema from a less ready rhymer than the one who wrote " Cursed be he who moves my bones !"   And it is the more inexcusable

PLATE XXIV.

CHURCH OF THE TRINITY WITH GARDEN AND FOUNTAIN IN FRONT.

when we reflect that there is not the least occasion for any mutilation of the kind, and that the most suitable trees for such places are those that would not require any alteration of the ground, and would flourish freely in a town atmosphere. The weeping willow, birch, ash, weeping elm, and a considerable variety of drooping and other deciduous trees, are above all others suited for this purpose, and might be planted without interfering with the stones in any way. Would the latter look any the worse for being shaded by a beautiful pendulous tree here and there? The fact is, town cemeteries may be made as beautiful as it is possible to make them with vegetation, by the use of deciduous trees and shrubs and a few well-tried evergreens; and instead of any clearance or levelling being required for the judicious placing of these, they will look all the better for being picturesquely grouped among the tombstones and other irregularities of the surface. When new gardens are made in connexion with a new church it matters not of course how the ground is moved, but it would be a great advantage if the churchwarden mind could get rid of the idea that before making a garden in a graveyard it is necessary to level the space and make it like any commonplace bit of ground. Instead of pursuing such a course they should procure a few pounds' worth of advice from a respectable landscape gardener acquainted with the subject, and say to him, " Embellish the spot without destroying its memorials or associations." If you want it levelled, mutilated, and planted with a few formal beds and shrubberies confide its execution to an intelligent navvy. In such graveyard gardens much temporary flower work should be avoided in consequence of the ceaseless care it requires, and all attention should be paid to the hardy and permanent ornaments fit for such a place. Among these are happily found numerous graceful and weeping subjects so suitable for cemeteries.

Our suburban cemeteries are often gardens, pleasantly green, and abounding with trees, weeping and otherwise; while in the country churchyard where—

" Scattered oft, the earliest of the year,
By hands unseen are showers of violets found,"

there is a quiet verdure which makes the spot sweet to look upon; but with the cemeteries of Paris it is very different. There human love is lavish in its testimony, but the result is ghastly to behold. The quantity of everlasting flowers or immortelles, that is there woven into wreaths, for placing on and about the tombs in the cemeteries, is something astounding. Next to seeing the contents of a hundred Morgues displayed, the great spread of decaying everlastings is the most ghastly sight. They hang them on the poor little wooden crosses, they pile them inside on the covered tombs, they stick them on the few green bushes, they sling them under little spans of glass placed purposely over many tombs to protect the immortelles from the weather, till in every part, and particularly the part where the second and third-class departed are buried, there is scarcely anything to be seen but everlastings in every stage of decay, the sight being most depressing to anybody used to green British churchyards. A considerable portion of each large Parisian cemetery seems made to be inhabited by ghouls. In addition to decomposing compositæ, there is no end of small crockeryware art, and countless little objects made in bead-work, and brought here by the survivors of the dead to hang on the little black crosses or tombs.

It is somewhat different in the portions devoted to the graves of those who could command money when they moved about on the surface, and such as passed on their way to the grave by the paths of fame or glory. In their case, a little chapel, a ponderous tomb, or something of the kind, usually protects for a little time the dust of particular individuals from mingling with the common clay of their poorer relatives, and affords shelter to the crosses of silver and little objects of art, and a little more permanence to the wreaths. But what a very wide difference between this portion and that in which the ground is not paid for in perpetuity! Here the dust is allowed to lie undisturbed, at all events till they want to make a railway through it, or the gardening taste of a future age directs the surface to be levelled and planted with horrid taste as a garden, as has been recently done in several cases in London, so that the

PLATE XXVI.

THE CEMETERY OF PÈRE LA CHAISE.

earth is not merely a deodorizing medium, as it would appear to be in other divisions.   In the select parts, in addition to small statuary, &c., you frequently see choice forced
flowers placed on the tombs, and one cold February day I
saw a dame, evidently a nurse or respectable servant, sitting
weeping by the costly tomb of a young woman buried that
day twelvemonth, which tomb she had almost covered with

Fɪɢ. 42.

View in the Cemetery Montmartre.

large bunches of white forced Lilac and beautiful buds of
Roses.

Next let us visit the wide spaces where the poorer people
bury their dead out of their sight, and you will see a most
business-like mode of sepulture.   A very wide trench, or
fosse, is cut, wide enough to hold two rows of coffins placed
across it, and 100 yards long or so.   Here they are rapidly
stowed in one after another, just as nursery labourers lay
in stock " by the heels," only much closer, because there is
no earth between the coffins, and wherever the coffins—
which are very like egg-boxes, only somewhat less substantial, happen to be short, so that a little space is left
between the two rows, those of children are placed in
lengthwise between them to economize space; the whole

being done exactly as a natty man would pack together turves or Mushroom-spawn bricks. This is the fosse commune, or grave of the humbler class of people, who cannot afford to pay for the ground. I am not certain what becomes of the remains of these poor people after the lapse of a short time, but by some means or other the ground is soon prepared for another crop. On this principle, " the rude forefathers of the hamlet sleep " but a very short time in their last bed, and there is a very wide difference indeed between " sickle and crown " in Père la Chaise.

One day, when in the Cemetery of Mont Parnasse, I saw them making a new road, the bottom being made with broken headstones, many of them bearing the date of 1860 and thereabouts. These had been placed on ground that had not been paid for in perpetuity, and were consequently

Fig. 43.

The Catacombs.

grubbed up when, as before described, they wanted to fill the trenches a second time. Sir Charles Lyell tells us about a graveyard being undermined by the sea on the eastern coast, and a stone inscribed to " perpetuate " the memory of somebody being knocked about by the waves on the beach—but I never fully knew what a poor, transient, weedy kind of grass is the flesh of the lords of creation till I became acquainted with Parisian cemeteries. A cutting thirteen or fourteen feet wide, with the earth thrown up in high banks at each side, a priest standing at one part near a slope formed by the slight covering thrown over the buried of that day, and, frequently, a little crowd of mourners and friends, bearing a coffin.   They hand it to the man in the bottom of the trench, who packs it beside the others without placing a particle of earth between ; the priest says a few words, and sprinkles a few drops of water on the coffin and clay ; some of the mourners weep, but are soon moved out of the way by another little crowd, with its dead, and so on till the long and wide trench is full.   They do not even take the trouble to throw a little earth against the last coffins put in, but simply put a rough board against them for the night.   Those places not paid for in perpetuity are com-pletely cleared out, dug up, and used again after a few years, the bones being doubtless sent to the catacombs. The wooden crosses, little headstones, and countless orna-ments are carted away, thrown in great heaps, the crosses and consumable parts being, I believe, sent to the hospitals as fuel.   The headstones from such a clearing (when not claimed in good time by their owners) go to make the drainage of a drive, or some similar purpose.   And yet these people, who cannot afford to pay for the ground in perpetuity, go on erecting inscribed headstones, and bringing often their little tokens of love, knowing well that a few years will sweep away these, and that afterwards they cannot even tell where is the dust of those that have been taken from them.   What an instance of human love and man's fugacity !   Let us hope that whatever else may be " taken from the French," we may never imitate them in their cemetery management.

The catacombs are simply old subterranean quarries stored with the skulls and bones of multitudes of men.   When some of the old and well-filled cemeteries of Paris were removed to make way for improvements, the bones, &c., were carted away at night, escorted by priests and torches, and shot down in these extensive burrowings.   Afterwards they were regularly arranged and packed, and these places now present the ap-pearance shown in the engraving.   These caves were origi-nally precisely like those in which is practised the extra-ordinary system of mushroom-culture described in another chapter.

CHAPTER VII.

THE BOULEVARDS.

PARIS is famous for its parks, its squares, and its gardens, but its noblest features, and those most worthy of imitation in other cities, are its magnificent open streets, avenues, and roads, called boulevards. There are people who regard these as needless, simply created to serve the designs of an astute autocrat, and only possible under similar rule; whereas the fact is, they are merely such means of communication as would be found in every city of the world, if cities were designed with any due regard to their being fitting and healthy dwelling-places for hosts of men.

Parks and gardens are excellent in their way, but they effect only a partial good if vast areas of densely-packed streets are unrelieved by green open spots where wholesome air may obtain a vantage ground in its ceaseless work of removing impurities. The slight good that is effected by fine parks here and there in or towards the outskirts of a city is as nothing compared with what may be carried out by so planning and planting streets and roads that the air in which the people work and sleep may day and night be comparatively pure and free, and the eye refreshed with green at almost every point.

Paris exhibits the noblest and most praiseworthy attempts yet seen to render an originally close and dirty city healthy and pleasant for man; and this has been chiefly effected by her vast system of boulevards—wide well-made open streets and roads bordered with trees, and excellent footways as wide as many of the old streets, or wider. They do not simply pass through the city in one or several important lines, but pierce it in every direction, and are designed upon a far-seeing and systematic plan, so that during the future existence of the city overcrowding of its parts must become almost an impossibility. Many visitors who stroll along the fashionable and crowded boulevards of central Paris, who see them running in all directions from the Arc de Triomphe and offering bold approaches to every important position, may yet have but a meagre idea of their vast extent in the backward and less known regions of the city. The elm-bordered Boulevards Sébastopol and St. Michel cut through Paris from north to south, running miles in a straight line, and on their way effectually opening up the old Latin and many other close quarters; but beyond their outer extremities and between the fortifications and the central districts still larger boulevards sweep round, wide enough to be planted with groves of trees and to permit the breeze to play freely through, no matter how high and thickly the buildings may be raised for years to come. Immediately within the fortifications there is a wide boulevard running round the city under various names for many miles, while from every circular open space—like the Place du Trône, Place du Trocadero, Place d'Italie, or Place de l'Etoile—they radiate like a star. In fact the whole of the space within the fortifications is netted over by them, and, instead of the outer and less frequented boulevards being narrower than the central ones, they are often much wider. In many instances these outer boulevards pass through parts but thinly or not at all populated, so that the buildings to which the future is sure to give rise cannot encroach upon the space necessary for the free circulation of air and traffic.

The architecture that borders the boulevards in the most important and populous districts has often been objected to, and with justice, as formal and not in any way attractive. But this cannot, except with the most thoughtless, pass for any objection to the creation of open, tree-embellished streets. The greenest and sweetest of gardens may be quickly rendered hideous by somebody with a taste for pottery, plaster, or geometrical twirlings on the ground, but this is clearly not the fault of the garden. The varied archi-

I

tecture that has of recent years sprung up in many of our leading streets is, with all its faults, infinitely to be preferred to the bald formality and monotony characteristic of the style adopted in the best streets of Paris.   With our street architecture, which has improved so much, and promises so much more, we might, if we could only obtain open, handsome, tree-enlivened streets, eventually produce a result of which we, and all interested in city improvement, might justly be prouder than the French are of their boulevards.

How far we are behind them at present, those can tell who know what has been done of late years in such cities as Rouen, Lyons, and Paris, and who are also acquainted with our own great, sooty, packed, and cheerless cities.   Are our cities and towns to remain a mere agglomeration of furrows —ruts which to the over-passing bird must seem an excellent contrivance for preventing foul vapours to escape from the abode of men—or are they to receive as much attention, as to laying out, as is bestowed upon the surroundings of a suburban villa?   At first sight there does not seem any reason why the places where men most congregate should be those from which all who can afford it escape as often as possible ; though, doubtless, in a country where the laws of supply and demand regulate everything and everybody in such a satisfactory way, and where political economy is so well understood, one would not have to travel far for reasons why things are right as they are.   But judging by results few will deny that the disposition of our cities is a disgrace to any civilized race.   Why, without touching at all upon the most crowded and filthy parts of London, one may see more in a walk from the Strand or Fleet-street to the Regent's Park than would suffice to make him exclaim, " What a miserable and disheartening accompaniment of all our boasted progress !"   Such a reeking mass of mismanagement as may be found from east to west and north to south, the world has probably never seen ; and yet London is the " richest city in the world !"   The wealth of it, compared to that of such towns as Rouen or Milan, is as Mont Blanc to Primrose Hill; yet either of these cities

PLATE XXVIII.

VIEW ON THE BOULEVARDS NEAR THE CHATEAU D'EAU.

would put the " centre of civilization" to shame as regards clean and well planned streets and promenades.

It is a city of commerce, and we cannot afford space or money to remodel it, say some; but apart altogether from questions of salubrity and appearance, imagine for a moment how much is lost from mere want of room even in our leading thoroughfares. In many cases they are almost impassable except to those used and compelled to force their way through them, while if the pressed pedestrian retires into a cab he may find himself brought quite to a standstill in some busy groove. Wide thoroughfares and free circulation would be found to agree as well with commerce on the banks of the Thames as on those of the Rhone at Lyons. All real improvements would result in a clear gain to the business of a city, as will doubtless be proved ere long by our truly worthy Thames embankment. But the space? Land is too dear! This is really not a great difficulty in London. There is no city which could be pierced with free, open roads and boulevards more cheaply and readily. In its very centre there are acres covered by shallow brick buildings, which have not cost, and do not pay, nearly so much as closely-packed, tall, stone houses in inferior parts of Paris, that are cut through every day almost as freely as if they were made of pasteboard. Regions like that of Tottenham-court-road, most important and well situated for business purposes, are covered by the veriest shanties, which are of comparatively little value. In such places houses to accommodate twice the number of persons might be built, and lodge them far more comfortably than at present, while the streets might be as wide again, and therefore have purer air and more light. Wide tree-planted avenues might lead from the embankment out towards the pleasingly diversified suburbs, and would act as veins of salubrity to the regions they traversed. The increase in the value of property along such main arteries would repay for the outlay. If land be really so valuable, why occupy it with such trifling and unprofitable buildings? The fact is, the objection as to space, which is usually urged as the greatest, is no objection at all. Half occupied and sometimes waste ground without the margins of the city, and square miles almost worse than waste within, attest this.

Of course such changes as I advocate would involve the adoption of the "flat" system, which some say our people have a great objection to.   But that there can be no real objection in this is proved by the fact that it is adopted by a non-gregarious people like the Scotch, as well as by the French.   In houses constructed in this way, and with all modern improvements and comforts never to be found in the miserable and fragile structures now everywhere to be seen, the additional warmth and dryness which would result from the thicker and better buildings necessary, could hardly fail to have a beneficial effect on the public health, while efficient ventilation would prove a still greater aid.   It is by no means clear that any less thorough mode of proceeding will prove a true remedy.   Our narrow streets, the want of anything like a generally recognised plan, and flimsy houses, are worthy only of a period when men first herded together for security, and not of the Victorian era.   No sprinkling about of disinfecting agents when danger becomes imminent, or pulling down of a few shops that have protruded themselves and their outhangings so far into the narrow street that they have become intolerable, even to those accustomed to dodge through the streets of London, will touch more than the surface-roots of the evil.   Most of such narrow, schemeless improvements as are now taking place will be more than counteracted by the vast growth of what even in Byron's time was "this enormous city's spreading spawn."

The change must be radical!   We want a plan with the Thames embankment for its backbone.   There is nothing to prevent us having the best embellishments seen in continental cities, minus their trees in tubs and paltrier features. But to have them it is indispensable that we first have breadth and room, that the street traffic may circulate without abrading them away.   Footways and roads, wide and open, are the first and greatest necessaries, and they ought to be planted with trees, which do better in London than in Paris.   No fancy gardening, no expense for vases, griffins,

PLATE XXVII.

THE PLACE DU CHATELET AT THE SOUTHERN END OF THE BOULEVARD SEBASTOPOL.

water-squirting—nothing whatsoever of that type—should be tolerated until the pure free air be enabled to search its way into the heart of the town, through open verdure-bordered roads; which indeed would induce it to ignore the boundary line that now so widely marks the difference between town and country.

To hope to attack the mass of disease and dirt that exists, without first giving men an opportunity of enjoying pure air and light, is in vain. These are the cheapest as well as the greatest of blessings; they are naturally the property of all; but civilized man completely annuls them by his muddling and stupid arrangement of our cities. To make them once more the property of all should be the aim of everybody who wishes well to his country. It should be one of the first and most important "questions for a reformed parliament." For what is the use of all our present efforts towards ameliorating the condition of the masses in our cities, if health and all its consequences be impossible in them? Of none indeed, except it be in perpetuating much of the misery and squalidness that occur amongst us by ministering to them.

The conditions complained of do not simply occur in central parts of London where land is very dear: far without the radius of the parks, the arrangements of streets are frequently quite as bad as in the poor central districts, and capital preparations are being made to secure a dozen years hence a suburban cordon of districts like St. Giles's. To experience the truth of this the reader has merely to walk from Kensington Gardens to Kew—not the most unpleasant stroll that could be selected in suburban London. In the course of his journey he will find in the least populated parts pleasant open roads, in some cases wider than a boulevard, and with useless spaces railed off, and spreads of gravel, wide as a princely avenue, before some low and isolated public-house; but the moment he arrives at a densely populated part, the dead rabbits, sheep, &c., thrust out from the shops into the few feet of crowded footway, oblige him to dodge so often among the dung-carts and omnibuses of the narrow, crowded street that, if he has ever seen even an approach

to a decently arranged city, he will be forced to admit that
suburbs of London, miles in extent, have received less
attention as to design than a cottager bestows upon his little
garden, or a designer of wall-paper on his rudest patterns.
From a like, or even a worse, condition our neighbours have
been delivered by their splendid boulevards, and in a very
short time.

The word boulevard, or boulevart as it is frequently spelt,
signifies in its primary sense a walk made upon the walls of
a fortified town, of which we have so good an example in
the ancient city of Chester: it is said to be derived from

Fig. 44.

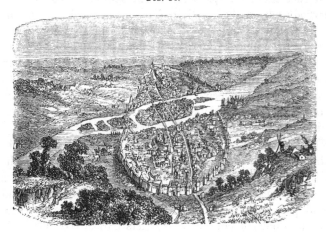

Paris seven hundred years ago.

the Low German " bullewerke," a word of similar meaning to
our own bulwarks.   Be this as it may, the Paris boulevards
proper, extending from the Madeleine to the Bastille,
undoubtedly occupy the site of the ancient wall built by
Etienne Marcel and Hugues Aubriot, to resist the incursions
of the English army which encamped round St. Denis. The
tables have long since been turned however, and instead of
entering Paris to plunder and slay, a huge undisciplined
English army is constantly passing through the gates of Paris
to be plundered by the shopkeepers, and take their chance

of being slain by one or the other of the numberless forms of dissipation to be met with in that gayest of all cities.

The first Paris boulevard was opened in 1670, in the reign of Louis XIV., and extended from the Porte St. Denis to the Bastille. It so pleased the Parisians that before the end of the following year the line was continued in the other direction from the Porte St. Denis to the Porte St. Honoré, which stood across the end of the Rue du Faubourg St. Honoré, where that street now joins the Rue Royale. A year or two after, the ancient Porte du Temple which stood in the boulevard of that name was

Fig. 45.

View on the old exterior Boulevards.

removed, and the road between the Porte St. Denis and the Bastille levelled and repaired, thus completing the original line of boulevards so often celebrated both in prose and in verse since the days of the Grand Monarque, and which have preserved their prestige, as *the* boulevards par excellence, over all those that have been since constructed.

The promenaders of the 17th century, on looking to the north of the capital, must have enjoyed an uninterrupted view of the open country beyond, dotted here and there with little villages or rich abbeys—the ancient fane of St. Denis lifting its grey towers in conscious superiority above

them all.   Looking south, he would see a somewhat sparsely
populated district between him and the Louvre, although
the Marais to the east was gradually extending itself
towards the fortifications.   The boulevards then under
Louis XIV., formed a long promenade some 150 feet wide,
and planted with rows of trees, beneath which the Parisians
could enjoy the double sight of the city and country in the
midst of rural silence and quiet.   The ancient buildings of
that epoch, from the Porte St. Honoré at one end to the
castellated Bastille, have long been swept away, and nothing
is left to remind one of the Grand Monarque but the
Portes St. Martin and St. Louis—both heavy masses of
classicalism—and a few names that have been bestowed on
the neighbouring streets.

The memory of the old fortifications is still preserved in
the Rue Basse des Remparts, which forms the north side
of the Boulevard de la Madeleine.   But it will not do to
linger over the recollections of the past when the present
has such pressing claims on our attention.   During the
following reigns the city gradually crept up to the boulevards, absorbing numerous convents, monasteries, and noble
domains in its progress.   The Revolution precipitated
matters by confiscating the remaining monastic and aristocratic lands in the neighbourhood, both within and without
the walls.   The boulevards soon became the favourite
resort of all that was noble, witty, or pretty in Paris.
Restaurants began to lift their heads above the small guin-
guettes that were first erected along the line, and house by
house, tree by tree, the boulevards gradually assumed their
present aspect.

The boulevards, par excellence, stretch from the Madeleine to the Place de la Bastille ; and a ride outside an omnibus from one point to the other will well afford a sight not
to be witnessed in any other part of Europe.   The roadway
and footpaths are more spacious than any in London.
The latter are usually thoroughly well asphalted, shaded
with rows of trees, and furnished with numerous seats.
In the more fashionable portion—that between the Boulevard de la Madeleine and the Porte St. Martin—nearly

PLATE XXIX.

THE BOULEVARD DU TEMPLE.

every other house is a restaurant, a café, or a theatre. Before every one of these are groups of little tables, at which pleasure-seekers from all parts of the world are seated laughing, talking, smoking, and drinking as if no such things as wars, revolutions, or financial panics ever existed.

The boulevards of Paris are, generally speaking, so very much alike that to describe them in detail is needless. The

Fig. 46.

Avenue Victoria, near the Hôtel de Ville.

illustrations will give a better idea of their actual appearance than any written description. From house to house they are usually, in the most frequented parts, over 100 feet wide, occasionally reaching between 130 and 140 feet, and even much wider than this in the outer boulevards, which are sometimes large enough for half a dozen lines of trees, in addition to very wide footways, and perhaps two minor as well as a wide central road, as in the Avenue de

la Grande Armée. The footways of the most frequented
boulevards are about twenty-six feet wide on each side, and
sometimes more.

But, notwithstanding their general similarity, there are a
few distinctive enough for special mention, and among
these none more so than the Boulevard Richard Lenoir,
which runs from the Place de la Bastille to the Rue du
Faubourg du Temple. This often escapes observation from
visitors, as the Boulevard Beaumarchais drains most of the
traffic from the Bastille to the fashionable boulevards; but

Fig. 47.

End view of the Boulevard Richard Lenoir.

it is one of the most remarkable in Paris, and more than
usually ornamental. It is nearly 2000 yards long, and is
in great part built over a canal. It was thought desirable
to cover a large portion of the canal, and to make a wide
boulevard over this huge bridge, in order to facilitate the
traffic and improve the appearance of the district. It
became necessary to have ventilating and lighting shafts
for the canal, and eighteen pairs of these openings occur
in the course of its length. These have been ingeniously
and tastefully hidden by eighteen little railed-in parterres.

In these the openings, which are wired over, are surrounded by a thick low hedge of Euonymus or some close evergreen, so that no opening of any kind is exposed to the passing observer. In the centre of each garden there is a long basin and a fountain, the whole being connected and surrounded by flowers and grass. Then on each side of these parterres there are very wide avenue footways, each shaded by two lines of Plane trees—a road being on each side of the parterres and tree avenues. For a considerable distance from the Château d'Eau, the flower-market that has its headquarters held there extends down amongst the little railed-

Fig. 48.

Place du Trône.

in parterres, and the effect is altogether very pretty and quite unique.

Every visitor to Paris must of course have seen the fashionable boulevards stretching from the Madeleine to the Place du Château d'Eau, but the great outer systems often escape observation. On the left bank of the Seine among the more remarkable of the exterior boulevards are those of St. Jacques, d'Italie, d'Enfer, Du Mont Parnasse, and the Avenue de Breteuil; on the right bank the Boulevards Pereire, des Batignolles, Clichy, Rochechouart, de la Chapelle, and de Belleville are amongst the most important, though, where there are so many of great extent and almost similar features, it is difficult to particularize.

Of avenues, the largest and most gardenesque is the Avenue de l'Impératrice, leading from the Arc de Triomphe to the Bois de Boulogne.

In order to put the centre of Paris in communication with the Bois de Boulogne by means of a wide direct road, an imperial decree ordered the Route départemental leading from the Rond Point de l'Etoile to the Porte Dauphine of the Bois de Boulogne, to be straightened. Half the expense was borne by the State, under the conditions that an iron railing of uniform design was to be constructed along the whole length of the road, that a strip of eleven yards in breadth be left for a garden between this railing and the houses on each side, and further, that no kind of trade or manufacture should be carried on in the houses adjoining. The avenue was made entirely through private lands which were acquired for the purpose. Its total length is 1300 yards; the width 130 yards. It consists of a central drive, seventeen yards wide, of two large side walks, each measuring thirteen yards wide, and of two strips of turf planted with choice trees and shrubs, including the whole of the species as yet naturalized in Paris, and lastly, of two footpaths running along the side of iron railings that separate the houses from the road. The total cost of the avenue amounted to over 20,000*l*., in addition to which the city of Paris expended a sum of 4000*l*. more on the flower-beds and plantations, for the enlargement of the Auteuil railway bridge, and for the general drainage of the ride.

The Avenue de l'Empereur, beginning at the Quai de Billy, opposite the Pont de l'Alma, and joining the Bois de Boulogne close by the Porte de la Muette, is another example of the great attention and expense devoted to avenues and boulevards in Paris during recent years. The portion of the avenue between the Porte de la Muette and the Place du Roi de Rome was laid down in 1862. The part included between the Rue du Petit Parc and the Place du Roi de Rome necessitated considerable excavations, many of them being as much as thirty-three feet in depth. Along

PLATE XXX.

THE BOULEVARD MONTMARTRE.

the whole of this section the owners of the adjacent property were obliged to set apart a strip of ground, thirty-three feet wide, enclosed by an iron railing of uniform character, and laid out as a pleasure-garden. They were also forbidden to let out any of the premises for trading or manufacturing purposes.

Fig. 49.

Avenue de Breteuil: the artesian well of Grenelle, and the Invalides.

The part between the Place du Roi de Rome and the Quai de Billy was begun and finished in 1866. The incline of 27 in 1000 that was obliged to be given to this portion of the road also necessitated large excavations of an average depth of six feet between the Boulevard du Roi de Rome and the Place Chaillot; whilst from this point to the Pont de l'Alma an embankment had to be made in some places

thirty-six feet high.   Between the Place de Chaillot and
the Pont de l'Alma that side of the road only which faces
towards Chaillot will be covered with houses.   The side
looking towards the Seine, which will form a terrace, supported by an immense wall, will give a fine view of the
river and the hills about Meudon.   The land on which the
terrace wall is built is formed of the alluvium of the Seine,
which rendered the work of construction most difficult,
owing to its want of firmness.   The method which seemed
to promise the greatest amount of safety, combined with
economy, was to spread the pressure of the vast mass over
a large extent of surface.   For this purpose a wide area
was formed of concrete, on which was erected a wall nearly
of the same size.   This wall was hollowed out on each side
by large spaces, forming on the front turned towards the
observer a series of vaults supporting a row of shrubs,
which allowed the eye to wander through them into the
neighbouring gardens.   On the other side where the embankment had been formed there were two rows of vaults,
in order that the weight of earth resting on them might be
added to that of the arcade itself, so as to counterbalance the
effect of the tendency of the embankment to throw the wall
outwards.

The portion of the wall standing to the right of the
premises that belong to the waterworks at Chaillot was too
low to render the same method of construction necessary.
However, in order to increase the resources of this establishment, and to shut out from the sight of the passers-by
the vast heaps of coal lying in the yards, the vaulting
of the wall was continued on a smaller scale so as to form a
footpath above and a series of coal cellars below.   This part
of the wall was surmounted by a railing covered with ivy so
as partially to hide the yard and buildings of the waterworks, which were somewhat awkwardly cut in two by the
new road.   This wall was built entirely in a kind of artificial
stone formed of compressed concrete.   The arcades that are
visible have millstone dressings, in order as much as possible
to vary the appearance of the wall, which is no less than 430
yards long.   The total cost of the construction of the Avenue

PLATE XXXI.

THE BOULEVARD ST. MICHEL.

de l'Empereur amounted to the large sum of 82,000*l*. From the preceding the kind of difficulties that have been overcome in carrying out such works in Paris of recent years will be tolerably apparent. A large work might be filled with like details.

Naturally the features of the boulevards which command most attention in this book are the trees. The advantage of having a full knowledge of the best boulevard and city trees is so great that a special chapter is devoted to them. It is truly surprising to see how well these are managed in Paris, and to what an enormous extent they are planted, as well in the centre of the city, on the boulevards, and along the river, as on the scores of miles of suburban boulevards, radiating avenues and roads, the sides of which one would think capable of supplying Paris with building ground for a dozen generations to come. The planting in all the London parks is as nothing compared to the avenue and boulevard planting in and around Paris. The trees are nearly all young, but very vigorous and promising. Every tree is trained and pruned so as to form a symmetrical straight-ascending head, with a clean stem. Every tree is protected by a slight cast-iron or stick basket ; it is staked when young, and when old if necessary, and nearly every tree is provided with a cast-iron grating six feet wide or so, which effectually prevents the ground from becoming hard about the trees in the most frequented thoroughfares, permits of any attention they may require when young, and of abundance of water being quickly absorbed in summer. The expense for these strong and wide gratings must be something immense, but assuredly the result that will be presented by the trees a few years hence will more than repay for all the outlay by the grateful shade and beauty they will afford the town in all its parts. As soon as a new road or boulevard is made in Paris, it is planted with trees ; and every one of the millions is as carefully trained and protected as a pet tree in an English nobleman's park.

The trees recently planted on the boulevards are placed at a uniform distance of between sixteen and eighteen feet. To plant so closely of course helps to furnish the streets with some shade almost directly; and as the trees are usually trained specially for boulevard planting some little effect is obtained at once; but there can be no doubt that it is too close a system of planting, as the trees cannot grow sufficiently when so much crowded. A better way would be to place them five or ten feet further apart, and plant, alternately with the trees destined to grace the boulevard eventually, some kind that grows very rapidly when young. This would help to furnish and freshen the avenue until the trees intended to permanently adorn it have been established and advanced a few years; and as soon as those of the free growing nursing kind have become large enough to deprive their neighbours of light they may be cut in vigorously, and finally removed altogether. Sometimes double ranks of trees are planted, but this is only wise where there are very wide boulevards. It is occasionally practised in avenues—like some of those that radiate from the Arc de Triomphe; but usually it has the effect of darkening the houses too much. Where, as is often the case in the outer boulevards, there is abundant room for a double or even treble line of trees to develope without disagreeably shading the houses, they should of course be planted. The trees are usually placed within three and a half or four feet of the edge of the footway, but there can be little doubt that it would be a better plan to keep them a few feet further from the road, and this would admit of giving them a larger body of soil. Generally in Paris they receive too little.

When the boulevard is marked out and levelled, if the soil is of bad quality, as is nearly always the case, trenches are dug in the footway, from one end of the boulevard to the other. The width of this trench is usually about six feet, and its depth four or five; and before filling it in, drain pipes are laid along the sides, made with lapped joints so that the roots shall not enter between them. The trench is then filled with good garden earth, raising it a little higher than the level, so as to allow for settling. In this ground the trees are planted at about six yards apart. They should be carefully chosen, with perfect roots, and moderately pruned. Formerly the stem was cut at about nine feet from the ground; but this had the bad effect of preventing the top of the tree from being straight, and the practice has been given up. The trees are next staked and tied with wire over a neat wad of straw, which prevents all injury to the stem. A protecting cage, neither heavy nor very expensive, is placed round the tree to prevent accidents; and if the weather be at all dry at the time of planting, the trees are copiously watered. As for the making of the roads and streets, it is admirable, as many readers may have learnt for themselves. When the repairing or making a road in Paris is finished, the surface is as level and crisp as the broad walk in the Regent's Park, so that the horses are spared much pain, and carriage movement greatly facilitated. Stones of about the same size as we spread on the roads are thrown down, and then comes the heavy steam or horse-drawn roller, making but a slight impression at first, as might be expected, and indeed it has to be passed over many times before the work is completed. All the time, or nearly all the time that this rolling is going on, a man stands at the side of the footway in charge of a hose on little wheels, and keeps swishing the stones with water, while others shake a little rough sand on them between the rollings; and so they wash and roll and grind day and night—the result being that the Parisian roads are as comfortable for locomotion as could be desired. But it would be a mistake to suppose that their system of road-making is otherwise superior to ours. If we took the trouble to grind down the rough and sharp stones used in repairing the streets, there would be little to complain of as regards the texture of our roads; and it would probably be impossible to find more perfect examples of roads than those in Hyde Park since the introduction of the steam rollers.

It is not an uncommon impression among us, that since his access to power the Emperor has most industriously employed himself in removing all the paving stones from Paris, so that they may not be used against him in case of an insurrection. This is an error; for, although the wretched old system of paving is being done away with—greatly to the relief of the ears of the inhabitants of the streets in which it existed—nearly all the important new boulevards have a considerable breadth on each side well paved for heavy waggon transit.   This prevents. the larger macadamized portion in the centre from being cut up by the heavy traffic of Parisian streets, and leaves it free for carriages and the lighter kinds of traffic.   In some of the older and narrower streets, in which there is not very much traffic, asphalte has taken the place of paving stones—making a road that is almost noiseless; but its chief use is in the formation of the wide and excellent footways that border all the new streets and boulevards.

Asphalte has long been used in Paris for two reasons: first, the supply of good paving stone, similar in quality to our York stone, is scarce, and the few quarries that do yield it are far distant from the capital; and secondly, the peculiar bituminous sandstone from which the asphalte pavement is made, is cheap and abundant.

Bad attempts at laying asphalte produce such very disagreeable results that the very name must be dreadful to some people; but in a sloppy climate the advantage of having in all weathers dry, smooth, and permanent footways, instead of cloggy, saturated gravel or mud, is so great that some account of the best system of laying this material cannot fail to be useful.   Some years ago asphalte produced a regular industrial fever, and pavements were made in all directions in Paris and London, of any material that at all resembled it.   Gas tar, wood tar, pitch, and all sorts of nastiness were ground up with stone, and laid down without proper preparation; the consequence of which was that a large number of failures took place, and asphalte pavements (at least in this country) were very soon completely tabooed by all good architects.

Bituminous limestone occurs naturally in many parts of the world, notably at Val de Travers, in the canton of Neufchâtel, Switzerland, and at Pyrimont, near Seyssel, a small town in the department of Ain, on the right bank of the Rhone.   The asphalte rock from Pyrimont consists of pure limestone impregnated with about 10 per cent. of fossil or natural bitumen. It may be asked how it is that ordinary tar or pitch of good quality mixed with pounded limestone does not answer the purpose of this natural combination; but it is found by experiment that, although natural bitumen differs but slightly in chemical composition from pitch and tar, it is much more elastic and durable. If made with tar, the resulting asphalte is sticky and soft in hot weather; if with pitch, it is too brittle, and soon cracks and splits.

In the natural asphaltic rock the bitumen is so intimately combined with the calcareous matter, that it not only resists the action of the air and water for a considerable time, but even that of some of the strong mineral acids. The ancients were in the constant habit of using natural bitumen instead of mortar; and there is a tradition that the stones of the Tower of Babel were cemented together with the same material as that forming the footways of the boulevards. The principal ingredients used in forming the mastic for the pavement is the dark brown bituminous limestone from Pyrimont, just described. The stone is first reduced to fine powder, and then mixed with a certain proportion of mineral bitumen, extracted previously from another portion of it.

When it is intended to be used for covering roofs, lining tanks, &c., no other addition is necessary; but if it is to be used for paving, a certain quantity of sea-grit is added. One specimen analysed by an English chemist yielded 29 of bitumen, 52 of limestone, and 19 of siliceous sand. The ingredients are exposed for some hours to a strong heat in large cauldrons, and kept constantly stirred by machinery. The mastic thus formed is made into blocks, measuring eighteen inches square by six inches deep, and weighing from 112lb. to 130lb. each. In this state they are sold ready for use, and are remelted on the spot where the asphalte has to be applied; for which purpose small portable furnaces fitted with cauldrons are employed. A pound weight of mineral bitumen is first put into the cauldron, and when melted 56lb. of the mastic are added, the whole being repeatedly stirred. When fully mixed, another 56lb. of mastic are stirred in, and so on until the cauldron is full.

K 2

When thoroughly melted—which may be told by the mastic
dropping freely off the stirrer, and by jets of light smoke
darting out of the mixture—it is conveyed quickly to the
spot where it is to be used, in heated iron buckets or
ladles.   The cauldron ought to be as close to the work as
possible, and in covering brick arches it should be hoisted
to the top of the building.   It must be clearly understood
that the only kind of bitumen to be used is that impreg-
nating the limestone itself.

In forming foot or carriage ways it is most important to
secure a good foundation by removing or ramming the soft
earth, and laying a course of concrete, care being taken to
allow the whole to dry before putting down the asphalte.
If this precaution is not attended to, the heat will convert
the moisture in the concrete into steam, and fill the asphalte
full of airholes and bubbles.   The thickness of the layer of
asphalte may be regulated by slips of wood arranged across
the pavement at a distance of 30 inches from each other—a
width quite sufficient for one man to work at at a time.   If
two men are employed, double the width may be taken, as
it is always better to have as few joints as possible.   The
work is levelled with a long curved wooden spatula, assisted
by a long straight ruler, which stretches across the layer of
asphalte, over which it is moved backwards and forwards,
the wooden gauges supporting its ends.   If the surface is
intended to be smooth, a mixture of equal parts of silver
sand and slate dust or plaster of Paris is sifted over it before
it has quite set, and rubbed down with a flat tool of wood.
If it is required to be rough, sharp grit is to be beaten in
with a heavy wooden block.   One portion of the pavement
being complete, it is best to proceed to lay the next but
one, leaving the intermediate space to be filled up afterwards, when the first layer is firm and cold, so as to insure
a good joint.

The thickness of asphalte for footways varies from
half an inch to an inch and a quarter, the former being
sufficient for common floors and courtyards, the latter for
carriage pavements.   A thickness of from half an inch to
five-eighths is sufficient for roofs and the coverings of arches,

and for lining tanks and ponds, and about half an inch for
the ground line of brickwork to prevent the damp from
rising. An asphalted surface admits of easy repair. By
placing hot mastic on the places requiring it, the faulty
part may be cut away and the edges cut square, when the
hot material will be found to adhere to them if they are
perfectly free from damp or moisture.

The great secret, then, in obtaining a perfect layer of
asphalte paving, dry, hard, elastic, warm, and durable, is
first to employ only the natural material, such as that from
Pyrimont-Seyssel; and secondly, to provide a firm, dry sub-
stratum of concrete for it to rest upon. For pavements,
terraces, &c., nothing can be better. It is always warm
and elastic to the tread; there are no joints to encourage
the accumulation of filth or the growth of weeds; and in
case of rain it dries in a few minutes. As laid down by
the Seyssel Asphalte Company, its durability is immense.
The whole of the quadrangle in Trafalgar-square has been
laid with asphalte since 1863, and yet there is no sign
of wear upon it, in spite of the enormous traffic. The
terrace at Bridgewater House is also an excellent specimen.
If the reader desires to see a really bad example, he has
only to examine the pavements of some of the metropolitan
stations of the Midland Railway to see what badly-laid
asphalte made of improper materials ·is capable of becoming, in the course of even a few months.

For roadways and carriage drives asphalte does not seem
so applicable. In dry weather it is all well enough, but
after rain, more especially if there is any mud about, it
becomes disagreeably slippery both for horses and foot pas-
sengers. For laying between the courses of brickwork to
prevent the damp from rising, it is unequalled, a layer, even
only one quarter of an inch thick, keeping all damp down
most effectually. It is especially fitted for this purpose in
the case of boat houses built by the sides of rivers or lakes.
For ornamental ponds and banks it is also excellent, but it
should be roughened for, say, a foot in depth, so as to hold
sufficient soil or mud to grow water plants and weeds, and
so entirely conceal its existence. As to its cost, in England

it is somewhere about 10*d.* per superficial foot for quantities not less than 700 feet. For roofs and terraces to
the same extent, the cost would be about 2*d.* extra.

During the last few years the preparation of the asphalte
in Paris has been much improved. Some years ago, when
a pavement was to be made with bitumen, a great nuisance
was experienced by the public during the operation. The
mastic was liquefied on the spot, and produced a nasty
smell and smoke, disagreeable and injurious; but now some
of these inconveniences have been done away with by a new
system, and asphalte is now laid down in the most expedi-
tious manner. It is prepared first in out-of-the-way places
devoted to the purpose, and the matter, ready for use and
liquefied, may be transported from these places to any
parts of the town without the least inconvenience in a semi-
cylindrical boiler, closed by iron doors, and moved about on
iron wheels as freely as a common cart. Under the boiler
is a fireplace, and the blaze, after having heated the two
sides of the boiler, passes out by a chimney placed at the
back of the machine. Means to keep the mastic in motion,
and prevent its burning by adhering to the sides of the
boiler, are secured by a simple mechanism easily worked
with the hand. These carriage boilers, full of liquid asphalte,
are driven from place to place with the greatest facility.
The boiler is emptied by the means of a pipe fixed to its
bottom, and the mastic is collected in a pail, and spread
on the surface to the thickness of three-quarters of an inch.

If the surface is not perfectly dry, the drying must be
accelerated with hot ashes, which are to be taken away
afterwards, or with a little spreading of quicklime in
powder. These operations are indispensable, as if the
mastic were laid on before the surface is dry, the heat
of it would dispel in steam the water underneath,
and that steam would produce blisters in the asphalte,
which would crack under the pressure of the feet, and
endanger the success of the operation. The workmen
place on the platform two iron bars of the same thickness
as the asphalte is to be, equally distanced from each other;
it is then laid down in a very warm state, and thick

enough to require some slight exertion of the operator to make it level. This operation done, a small quantity of fine gravel must be spread over the asphalte while hot, and slightly beaten down to penetrate in it. This gives a greater hardness and solidity to the footway, and insures its lasting for a very long time.

The roads before spoken of are made of the powdered and not liquid asphalte. The surface of the roadway must be beaten down very hard, and covered with a thickness of about three inches of concrete, well beaten down and dry. If the dryness is very necessary in the making of a pavement, this condition is of a greater importance for the road, as, if the powder were spread on a wet surface, the steam caused by the heat would produce a great number of little fissures, the elasticity would be destroyed, and the road would be useless after a few months' use. The concrete well dried, the powder (hot) must be spread about three inches thick; and then well levelled and beaten. The sides must be done first, and pressed down with a rectangular iron pestle eight or nine inches in length and two or two and a half inches in width. When the sides are done, proceed with the middle. The pestles used in pressing it are made of cast-iron, circular, and about eight inches in diameter. The pestles of either form are heated and used quite hot, so as to compress the asphalte into a hard smooth mass.

When the crust of asphalte is brought to the thickness required, and is sufficiently smoothed and beaten hard, they spread with a sieve a little quantity of very fine powder to fill all the unevenness, and again smooth the whole with a flat piece of hot iron. The compression is completed by the employment of two cast-iron rollers, one of 4000lb. weight and the other of 3000lb. Sometimes three of these rollers are employed, the intermediate one being about 1500lb. or 1600lb. in weight. This rolling is not always necessary, and in many cases the beating down with pestles is sufficient. The roads thus made, completely noiseless and lasting a long time, have been adopted with the greatest success by the city of Paris, and are supplanting the paving stones, macadamizing, and other pavements, in narrow streets where there is

not much traffic.   Some beautiful smooth roads through
the Luxembourg gardens have been made of this powdered
asphalte, and without the use of heavy rollers, the hot
smoothing irons only being used.

Bathing.—With the boulevards one naturally associates
the quays, planted in every available spot with trees, and in
Paris the public swimming baths are all on the silent boulevard.   However, the Seine at Paris is not a noble river, and
the ugliest things to be seen from its banks in summer are
the floating baths, which in some places half cover its surface.
But public bathing is a matter of the highest importance, and

FIG. 50.

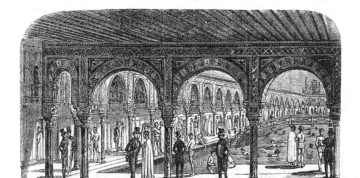

Interior of floating bath on the Seine.

it is perhaps better to have floating baths on the river than
tolerate the astounding exhibition of naked humanity which
may be witnessed on the Serpentine on any warm summer
evening.   My friend Mr. Gibson, of Battersea Park, thinks
these floating baths would be a desirable improvement on
our river, and from its size they need not produce such an
ugly effect as at Paris.   With a clean river, a constant current of fresh water is a great advantage of these baths; and
then they are not costly, and are closed in from public view.
But whatever may be thought upon this point, it is certain
that there is no question connected with the healthful exer-

PLATE XXXII.

THE LOUVRE, INSTITUT, QUAYS PLANTED WITH PLANE TREES, THE RIVER AND FLOATING BATHS.

cise of the people that has been more neglected than that of public bathing.

Everybody knows with what alacrity the cockney "takes to the water" in summer, although indeed he can, as a rule, only do so under great difficulties. The pluck that must be required to venture to bathe amid such an assemblage as that in Hyde Park must be of itself considerable, and yet the enormous crowds that practise it here show us to what an extent decent cleanly bathing would be taken advantage of by the working population of London. If it were provided it would prove one of the greatest boons that could be conferred on them ; and surely no great good could be so cheaply effected as that of providing proper bathing-places in all our parks and open spaces. The benefit to be got by the regular practice of bathing by our working men during the summer months could not be equalled by any other exercise or recreation. It is a good of which the advantages have not to be pointed out to the people ; and every one of our parks offers capital positions, in which inexpensive bathing-places might be made. Bathing-places should always be introduced in a quiet and somewhat retired part of a park or public garden. They should be surrounded by plantations sufficient to thoroughly conceal the bathing from all but the bathers. They should be made of a convenient depth for swimming purposes, and, above all things, should have a clean level bottom; for a sticky, muddy bottom, such as is likely to occur in some places about London, is very objectionable ; and in making a swimming pond it would not be difficult to provide against this. They should be surrounded by a diversified plantation of trees and shrubs, with the taller growing subjects kept somewhat back, and with an inner edge of dense dwarf shrubs. The free-growing and smoke-enduring evergreens, such as Box and Aucuba, should be extensively used around these bathing-places, which should also have a very wide marginal walk of clean gravel, and long seats in recesses of the shrubbery borders.

It would be an excellent plan if roomy sheds were also erected near the water's edge, but slightly thrown back and concealed by vegetation. These might be well utilized in

the winter as a working place for the park men, or any
other labourers employed about.   In it they could make
brooms, prepare wood, paint hurdles, make labels and pegs,
and many other things that are in constant use.   Of course
this should only be done in bad weather.   In all places
where a number of workmen are employed in winter, there
is generally a difficulty in providing men with work unless
there are large sheds : such to the sagacious manager prove
a great boon.   They might also be advantageously used as
winter storehouses for seats, boats, and other things which
must be used in our recreation grounds, if we are to afford
the people sufficient amusement and attraction therein.
But their chief use would be in making it possible for
people to bathe at all times.   How many summer, spring,
and autumn days are there on which bathing would be a
delight, but when showers of rain forbid it in the open air !
Of course men cannot strip and bathe and then get into
wet clothes ; and, like other exercises, bathing, if not
regularly practised, is not productive of much good.   Those
sheds would afford a place where the clothes could be kept
dry, and then rain, light or heavy, would not produce any
difference—a swim would be as enjoyable in the heaviest
of rains as at any other time.   Partial bathing, such as that
practised in the Serpentine during the mornings and evenings, merely meets the wants of a few persistent morning
bathers, and a host of the roughest of the great unwashed
in the hot summer evenings.

CHAPTER VIII.

THE JARDIN FLEURISTE AND OTHER PUBLIC NURSERIES OF THE CITY OF PARIS.

In its public nurseries Paris possesses a very useful aid which we have not in this country. With us each park or garden produces or purchases its own supplies; in Paris all the gardens of the city are furnished from its nurseries. It should be observed that in Paris there are two sets of public gardens—those of the city comprising the boulevards, squares, parks, church gardens, and so on, and those of the State, the gardens of the Luxembourg, Tuileries, &c. All are equally open to the public—all arranged with a view to its pleasure and convenience; but in the case of the State gardens each supplies its own stock. What we have to deal with now is the manufactory, so to speak, for the vast array of gardens and open spaces made during recent years. At one time the old State gardens were by far the most important in Paris; now they are quite eclipsed by those created specially for the city and its people, and not merely as the surroundings of a palace or the pleasure-gardens of princes. Considering that the whole is of such recent growth, the success of the arrangements is surprising. In commencing to improve the town by means of public gardens, there can be no doubt that it was a wise step to begin with central nurseries or plant manufactories, from whence all those gardens could be supplied.

The advantage of having public nurseries of this kind to supply the parks, gardens, and squares of a great city are so great, that it is surprising they have not been already adopted with us. Not only could the necessary trees, shrubs, and flowers be procured much more cheaply, but a far greater selection of choice subjects would be at the disposal of the planters. By selecting ground favourable to each class of plants, shrubs, or trees, the whole of the subjects in that particular section could be grown to as great perfection as by any nurseryman—could be produced at a far cheaper rate than they could be bought; and the necessity of searching for, bargaining, and selecting would be done away with, the planter having merely to indicate the subjects required. They could be quickly despatched to any given point in vans constructed for the purpose. In addition to these advantages, a small portion of each nursery might be devoted to an experimental ground to test newly introduced or imperfectly known plants; and in this respect each would be of valuable aid, not only to the State, but also to the general public. With our parks and crown lands in which to select positions, the establishment of such gardens would not be expensive, and would in a very few years save the first cost of their construction. Our large nurserymen would feel a pleasure in contributing their novelties and rarities, as they now do to our botanic gardens, and a system of exchange might be arranged between them to the advantage of both public, private, and commercial establishments.

The present system is too bad to last. We have, in and around London and our other great cities, numerous public parks and gardens, and it is to be hoped their number will go on increasing from year to year. Let us suppose that the superintendent or designer of a new public park or garden wants many thousand trees and shrubs for its embellishment. He has to obtain them wherever he can, and as the nurseries are arranged chiefly or solely for private use, most probably there will be great difficulty in getting some things even at a high rate. For example, a very important item in town gardening consists of trees for park and avenue planting. If at the present moment we wished to plant an avenue of Plane trees, of suitable size and properly prepared for the purpose, we should no doubt have to send to the Continent for them, as in our own nurseries they are not prepared for street planting; in which case they would cost much more than if bought in this country, and be in far worse condition for the purpose than if they had been grown at home. The Planes recently placed on the Thames embankment have been imported from the Continent, and of course there would not have been the slightest occasion for this if we possessed the kind of establishment I suggest, and of which the necessity must be seen by every reader. In Paris there is a great central establishment at Passy where all the tender plants are grown and increased, and there are nurseries specially devoted to the production of city trees and shrubs, in which the most suitable kinds are grown, and grown exactly to the size and shape in which they are best suited for being placed on the boulevards, or in the parks or gardens. The cost of each plant or tree is in this case a mere trifle; in our own the plantation of even a very small park, or one boulevard, would amount to a very considerable sum. To pay a guinea apiece for specimens that we could produce for a few shillings, and a shilling or two each for common stuff that we could grow for a few pence, is to follow a plan whereby our public gardening, and consequently the health and beauty of our cities, are considerably retarded.

The Jardin Fleuriste of the city of Paris is situated in the Avenue d'Eylau, close to the Porte de la Muette, leading to the principal promenade of the Bois de Boulogne, and should be seen by every visitor interested either in public or private gardening. It is the depot for all the tender plants used in the decoration of the parks, gardens, and squares of the city. Entering from the Avenue d'Eylau, the first objects of interest that meet the eye are collections of handsome plants growing in the open air on a small lawn amidst the glass-houses with which the place is nearly covered. My object is not to describe the garden in detail, but simply here as elsewhere to point out its most instructive features. To me the most interesting and valuable group planted on this lawn is a number of hardy Bamboos, proving clearly that in our latitudes we may enjoy the peculiar grace and verdure of these giant grasses, and by planting them highly improve the appearance of our gardens and pleasure-grounds, especially in places under the mild influences of the sea and in the west and south of England and Ireland. As the family is alluded to in another chapter, I will say no more of it here.

On the grass here during the past summer might be seen one of the most magnificent of all pea flowers, Clianthus Dampieri, flowering very freely in the open air, although we find it so difficult to grow even in our greenhouses. It was sown in February and planted out early in June as a tuft of several plants, isolated on the grass, but rooted in peat soil. The shoots grew to more than two feet in length, and began to unfold gorgeous blooms at their apex about the beginning of August, continuing to do so till the commencement of October. As an isolated group upon the grass, I need not say it was very fine; and I believe the same success could be obtained in mild parts of this country, and in many places against the low, warm walls of glass houses, &c. It should be raised as a greenhouse annual and planted out about the first week in June in peaty soil. Some may not be aware that it is infinitely more beautiful than the old brilliant and popular C. puniceus, though, unlike that, difficult to cultivate and impatient under the most skilful treatment in houses. There are usually many fine groups of Yuccas, Musas, Cannas, various new plants, and other objects of interest on this little lawn which will well repay a careful examination. The most remarkable of the novelties of the past season was Dimorphanthus manchuricus, a remarkably handsome plant, reminding one of Aralia japonica. A plant of it a few inches high put out at Passy in July, 1868, had leaves a yard long and thirty-four inches across by the middle of September. It will prove of the highest value in the ornamental garden.

The first great group of glass houses are span-roofed. The interior arrangements made in them for the convenience of the workmen and for the preservation of the plants in winter are most admirable, and should be adopted by us in all similar instances. We build more hot-houses than any other people, construct them better, and furnish them better; and therefore it is desirable that in disposing them in relation to each other we should employ the most economical and convenient plan. Everybody knows how often they are scattered about without any connexion with each other, and the consequent additional expense and trouble. But, even where the errors of the scattering system are guarded against, there is seldom an effective means of communicating from one to the other without going in the open air. We all know how disagreeable it is to pass from a moist stove to frosty air—from wet gusts to damp greenhouses; it is dangerous to tender plants that often have to undergo it unclothed; nor can it be otherwise than injurious to the

FIG. 51.

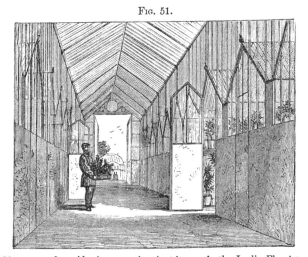

Glass-covered corridor between the plant-houses in the Jardin Fleuriste.

health of those employed in such structures. All these inconveniences are got rid of by the very simple plan adopted in the case of the group of houses, the arrange. ment of which the following diagram may serve to explain. The plant houses diverge on each side of a glass-covered passage, and there is no necessity for taking the plants into the open air in winter, or for the men who work in the houses to undergo any change of temperature for hours at a time. The houses are so closely arranged together, that heating them becomes much less difficult than when they are separated. The advantages of the plan are so great that I should strongly advise everybody building a batch

of houses for growing or storing plants to adopt no other.
For graperies with the borders outside it would not be so
suitable ; but where good borders are made inside it would
answer well; or the vineries or peach-houses might form
the outer four houses of each block, leaving the plant-
houses, forcing-houses, &c., inside.

Fig. 52.

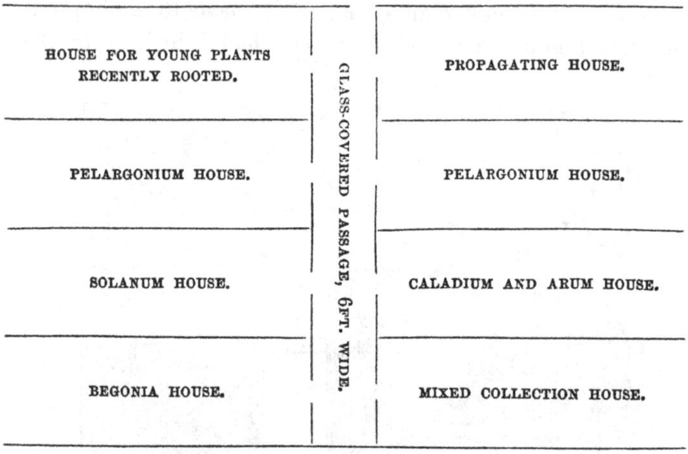

HOUSE FOR YOUNG PLANTS RECENTLY ROOTED.	GLASS-COVERED PASSAGE, 6FT. WIDE.	PROPAGATING HOUSE.
PELARGONIUM HOUSE.		PELARGONIUM HOUSE.
SOLANUM HOUSE.		CALADIUM AND ARUM HOUSE.
BEGONIA HOUSE.		MIXED COLLECTION HOUSE.

Plan showing the arrangement of glass houses in the Jardin Fleuriste.

Plants may be grouped in the passage, where narrow, in
half-oval groups between each door.   In large places, where
money is not an object, and where the houses on each side
would be filled with very ornamental specimen plants, it
would be a capital plan to make the central passage as wide
as one of the houses.   Beds may be placed between the
doors, in winter garden fashion, and climbers run up the
roof, thus converting the passage into a most agreeable
promenade.   With the better kinds of climbers depending
from the roof; a few belts of Oranges and Camellias,
and some palms and fine-leaved plants here and there,
to lend the scene grace and character, I can fancy
nothing more agreeable in the way of winter garden
or conservatory, particularly as the varied contents of
the houses on each side could be seen through the glass

ends and doors from the promenade. A wide gutter separates the roof of one house from that of its fellow—forming a passage along which men can freely move to arrange shading, ventilation, or repairs.   It will be seen at a glance that easy communication between all parts of the range is secured, that the plants just rooted in the propagating house have merely to be carried across the passage to the house devoted to their further de-velopment. The plan is capable of adap-tation in various ways, to houses either large or small.

One of the houses in the block just referred to is the largest and most perfect propagat-ing house I have ever seen—being more than eighty feet long and twenty-four feet wide. From this house im-mense quantities of plants are turned out in the course of a year, many of them being large-leaved Ficuses and plants that are difficult to strike, as well as Be-gonias, bedding and free - rooting plants. It contains three cen-tral and two side beds; the central pits are well elevated, and every space is in active work, the whole presenting a most imposing array of large bell-glasses.

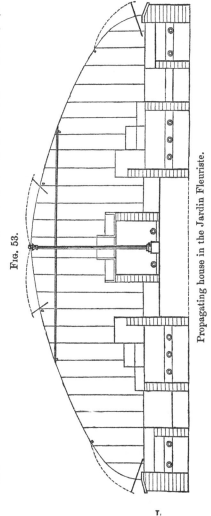

FIG. 53.

Propagating house in the Jardin Fleuriste.

T.

The propagating which seemed most successful, is carried out on a different plan to ours.

FIG. 54.

Propagating pot used in the Jardin Fleuriste : full size.

No pans are used in this house, but very small pots, a shade larger than a thimble : in each a cutting is placed, the little pots are placed in the tan, and covered with large circular bell-glasses, as shown by fig. 55. The greater part of the house is occupied with these, all being of the same size. But there are some special arrangements for propagating the more difficult subjects, and among them may be noticed what appeared to be an improvement—the bell-glasses, which are somewhat of the ordinary shape, being provided with an aperture at the top of about two inches in diameter, into which a piece of sponge is squeezed to absorb the moisture from the inside. Nothing could be more business-like than the arrangements for propagating in this house. We will next glance at a few of the more re-markable collections and structures.

FIG. 55.

Small cutting-pots under bell-glass.

Imagine yourself prepared to visit a propagating establish-ment, and then finding yourself ushered into a grand con-servatory of Camellias—a second being in connexion with it filled with Aralias, Yuccas, Beaucarneas, tree ferns, Nicotianas, Dasylirions, Dracænas, and a host of such plants, all in fine condition and well arranged ; and another, on the other side, containing healthy palms in vast numbers. These are arranged in three longitudinal beds, while all along the sides of the house is a belt of the smaller and younger kinds, plunged in tan to give them a little encouragement. To look along the pathway between these long beds is like glancing into a fresh young tropical palm grove, in such perfect health are the plants. When it is considered that

many other great houses are in the garden, besides a large
field of pits and frames, the reader will agree that it would
be out of the question to examine each subject, particularly
when it is stated that there are nearly 400 kinds of palms
alone in this establishment.   Though it is essentially a
business garden, and one in which an almost innumerable host of plants have to be annually developed, no
slovenliness of arrangement or culture is apparent in any
part.

Seldom indeed do we see such efficient economy of space
in gardens as is the rule in these houses. Under the benches
are packed quantities of Caladiums, Fuchsias, Cannas, and
other plants that may be efficiently preserved in such places
in winter; and even after the great Arums, &c., are potted
off in spring, they are placed underneath for a short time,
every available inch being taken advantage
of.   Some of the houses are large lean-to's,
and instead of the back wall being left naked,
or with one shelf placed against it at the
top, there is a series of shelves one above
another, six altogether, and on these a
multitude of plants are accommodated—
Coleuses, &c., in the warm houses; Lan-
tanas, and the like, in the cool.   They
keep well on these during the winter, and,
if drawn a little or discoloured, the mischief
is soon counteracted by a sojourn of a few
weeks in the frames in spring. In the large
span-roofed curvilinear houses, with a narrow passage through the centre, there is
a series of shelves affixed to irons on each
side of the central pathway, and on these
great numbers of plants are stored, so that
every space is taken advantage of without
in the slightest degree interfering with the
health of the plants, which is truly admirable.   But doubtless it is necessary thus to
economize space, for the enormous number
of nearly three million of plants is annually furnished by this

FIG. 56.

Shelves for storing
bedding plants along
the central passages
of the span-roofed
houses in the Jardin
Fleuriste.

establishment for the embellishment of Paris and its environs. They are raised at a very cheap rate—less than a penny each. It should be observed that many of the plants are such as would be fit to embellish any exhibition, numbers of them being palms and fine-leaved plants, while of course the least valuable are simply bedding plants, from Nierembergias to Pelargoniums, of which last 400,000 plants are sent out annually.

If neither houses nor plants were seen, the potting-shed would tell of extraordinary operations, for in the centre there is a great wide bench, around which sixty men can work. Ordinary bedding plants are kept here in an unusually economical manner. A large space of ground is covered by parallel lines of rough and rather shallow small wooden frames, simply and cheaply made—in fact, such as the rudest workmen could put together during wet weather. The frames are rather closely placed; and the pathways between, and indeed all the spaces around them, are filled up with leaves and mossy rakings from the adjacent Bois de Boulogne. These are nearly or quite piled up to the edge of the frames, and of course keep the plants warm through the winter. In winter the floor of the frames is low; in spring, by putting in a quantity of the well consolidated leafy stuff before named, it is raised so as to bring the foliage of the plants right up to the glass. All the material is removed from between the frames in summer. Many of these frames are furnished with iron sashes, so that only the rough cheap framework is exposed to the decaying influences of the weather. The large quantity of leaves and moss thus decomposed is preserved for potting purposes, making of course excellent leaf-mould.

A number of houses that have lately been erected at La Muette materially encroach upon the space occupied by the rough framing just alluded to, which they are destined eventually to replace. These houses are especially intended for bedding plants, and are so well adapted for that end that some details about them may be useful. They have been designed on an excellent plan for the culture of such plants, the raising of seedlings, and for the growth of seedling palms, and all dwarf and young plants. I have seen a good many houses devoted to similar purposes in public, private, or commercial gardens in all parts of these islands, but never any so well-arranged as those in the Jardin Fleuriste. They are low, and rather narrow, so that all operations may be conducted from the central pathway. The sashes are cheaply made of thin iron, and the roof consists of one sash at each side. Many of the iron sashes of the old frames were utilized in the building of the houses.

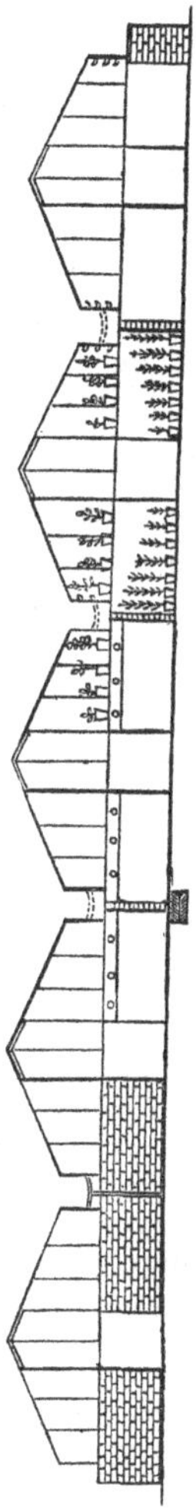

As you pass along by the ends of these plant houses you may see a bench about a hundred feet long, filled completely with the deeply dyed Alternantheras—a sheet of colour; the next devoted to young palms, as green and vigorous as if in their native wilds; another devoted to young Dracænas and fine-leaved plants generally; and so on. The benches are of slate, and the plants are held well up to the glass, while quantities of subjects in the way of Cannas and Dahlias may be stored beneath, as shown in the engraving. We generally prefer wooden houses, but any horticulturist who has seen the plants in this low range at Passy will agree with me that no plants could be in finer health or condition; while the very permanent nature of the structure is a great gain, inasmuch as a wooden series of the same character would require a complete overhaul in the course of a dozen, and perhaps reconstruction at the end of twenty years.

A mode of protecting these houses from frost by means of wooden shutters, each about the size of the sash of the house, is deserving of notice. As will be seen by the engraving,

Fig. 57.

End view of new range of bedding-plant houses in the Jardin Fleuriste.

the gutters, strongly lined with zinc, are wide, so that men can run along them with the greatest ease to protect or shade the houses. The shutters are not taken from between the houses every day, but simply left in piles of ten or so over some unoccupied spot, or if the range happens to be completely filled, each pile is shifted every day so as to prevent the plants beneath from suffering. The facility and simplicity with which these houses may, in a few minutes, be thus encased in wood to meet a very severe frost, and without the least untidiness of any kind, are admirable. However, matters are so arranged in the houses that they could dispense entirely with this precaution, which is noticed merely from its adaptability to many places where a great number of bedding plants have to be kept, and where the means of heating sufficiently to keep out very severe frosts are not forthcoming. The ground plan of the range is nearly the same as that already described, so that

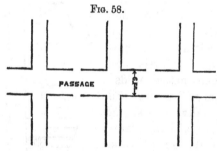

Fig. 58.

PASSAGE

Portion of ground plan of the bedding-plant houses in the Jardin Fleuriste.

the men at work in any of the eighteen houses of the block already completed, may pass and convey plants from one to the other without passing through the open air. Thus the comfort of the men and the health of the plant are both secured. Already nine houses are arranged on each side of the central passage, and it is proposed to continue the arrangement till all the ground previously devoted to framing is covered with this class of house. The visitor, entering at the outer end and continuing his way through any of the houses, would at its further end meet with the covered way running at right angles to it, through which he could enter any of the other houses he wished to see without again exposing himself or opening any doors to chill the plants in winter, or running the draughty gauntlet, as he usually has to do where houses are arranged in the ordinary scattered way. Moreover, as in many cases, one long house is devoted to a particular species or variety in much request, the visitor or superintendents may see the state of the stock by simply traversing the central passage, and looking through the glass dividing it from the houses.

But though the ordinary dwarf bedding plants are preserved in vast quantities both in the rough frames and the houses, these are not the cheapest ways in which they manage such things here, as we shall presently see. Many have heard of the graceful use made of the Cannas in Parisian gardening. These are preserved in a most efficient way in caves under the garden. When the stone is taken out of the ground for building purposes, a rough propping column is left here and there, and thus dark and spacious caves of equable temperature are left underground. They are in this case about seven feet high, and are used for storing plants that may be well preserved without light in the winter. You descend by a sloping tan-covered passage, and most likely you will imagine yourself in a large potato store immediately you get down, as heaps of different kinds of Canna, and those that are by no means common with us, are in winter spread upon the floor a yard or more deep, and twenty feet long. The tubers of some of the large varieties are from five to ten inches long, and the men turn them over just as they would the contents of a series of potato-pits.

Here too in wide masses against the wall are arrayed quantities of Aralia papyrifera, the handsome and much grown species so useful for subtropical gardening. It seems in a perfectly firm and safe condition, growing in this dark or rather gas-lighted atmosphere, and sends out long blanched leaves of a delicate lemon colour, which will of course soon acquire a healthy green when the plants are placed in the open air. Thus they preserve Aralia papyrifera in all sizes, and this fine thing is turned out for garden embellishment almost as cheap as wall-flowers. Of course analogous protection could be given to such things in many English gardens where space may be limited, and much expense out of the question. In these caves were also preserved Brugmansias, American and other Agaves, Dahlias, Fuchsias, &c., and it seemed to me about the best possible place for storing such plants.

The quantities in which you see rare things and new bedding plants here are surprising. Houses, eighty and a hundred feet long, are filled with one variety; and others of equal size are devoted to the raising of seedling palms, &c.,

Fig. 59.

Caves under the Jardin Fleuriste, used for storing large quantities of tender plants in winter.

in large quantities. If a plant be considered worthy of attention at all it is propagated by the thousand; 30,000 being the opening quantity for a new thing of any promise. During the past autumn 50,000 cuttings of one kind of Fuchsia were inserted in one week. Dracænas are grown here more abundantly than variegated Pelargoniums in many a large English bedding garden, and the Jardin Fleuriste is believed to possess the finest collection of them in existence. In one house a specimen of each kind has been recently planted out for trial in the central pit, and among them are many handsome kinds worthy of extensive use with us.

It is a favourite plan here to devote a house to a special subject. Thus there is a large and fine span-roofed stove for Ficuses; a house for the collection of Bananas, with a line of thirty healthy plants of Musa Ensete forming its backbone, so to speak; a very large and high curvilinear stove for the great collection of Solanums; special houses for Arums, Caladiums, &c.; and a winter garden about 120 feet long by 40 wide, well stored with a healthy stock of the usual conservatory plants, with here and there fine-leaved things like Phormium tenax, a very effective plant when well grown in pots and tubs, and of which they have here thousands of plants of various sizes. Of course all this vast collection cannot be and is not used for summer decoration. It is employed for the decoration of the Hôtel de Ville, where 10,000 plants are sometimes required upon a single occasion. The boilers of some of the smaller houses are heated by gas, and in this way a very equable temperature is preserved.

It may give some approximate idea of the collection, when it is stated that there are in cultivation nearly twenty species of Banana, about fifty kinds of Aralia, forty of Anthurium, fifteen of Pothos, thirty of Philodendron, nearly one hundred and twenty of Canna, eighteen of Zamia, and more than one hundred and ten of Ficus, while families better known and more popular are counted by hundreds!

Although the place is chiefly devoted to tender plants, and most of the dwarf hardy subjects are grown in the nursery in the Bois de Vincennes, there is, nevertheless, some interest taken in hardy plants, seeing that a part of the garden is devoted to one of the most extensive collections of Tulips in existence.

It is a regular practice in this and other new public gardens in France to plant out a sample of their stock of tender flower-garden plants each year for comparison. In the parks, squares, &c., they of course have opportunities of seeing how they thrive, but the object is to test them all growing on the one spot and under the same conditions. Thus,

you see all the kinds of Canna planted out in one place, all
the varieties of Pelargonium in another, and so on.   It is
a good practice, but it is needless to repeat it year after
year to a large extent.   If you have thirty species of
Solanum planted out for several seasons in succession, you
must know all that you want to know respecting their comparative value, and the practice here of planting out every
year old kinds time after time is useless.   All that is necessary is to test the new additions, and in some cases it
may be desirable to plant the old ones by them for comparison, but to plant out annually a vast collection from a
well-known family is quite unnecessary.

Large, light, and well-made spring vans are used for
transporting the stock of flowers from the Jardin Fleuriste
to the parks and gardens, and from one nursery to the other.
They are about twelve feet long, and a little over six feet
wide.   By a simple arrangement each van is made to do
the work of two—a second floor of strong shutters, hinged
two and two together, being placed at the height of a foot
above the lower floor of the van.   The shorter plants are
stowed underneath, those on the upper floor may be as tall
as you like; but as the stock removed in this way usually
consists of dwarf subjects, one serves as well as the other.
By means of this plan 2000 plants, each in single pots, are
removed at a time.   The contrivance is merely such as
common sense would suggest ; yet for want of a little such
common sense how much labour is wasted ! How frequently,
for example, do we see in country places two men attached
to a handbarrow dragging about plants !   Of course it is
as unnecessary as it is laborious for the men.   There is often
more fuss and labour over transporting the summer flowers
of a country place from the propagating houses to the flower-
garden than occurs with the several millions of plants furnished yearly by the city of Paris, and all for the want of a
few simply-contrived spring barrows. Not to adopt simple facilities of this kind in our public gardens is sheer mismanagement.

There are also vans of peculiar make for conveying ornamental plants to the Hôtel de Ville.   Those used in winter
are furnished with a little stove with flat hot-water pipes

passing round the interior, so that, while space is not cur-
tailed, the van is efficiently heated, and tender plants can
be conveyed by it in safety in the depth of winter.

Students of all nations are admitted to this establishment.
They must be eighteen years of age, and must have spent
some time in practical horticulture. Their pay is sixty francs
per month during the first three months, seventy during
the second, and after that eighty or eighty-five francs per
month, after which they are paid according to capacity
and intelligence. They are changed from section to section
of the establishment, so as to study with profit each kind
of culture. An extensive botanical library has lately been
added for the use of the officers and students of this establishment, and is now being catalogued and arranged. It contains nearly all the standard English books on horticulture;
indeed quite half the books are English.

Attached to the Jardin Fleuriste are a forge, a carpenter's
shop, a glazier's and painter's shop, stables, and other offices.
These are of course indispensable where economy is necessary; and saving money is a consideration even for the
city of Paris at present. The mode of glazing with several
strips of lead-paper laid one over the other, as practised
here, is too expensive to be recommended: it costs as much
as the glass itself, and after all peels off after a time. It is
known as the couvre-joint métallique of Celard, 16, Rue du
Faubourg du Temple.

*The Public Nurseries for Trees, Shrubs, and Hardy
Flowers.*—The nursery for trees for the boulevards is situated
at Petit Bry, near Nogent-sur-Marne—a somewhat out-of-
the-way place. The nearest railway station to the nursery
is that of Nogent-sur-Marne, on the Strasbourg line. It
consists of nearly forty-five acres, entirely devoted to the
raising of the commoner and more useful kinds of trees for
avenue and boulevard planting. On entering it the first
peculiarity that strikes the visitor is, that the whole of the
surface of the ground is thrown into ridges nearly six
feet in width, on the apex of which the trees are planted.
This arrangement is adopted in consequence of the ground
being occasionally flooded by the river Marne, which is close

by, and the trees being injured by the water being frozen above the base of their stems. But the necessity of taking this precaution resulted in an advantage, as the trees being planted on the apex of these ridges, and with the collar of each, say, a foot above the level, make their roots much nearer home, so to speak, and thus their transplanting is rendered much more easy. When the time comes for removing them the workmen begin at one end and turn them out quite rapidly, all with close bundles of roots. The whole surface of the nursery is thus treated. The trees are a little more than a yard apart in the lines, which are, as may be inferred from what was before said, within a few inches of six feet from each other.

The kinds mostly used are the Western Plane, the Horse-chestnut, the large-leaved Elm, the Ailantus glandulosa, Planeras, and Lombardy poplars—the last, however, are not used for avenue or street planting. Other kinds used on a smaller scale than these—the Paulownia, for example, are grown at Longchamps. These trees, the names of which are put down in the order of their importance, are all trained straight, and sent from hence to the boulevards for planting as far as possible of an equal size. The rule is to send them out with a clean stem nearly ten feet high, and about eight inches in circumference. The portion above the ten feet of clear stem is not of so much consequence and may vary, but if the trees when taken up for planting do not present the length of clean stem considered necessary, the lower branches are cut away till it is attained. Of course the trees are so pruned when young that straightness of stem is obtained. To arrive at the necessary size and fitness the Plane requires five years, the Horse-chestnut ten, the Ailantus four, the Elm and Planera about five years each. The Elm and Planera are the only trees that require support in training them into the necessary form, for which purpose stakes from fifteen to eighteen feet high are used. The whole place is surrounded by a hedge of Tamarix tetrandra, which is cut down occasionally, and the shoots sent to the Jardin Fleuriste for stakes for house plants and the like. This nursery is well kept and managed, and has a large stock of street trees.

The nursery for shrubs is very pleasantly situated near the racecourse of Longchamps in the Bois de Boulogne, and is somewhat more than twelve acres in extent. I found it in excellent keeping, and with a good stock both of well-known and rarely used subjects. Roses and all kinds of shrubs and hardy climbers are grown here, as well as nearly every description of low tree. The superintendent considered the Caucasian Laurel (Cerasus caucasius) the hardiest and best of any he had tried. There were good stocks of those fine hardy Aralias—spinosa and japonica : they should be everywhere employed for the sake of their large and handsome leaves. A good many subjects were out for trial as to their hardiness, among them an extensive collection of Japanese plants. Melia Azederach was in a healthy condition after passing a sharp winter in the open air. From this nursery all the shrubs of the various parks, squares, and gardens of the town are supplied.

The nursery for herbaceous plants is situated in the Bois de Vincennes, and consists of nearly twenty acres of sandy ground just outside the fortifications, near the Porte Picpus and Lac Daumesnil. There were here, at the time of my visit, five or six acres of Chrysanthemums, prepared for bedding in the various parks as soon as the frost had cleared them of their summer occupants. There were also large stocks of the flowers used to replace the Chrysanthemum and decorate the gardens in spring. The stock of spring flowers is an unvaried one, and leaves much to be desired. Where there is so much ground devoted to a specialty it ought to be well done ; and it will be a pity if with so much improvement in other ways a large stock of all the really ornamental hardy flowers is not formed. The public gardens cannot fail to have a great influence on all visitors to Paris, and it would be conferring a very general benefit if, instead of depending so much on plants requiring expensive stoves and ceaseless trouble for their preservation, the chief gardeners of the city showed what may be done with the hardy plants belonging to our own and similar climates. At present their collections of herbaceous plants and spring flowers consist of quantities of common and not always first-class kinds.   They have, for instance, very few Tritomas in the Vincennes Nursery, and none at all in the parks, though they are perhaps the most useful and attractive of all autumn flowers.   It is, however, only fair to state that the nursery stock was killed in the winter of 1867. But when groups of these plants are established in the parks or gardens there should be little difficulty in preserving them by placing leaves over the roots in winter.

The nursery for the Pines and Rhododendrons is also in the Bois de Boulogne, near Auteuil, occupying somewhere about the same space as the one previously described.   The climate of Paris is not so favourable to the growth of coniferous trees as that of England, and consequently to the English visitor the Auteuil garden does not look so attractive as that at Longchamps, but it is well stocked, and serves its purpose admirably.   The American plants are mostly grown in the slight shade afforded by thin hedges of Arborvitæ.   The ivy used for making the edgings, which are so much admired in Paris gardens, and for every other purpose for which the plant is employed, is grown here.   Cuttings are first put in in handfuls, so close that the stems touch each other.   After a year or so they are transferred singly into four or six-inch pots, and plunged below the rims into the sandy soil.   They are used for forming the edgings at the age of two or three years.   Galvanized wire is extensively used here for the purpose of supporting plants that are usually staked.   Stretched tightly in parallel lines at about the height the line of plants requires it forms a neater, handier, and cheaper support than ordinary wooden stakes, which are so liable to decay and shake about.

CHAPTER IX.

TREES FOR CITY PARKS, AVENUES, GARDENS, STREETS, ETC.

It is a very popular but utterly erroneous notion to suppose that " trees will not succeed in London." On the Continent people are accustomed to see wide open streets and road-ways embellished with trees that are properly planted, well cared for and flourishing. They naturally at once compare these verdant avenues with our own streets, in which trees are never planted at all, or where, if they are, no care is taken of them, and at once jump to the conclusion that there is " something in the air." If you tell them that trees may be grown better in London than in Paris they will stare at you in incredulous amazement ; but such is nevertheless the fact. In August last a correspondent of the *Pall Mall Gazette* wrote several letters to that journal against planting trees in London, the following extract from which affords an excellent example of the notions almost universally held upon this subject.

" When people propose to decorate London, the first con-sideration should be what will suit the climate. It cannot be too often repeated that our city atmosphere is fatally inimical to delicacy of architecture, and quite equally so to delicacy of vegetation. Our skies will rain soot continually, and moisture therewithal to make the soot adhere ; the soot will insinuate itself amidst fairy tracery of stone, and clog the pores of beautiful trees and shrubs ; and it is an utter waste of art and money to disregard these inevitable conditions of the question. It is very childish to tickle our fancy by providing for a momentary admiration of things which a short time will make hideous, and then, when the inevitable has taken place, contenting ourselves with a shrug of the shoulders and ' what a pity !' Few young trees will really flourish in the climate of our modern London ; the case was perhaps different some years ago.　But from any general planting of trees in London, especially in leading thoroughfares, however wide, I cannot expect agreeable results.　I cannot dismiss from my mind that mournful spectacle I have so often witnessed with depressed spirits, all through the after-summer (as Germans call it) and the autumn, of gloomy civic avenues, every trunk black and filthy, with all its fur-rows clogged with soot, the branches showing symptoms of speedy decay, the scanty withered foliage distilling a drizzle of mingled smoke and moisture.—MISODENDROS."

These opinions are as erroneous as they are emphatic ; yet it is not to be wondered at that similar ones are enter-tained by the general public, when we find those who ought to know equally ignorant on the subject.　Not long ago, I was walking up Regent Street with a landscape gardener who had mostly worked in pure air, and he almost ridiculed my statement that trees could be grown in perfect health and beauty in London.　I felt it was useless to argue with him, but remembering the splendid Planes in Berkeley Square, a few minutes enabled me, through his own eyes, to cure him for ever of the erroneous opinion that trees cannot be grown well in London.　It is the custom in Paris and other continental cities to plant trees with care, to provide them with good soil, to spend a great deal of money in attend-ing to them and watering them, and yet neither in all Paris, nor in any continental city with which I am acquainted, can such noble examples as these be found.

But some may say, An open square at the West-end of town may do that which the smoky, densely-packed city will not. If these persons, who are evidently not yet acquainted with Stationers' Hall Court, will inquire for that narrow enclosure the next time they are passing near Paternoster Row—or St. Paul's Cathedral, to select a more conspicuous land-mark—they will find in it a noble Plane tree looking as happy as if it were in its native forest.　It grows in what to a tree is practically a brick well, and yet to stand under it in summer and look up the bole towards the top of the tree, is to get a glimpse of tree-beauty of the most refreshing kind. I could point out to friend " Misodendros " numerous places in the heart of London where trees flourish in the most satisfactory way. It must be borne in mind, however, that everything depends upon the kind of tree selected for the purpose ; for even our best landscape gardeners make a sad mistake by obstinately persisting in planting evergreens, which as a class are totally unfit for town cultivation. Even when moderately healthy these trees are generally so coated with smut that they entirely lack that polished and refreshing verdure which is so characteristic of evergreens grown in fresh and pure air.

In winter the atmosphere of London, and of many of our great cities and towns, is contaminated by certain minute bodies vulgarly termed " blacks." They belong neither to the vegetable nor to the animal kingdom, but they exercise a powerful influence on the former, for they are deadly enemies to all breathing surfaces ; and though they have not caused civil war and bloodshed like their human namesakes, yet if the tale of death were carefully summed, no doubt our " natives" would head the fatal list by a long way. How then fares it with the vegetation subjected to their pernicious influence ? Thousands of expiring and leafless shrubs furnish a reply. The once handsome and healthy Araucarias planted in front of Tattersall's at Knightsbridge, and now draped with filth and soot; the young pines and evergreens planted annually in the Regent's Park Botanic Gardens, only to dwindle and die ; the handsome Hollies, Yews, and other shrubs planted in the new avenue gardens in the same park ; the fine and costly evergreens in the Royal Horticultural Gardens at Kensington when it was first laid out,—these and many other cases that I could enumerate, were it necessary, answer the question, and tell the same tale of how they were deprived of life by the vile atmosphere.

Not so with the deciduous tree or shrub ; nor with those beautiful rosaceous bushes which are the glory of the grove in all temperate climes. After summer's " fitful fever they sleep well," and when " people return to town," and the flues begin to vomit forth poison and smut, they cast away

M

the leaves that have done their work, and with them the
filth of the year; and, so to speak, retiring within themselves
they remain till the winter is past, safer from deadly vapours
than the Esquimaux in his snow-hut is from cold. The
consequence is, they grow nearly as well in London as in
the country. When the fires begin to go out in spring,
and the air of towns becomes more free from evil humours,
they burst out into leaf and beauty—clouds of light, fresh,
budding green. What sight on earth can surpass the
bursting into leaf of deciduous trees in temperate and
northern climes ? We should see few things more beautiful,
nothing more magical, even if it were possible to pass high
over earth, like the swallow in its migration from the wolfish
north, with its pines and weird heaths, to the south with its
Vines and Oranges, and to the tropics with its Palms and
giant Bamboos. No charm of tropical or other climes sur-
passes the freshness and joy of an English spring. Why
then should we not take advantage of the fact, and make
our city springs more English still, by developing chiefly
those plants which flourish as well in towns as out of
them, instead of everlastingly purchasing evergreens which
are doomed to perish sooner or later? I have repeatedly
noticed that Peaches, Almonds, the double Cherries, and
the numerous exquisite trees and shrubs allied to them,
flourish and attain the same perfect shape in towns and
cities that they do in the country, while beside them valuable evergreens are but the ghosts of what evergreens
should be.

Supposing for a moment that evergreen trees and shrubs
throve as well within city influences as deciduous ones do,
it would even then be a questionable practice to use them
extensively, because they do not gladden us with that floral
beauty which deciduous trees are wont to put on; their
verdure in the parks and open spaces goes for little in
winter, as at that time people seldom frequent these places.
They do not keep time with our suns and seasons; and they
are not so beautiful, because not so changeful, as the deciduous kinds. They do not flower or fruit conspicuously,
as many deciduous plants do, and they tend to preserve a

disagreeable moisture round us in winter, which we are certainly better without.

I am certain that if the expense and trouble taken to plant evergreens in cities were devoted to the best of our deciduous trees and shrubs, a beauty would result to which towns are at present strangers. Even in parks and places

FIG. 60.

Sophora japonica var. pendula. One of the many deciduous trees of which the wintry aspect is preferable to that presented by smutty and half-dead " evergreens."

where one would be led to expect a tolerable display of fine flowering deciduous trees, the shrubbery vegetation is so intolerably poor and monotonous—so devoid of variety and interest—that it is not surprising that town planters fall back on evergreens and plant little else round their churches and in their squares. A fine double Cherry, pyramidal in outline, and hung with snowy bouquets, seen against one of

M 2

our dark churches, would be more beautiful than all the
evergreens within four miles of Charing Cross, and yet it is
only one out of a host of flowering trees belonging to
temperate or northern climes, nearly all of them far more
presentable objects even when leafless than the debilitated
soot-varnished " evergreens" which we now select for town
planting.   Even the Pear and Apple and the Hawthorn
families would furnish a grand array of beauty ; but any one
who examines the list of our deciduous trees and shrubs,
from the tall Acacias to the dwarf early-opening Daphnes,
may find a selection which, judiciously arranged, would
create a greater attraction in town gardens than has yet
been seen.   All who know the amount of beauty to be
found among deciduous trees will have no difficulty in
imagining how attractive our parks could be made by tastefully grouping and cultivating other flowering trees of equal
or nearly equal merit.   All those mentioned above thrive
well on the London clay, and indeed the same is true of the
majority of deciduous shrubs.   It would be a great benefit to
city gardening if landscape gardeners were to be cooped up
in town for a few weeks in the dead of winter, instead of
being permitted to run about the pleasant country : they
might then consider our wants more than they do.   Meantime, I strongly advise city planters to pay nearly all attention
to deciduous vegetation, promising them that their efforts
will not be thrown away, as they too often are at present.
They would then find that planting trees is not a " sheer waste
of art and money," but one of the most praiseworthy modes
of rendering our great unwholesome and ugly human hives
healthy, habitable, and cheerful.

    Although so deficient in street trees proper, one of the
best and most distinctive features of the suburbs of our
English cities is that resulting from the practice of placing
little gardens between the house and the road : it is the
absence of these which gives such a hard, uninviting, and,
to an English eye, hungry look to the unplanted streets of
many towns on the Continent. Although the space is small,
a line of trees is usually planted immediately inside the
wall, which line sometimes acts as a screen, but is generally

ineffective for that purpose. Considering the space at our disposal, and the fact that strong-growing trees prevent, to a great extent, smaller, more useful, and prettier subjects from being grown in these gardens, the right method to adopt would be never to plant anything stronger in them than dwarf trees. We certainly have a considerable gain in the large number of streets and suburban roads, where little gardens run along in front of the houses, affording greater breadth, and a little repose between the house and its inmates and the hard and dusty street. It is a pity, however, that when neighbourhoods become populous, crowded shops are built upon these gardens, additional rooms being eventually placed over them, thus narrowing instead of widening the street at the very time when more space is required for increased traffic. The Marylebone-road is an example of this kind.

The street gardens of London, and our other large cities, are true British institutions which I hope will never become unpopular; but go on increasing in favour. The following selection of city trees has therefore been made with a view to their improvement, as well as to the requirements of street or boulevard planting, in which branch continental cities are now before us.

A selection of the best trees and shrubs for cities.

The best of all trees for European cities is the Western Plane (Platanus occidentalis). I have seen it in many places in towns, from the heart of the city of London to the shores of the lakes of Northern Italy; in the town gardens of central France, in the fine old cities in La Belle Touraine, and in Anjou, where the Camellia and Azalea grow luxuriantly in the open air; in Brittany, where the glossy evergreen Magnolia becomes a tree bearing huge waxen flowers as big as plates; in the numerous new boulevards of Paris; and everywhere it is by far the noblest city tree; but in no place are there finer individual specimens of it than in London, although receiving no such attention as they do elsewhere.

Looking for a moment at the Plane in a wild and culti-vated state in the pure air, we find it second to none. It possesses the hardiness of a North American Indian, and the massive and noble port of " Daniel Boon, backwoods-man, of Kentucky." Found in a wild state over a vast portion of the North American continent, in its fullest per-fection, along the great rivers which fall into Chesapeake Bay, and in the fertile valleys of the West,—where it is constantly found to be the loftiest tree of the United States, it becomes one of the noblest trees in British parks and woods, and reminds us of its native land of great trees, rivers, woods, and prairies. Fairly and roundly de-veloped specimens have in summer almost the grace of a weeping tree. In winter the branches retain this character, but also present a rugged Gothic picturesqueness, which makes them highly agreeable to look upon, while the pendulous seed-vessels and striking column-like bole add to their at-tractions in the wintry season, when the trees are at rest and safe from the evil effects of smoke. To these advan-tages may be added the one that large specimens may be transplanted with safety—a very desirable point in a city tree. I could point to many parts of London where what is here stated of the advantages of this tree could be seen in a moment—from north-western squares near smoky King's Cross, to the western and southern parts of town, with a drier soil and better air, and even to the very heart of the city, where it appears to do as well as anywhere else.

As we are now almost commencing street and city garden-ing it is most desirable that we should have no failures—that things of this kind should be done so as to satisfy all. Places like the Thames Embankment should command the finest tree :

" We needs must love the noblest when we see it,"

says Tennyson ; and amongst town trees it must be the Plane. I can imagine nothing more calculated to bring town-garden-ing into disrepute than such a specimen of planting as that in the Mall in St. James's Park. Had the Plane been planted there it would have made a noble avenue—the Elm now forms a miserable one.

Next to the Plane, the Horse-chestnut seems to offer the greatest advantages. It has not indeed the stature and beauty of frame of the Plane, nor does it attain as large a size and as perfect health in cities, but it possesses great claims from its fine foliage, large sweet silvery spikes of bloom, and proved capacity of growing well as a town tree, even where the ground is hard and root-room scarce. If Paris is seen to greatest advantage when her groves of this tree are piled up with little pyramids of flowers, what might we not expect from the fact that it does even better with us? The avenue of Horse-chestnuts in the Regent's Park is a case in point. In Paris during the past year they lost their leaves rather early and became too rusty to be agreeable to the eye; but on coming to London at the beginning of August I found them in a green and healthy state after one of the most trying summers we have ever had. While selecting picturesque trees for towns that are not liable to suffer from disease or insect pests, we must also encourage variety as much as possible. The Horse-chestnut would be worth growing for the sake of its foliage alone, but when the additional charm of its superb inflorescence is taken into account there can be no hesitation in placing it among the most eligible of town trees.

The common Robinia or Locust tree has been so long and extensively tried that we need have no more doubt about it. It will never justify the reputation that Cobbett gave it, but I know of no tree which maintains such a depth of sweet verdure and freshness by the sides of the dustiest roads and in the most unlikely places. No drought seems to touch it; no heat renders it rusty-looking or fatigued. Few other trees stood the heat of the summer of 1868 so well, and after the drought was ended it looked as if it had just passed through a showery month of June. It is worthy of being much more extensively used as a park and square tree; it is also good for street use, not growing too large, and is the best of all known trees for planting in the front of a suburban house or villa, or in any position where a pleasant object is required to refresh the eye at all times. Compared to the Lime which is so often planted before London houses, it is as gold to pewter.

At first sight there seems little reason why the somewhat despised and roughly treated Robinia, or Acacia, as it is sometimes called, should come in after such stately and noble trees as the Plane and the Chestnut; but, taking the varieties as well as the original tree into consideration, I have no hesitation in giving it this rank, knowing it to be as well adapted for the smallest town garden as for the largest public park. Naturally it is not such a strong-growing tree as the Lime, while it may be cut in to keep it neater than it usually grows.

To many lines of suburban houses a thin line of trees is a great improvement, and forms the only species of garden embellishment of which they are capable. The qualities necessary in such trees are perfect hardiness, healthy constitution, and size and habits suited to the positions for which we require them. I know of no tree that combines these better than the spineless round-headed variety of the Robinia (R. inermis), and it is a very elegant object all through the season. It is, to be sure, somewhat dearer than the Lime and such trees; but the difference in appearance is such that nobody would refuse the difference in money, even for the improved appearance of the trees during a single year. It is usually grafted on straight stems, six feet to eight feet high, which support the umbrella-like heads and their mass of graceful, healthy green leaves. With a little cutting in now and then, they never become an inch too high.

Perhaps the most beautiful and appropriate city trees I have ever seen are those formed by the round-headed Robinias in the cities of Northern Italy; their grace, dense and grateful shade, and deep verdure being perfect. I measured several thirty feet in diameter of head, and with a bole a foot or more through, the heads being picturesque and somewhat irregular from age, while preserving their compactness and valuable shading properties. It would be impossible to find a greater advance upon the hideous lines of clipped Limes so common in France than is presented by these trees at Novara and other cities and towns in North Italy. But as we have no proof that as good a result could be obtained in our English streets, we must turn to those trees that we have already tested thoroughly.

We will next deal with the Lime, its bad and not its good qualities placing it so high in my list; for, while planted more abundantly than any other city tree, it is by far the worst that I am acquainted with, and the extensive use of it in our streets is the most blameworthy of bad practices in either town or country planting. I am speaking of the aspect of the tree as displayed in cities. I have seen the tall lines of Lime trees in the Jardin des Plantes at Paris fall into the sere and yellow leaf before one had time to admire the pleasing soft green which they display when their leaf blades are first rolled out; and on coming back to London, at the end of July, 1868, I found the Limes the most miserable and mangy-looking trees anywhere to be seen. And all this in the midst of the summer when we are most oppressed with heat! The withered, burnt, insect-covered leaves rustle lifeless upon the trees, hoarsely whispering the death of the year in our ears before we have half enjoyed the summer. In many cases they have perished prematurely and unnaturally, and have even lost the power to fall off the tree, but remain rustling on the branches, giving the ear as well as the eye a foretaste of winter three or four months before the proper time.

Can anything be more unwise than to persist in planting such a tree as largely as we do when there are dozens of deciduous trees that will do all that a Lime does at the best of times, and that present no such objectionable features as those alluded to at that season when a tree ought to be full of life and beauty? Our winter, the period when our deciduous trees must be devoid of leaves, is long enough without making it needlessly so by lining every street with the Lime. In the parks this tree may sometimes be planted, but never in streets, quays, or boulevards. Apart from its presenting a diseased appearance for more than half the time that it ought to be full of green life, the Lime grows much too large for the little front gardens where it is so abundantly employed, and will soon keep away from the house it is planted to adorn a large portion of the light and sun that we grumble so much at being deprived of. If, in order to obviate this, we cut it in periodically, it becomes an object that every person of taste should abhor and cut down as soon as possible, to say nothing of the labour and expense of this periodical mutilation. And all this is in face of the fact that we have several handsome trees that do infinitely better and without clipping. In many cases in towns it would be better, instead of planting a coarse tree of any kind, to cover the railings with the Irish ivy, as the French do; and then, no light being intercepted, it would be possible to have something worth looking at in the little garden, and the heavier rushes of dust would be kept out by the dense covering of ivy, which would moreover look green at all seasons. Dwarf shrubs more suited to the size of the place and plenty of flowers might then be grown with success. Every Lime tree in every small garden in London should be cut down.

The Elm is a tree much used in the London parks, and sometimes seen of fine dimensions, but occasionally it is much diseased when used as an avenue tree—for example, in the Mall, in St. James's Park, where the effect of the avenue planting is as bad as it can be. And all this from not having selected a good kind of tree at first; indeed it is so bad that there need be little surprise at our not yet having attempted street-planting. A few Plane trees near the Buckingham Palace end of the Mall almost save it from looking absolutely hideous from that point of view. The effect of the Elms in Rotten Row, though much better, is not nearly so good as may be produced in like positions by using other trees. The variety chosen has a good deal to do with it—the long Boulevard St. Michel, in Paris, planted with the large-leaved Elm, is quite a success. However, looking to the gross insect enemy of the Elm, and its aspect when planted, as we plant it in avenues, it is not a desirable tree for this purpose, though indispensable for grouping in the parks.

Paulownia imperialis is a very noble subject for town gardens, especially so for those on a dry soil like Paris, and where a good shading medium is wanted. It might well replace several of the miles of poor clipped Elms and Limes about Paris, and around numbers of French country houses. It has been well employed in Paris, as may be seen on the north side of the Boulevard Bonne Nouvelle. During the first days of September, 1868—days as hot as they were in July—I examined these trees, and found them as green and as fresh as could be desired, when Chestnuts, Limes, Elms, and all around looked as if they had passed a few moments over a brisk fire. The stems of the trees were straight, and just about the right elevation for a shade-giving tree, and the heads spread out flat, so as to give complete shade without betraying an awkward tendency to rise too high, so as to require clipping to prevent them from keeping the light from the upper windows. The large leaves were quite fresh—a sufficient proof after such a season as the past that it is one of the very best trees for city planting. The Paulownia, the Ailantus, and the Plane seem to preserve a freshness and vigour no matter how great the heat and abundant the dust. The forest trees of our own latitudes do not do this, and fall into the sere and withered leaf while their companions from Japan and America are in the greenest health. It is not easy to imagine a greater improvement than that which would be effected by planting this tree where a low and yet good shade is required. It is also worthy of attention as a town garden tree, and for similar reasons.

Ailantus glandulosa—sometimes called the "Tree of Heaven," and by the French Vernis du Japon—is a town tree of great excellence. When in a young state it is graceful from its long pinnate leaves—when old and well-grown it becomes a noble forest tree. But the qualities that will above all others recommend it to the town planter are its perfect health and freshness, under all circumstances, in towns. Dust, foul air, or drought seem to have little or no effect upon it. For parks and avenues it is indispensable, as it perfectly retains its foliage long after our own deciduous trees have been scorched by drought and dust. It seems to do equally well on all soils, having a constitution and a leathery texture which seem perfectly indifferent to any vicissitude of climate witnessed in these latitudes.

Everywhere in cities that beautiful and distinct tree, the

Lombardy poplar, retains its glossy health and vigour, proving its claims to be far more abundantly used than at present. Avenues of this tree would tell as well in some positions in cities as single specimens and groups of it do in the landscape. The drip of trees is sometimes objected to : this erect and close growing kind would seem to offer itself for rather narrow streets and positions, where a spreading habit or drip might be an inconvenience at any time. I do not recommend its use on an extensive scale; but it is so handsome and distinct in outline, stands drought, dust, and bad soil so admirably, that we are bound to recognise its merits far more than at present, and there are many positions in London in which it would be highly appropriate. Of other tapering columnar or fastigiate trees, the pyramidal variety of the Oak and the tapering variety of the Robinia do particularly well in the parks ; and the last is deserving of recommendation for the town garden.

Of weeping trees, in addition to the long-proved and indispensable Weeping Willow—which it is needless to recommend here, as most people will have noticed its bewitching outlines in some of our parks—we have the weeping Birch, Ash, Beech, and Elm, in all cases in perfect health in the parks. There is one tree of those above-mentioned which deserves to be much better known—the weeping variety of the large-leaved Elm (Ulmus montana pendula). This is a tree of much beauty and character, and it does not seem in the least to suffer from the atmosphere of London. It is a weeping tree of the first order : its foliage is massive, shade dense, and outline most picturesque when thickly clothed in summer—the backbone, so to speak, of each widespreading branch being seen just glistening above the dense mass of leaves, in consequence of none of the branchlets showing above their support. They are all of the true drooping tendency. It is a hardier and better constitutioned tree than the Weeping Willow, and never grows too high for a London or any other town garden; however, it spreads too wide for those of the smallest pattern. In all courtyards or open gravelled spaces, in little squares wherever a shady tree is desired, it is invaluable. To form a shady bower

there is nothing to surpass it. Should anybody doubt this,
I refer him to the specimen of this tree on the lawn of the
Botanic Gardens in the Regent's Park. I have also seen it
in perfect health in small squares in less airy parts of
London than that just named.

A very charming town tree is the weeping variety of
Sophora japonica (S. j. pendula). This is perfect as regards
size, not spreading so wide as the weeping Elm, Ash, or
Willow, yet quite as graceful as any of them. It is always
densely green, no matter how hot the season, and enjoys a
poor sandy soil, and the dry conditions from which, from
overdrainage and other causes, town trees are liable to suffer.
Bear in mind, however, that this and all weeping trees should
only be used where they are not likely to suffer from mutilation of any kind, and where their character and grace may
be seen and enjoyed.

The numerous free-growing trees of the Rose order, from
the Chinese Pear and the Almonds that illuminate our groves
with masses of light rosy flowers in earliest spring, to the
dwarf double Prunus, all grow healthfully in our parks;
and though unfit for street-planting, are worthy of the
highest attention both for small gardens in towns, and for
squares, public gardens, and parks. It would take a very
long list to enumerate all the really handsome members of
this family now almost entirely neglected in comparison to
their merits. They alone are almost capable of saving us
from the aspect of the soot-covered objects courteously
termed evergreens; and in spring, when all the world bursts
out in leaf, they are almost typical of London seasons.
From sooty-brown sticks they would "spread out their little
hands into the ray," quickly become clouds of virgin green,
afterwards great bouquets of flowers, white, pink, and
rose, would give shade and verdure as well as other trees
in summer, and in the fulness of time become covered
with gay fruits. Let us, for example, look at the Hawthorn
family, known popularly through one of its members, the
common May, the admired of everybody. In the Phœnix
Park in Dublin there may be seen many thousand quaint,
gnarled, indigenous plants of it—old fellows that must have

first sprung from their tiny stone many generations ago.
During the flowering season the whole park (about 1760
acres in extent) is perfumed with them, and nothing can
be more agreeable than a stroll there of a May afternoon.

Of course the Phœnix Park is practically in the country,
but the group near the museum in the Botanic Gardens in
the Regent's Park proves that the hawthorn family will
succeed perfectly on what is by far the worst kind of soil,
and the worst enemy with which the London gardener has
to contend — the deep bed of clay on the north side of
the city.   Generally people regard the Hawthorn as a thing
apart from all others, and know little of the varied beauty
of the family ; but the fact is, it furnishes a greater number
of hardy ornamental dwarf trees than any other known to
us.   They are not only pretty and fragrant in flower, but
the aspect of the fruit in autumn—borne in showers of
bright red, yellow, black, and scarlet—is of itself a recommendation which should entitle them to general cultivation,
even if the bloom and fragrance were of that obscure type
which never attracts the attention of any but a botanist—
a type too common among our popular trees and shrubs.

Of the common Hawthorn alone, the double and pink
and other varieties are capable of a rich display of beauty in
spring ; and the fruit, too, varies in a remarkable manner.
Varieties are grown with black fruit (Cratægus Oliveriana),
yellow fruit (C. aurea), woolly fruit (C. eriocarpa), and white
fruit (C. leucocarpa).   C. præcox, the early flowering, is the
Glastonbury thorn ; C. punicea flore pleno, the double
scarlet-flowered ; and there are many other varieties of the
common haw.   In fact, the plant would serve as well as
many to illustrate the variation of which a species is capable,
if such were wanted by a Darwinian.

If a single species displays so much ornament and
diversity, how much more may we expect from its numerous
congeners, all of which are hardy?   To have a full idea of
their value it is necessary to visit places where a collection
is grown not only in the early summer-flowering season, but
also in the autumn.   The brilliancy and profusion of the
fruit—some of them many times larger than the common

one, some of an agreeable acid flavour, and others like miniature apples, both in shape and taste—are quite refreshing amidst evergreens and common trees which never produce a noticeable fruit or flower. Some are as large as marbles, others more pyriform in shape, but large and eatable; such indeed as I should be very glad of if I were cast ashore on some desolate isle, like old Byron, and such as would have been a godsend to poor Burke and Wills and their party, who lived upon the tiny and miserable Nardoo fruit. Where the feeding and attracting of the feathered tribes is a consideration, there is nothing to equal these exotic thorns. Among the best kinds are the following :— C. coccinea, and its varieties, corallina and maxima ; C. nigra ; C. crus-galli, and its varieties, splendens, pyracanthifolia, and salicifolia ; C. punctata, and its varieties ; C. macrantha ; C. Azarolus ; C. obtusata, a variety of the common species which grows seventy feet high ; C. Douglasi, a purpleberried North American kind, named after the famous and unfortunate plant-collector Douglas, who sent us home the noble Douglas fir and a host of valuable American plants ; C. Orientalis, C. Leeana, C. Aronia, berries yellow ; and C. tanacetifolia, a native of Greece, and its German variety glabra ; but almost all the species are worth growing. The well-known evergreen species, C. pyracantha, so extensively used for training against houses and walls, will not do for association with these ; but it is of course valuable for the embellishment of the walls of the town garden.

It should be observed that the above species flowering at various times, and some of them a good deal later than the May and its numerous varieties, prolong the bloom of the family for a considerable period. They are more suited for grouping in the irregular and diversified parts of parks and public gardens than anything else, and may also be used with good effect in squares ; avoiding, however, the very common error of putting all our native and hardy shrubs roughly in under the shade of big trees, &c. Numerous subjects are never seen to present their native charms in consequence of being overcrowded, or overshadowed, or robbed at the root by heavy-feeding neighbours. If a thing be worth planting at all it is worth planting well; and the rule in our squares, and too often in our parks and gardens, is excessive crowding and little or no attempt at the fair and full development of individual plants, be they costly exotics or merely "common" wild Roses or Hawthorns.

Of tribes that may be associated with the Hawthorns there are the Cotoneasters—the freely flowering and fruiting deciduous species; the Almonds and Peaches, double and single; the various double Cherries and Plums; Amelanchiers (Snowy Mespilus); the Bird Cherry and the Weeping Cherry; the Judas tree; the Quinces and Medlars (particularly Mespilus Smithii); the varieties of the Scotch and common Laburnums; the Daphnes, the Deutzias, the various kinds of Lilac, and numerous other rather dwarf shrubs for the embellishment of the margins of groups, &c.; the various kinds of Pyrus from the great P. vestita to the handsome Chinese Pear and Japan Quince; the Rose Acacia; not to mention many other useful species.

The common Stag's Horn Sumach succeeds so well in the small town garden that it deserves a word of praise. It does not grow so gross as to require clipping, and retains its verdure without taint long after that miserable town tree the Lime has parted with it; but it is apt to produce suckers too abundantly. Amongst deciduous flowering shrubs, the Althea would seem to be the king. With attention it should form a telling object in all parts where the bottom is dry. By "attention," I mean planting it so as to develope it into a specimen, and not thrusting it promiscuously amongst rough and mixed shrubs, which may obscure it from the sun or unduly rob it at the root.

Liriodendron tulipifera, the Tulip tree, seems perfectly at home in city parks or gardens, and being a handsome and distinct tree in every way deserves to be planted largely in such places. Sophora japonica forms a grand tree in the neighbourhood of London, and has the valuable property of never seeming to suffer from drought, no matter how dry the soil, but retains its verdure to the end of the season. It therefore merits an important place in all our parks and squares, especially those with a dry gravelly, sandy, or light bottom.

Among various trees thoughtlessly recommended for London planting in the journals of the past year was the Copper Beech. I trust nobody will ever use that as a town tree; not at least until we have too much green in our cities, instead of square miles of dull brick without a verdant spot, as at present. When we become sufficiently Gallicized to establish a Morgue, a few dark and gloomy-looking Copper Beeches might appropriately adorn its neighbourhood. Nobody can object to grouping this tree here and there in the parks, or to the use of the Copper Beech in an isolated manner among other trees, but to talk of planting avenues of these trees is harrowing. There seems a sort of purgatorial ingenuity about this recommendation. We are coppery enough in all conscience; and though a line of rusted Limes relieved by Coppery Beeches might suit a nation of very strict ritualistic tendencies, anxious to find even an additional pang among their trees—surely the most noble, stately, and useful objects that nature has given for the embellishment of the earth—I trust such a peace-destroying combination will never be seen in my time. I would punish the writer of it by shutting him up in a London house of a hot August day, and let Copper Beeches and hideous Limes be the only things on which he could refresh his eyes—a dreadful punishment for anybody with a nervous system and a slight knowledge of trees.

Evergreens, as has been frequently pointed out, are as a class better avoided; and yet there are for city gardens some kinds which seem to flourish disregarding smoke. Of these the Japan Privet (Ligustrum japonicum) is worthy of a front place. The beauty and utility of the Japan Privet is insufficiently known to the town-gardener, though it is extensively planted by the judicious landscape gardener and planter. Large in leaf almost as a goodly Orange, producing flowers almost as large as the white lilac, and very sweet, it possesses first-class attractions as an ornamental shrub; but it is to its value as a London plant

N

that I would call the attention of the town-gardening portion of my readers. We all know how difficult it is to get any sort of an evergreen to grow well in towns; those with the best character for good behaviour within the vile influences of smoke are too apt to become hopelessly deciduous. The Japan Privet may be tried with safety in a back garden, far into what Cobbett called the great " wen." Having the advantage of flowering so sweetly and freely in addition to being a shrub with comely leaves and good habit, I am sure those who so plant it will not be disappointed. The remaining kinds are mostly those that are well known and frequently used—the Aucuba, Holly, in great variety, Box, and Rhododendrons, Ivies, Berberises, particularly Darwini, and the common Laurel —often cut off, however, and not so good or hardy as the Aucuba; the Caucasian Laurel, better and hardier than the common kind, Euonymus japonicus, Mahonias, and several kinds of Yuccas.

All these are known to do very fairly if properly planted in pretty good positions. Considering how excellently the common Aucuba grows in our towns, we may look forward with much hope to what may be done with the numerous new and fine kinds as soon as they are common enough to be tried extensively in city gardening. But the town-planter cannot be too often cautioned against the over use of evergreens—there is scarcely a suburb in which thousands of pounds worth of them are not to be at any time seen in a dying state ! Even the kinds above enumerated are often seen to languish and die after a year or two in a west central garden, like Mecklenburg-square, where the deciduous trees are fine enough to freshen the heart of a North American Indian should he happen to pass by.

Anxious to promote as far as possible permanent and noble rather than fleeting and mean styles of park decoration, I venture to add the names of some fine distinct trees that deserve to be more widely planted in our city parks and gardens, if only to vary the monotony caused by the profuse planting of well known kinds. The Oaks offer an example of the arboreal riches within the reach of planters, and the American Oaks especially cannot be overpraised. Of course they are not recommended here for street or boulevard planting, but for parks and open spaces where the dust is a little subdued and where they may have plenty of root-room.

The following species of Oak are well worthy of attention in our parks :—Quercus ambigua—fine foliage ; the best of all the American oaks for quick growth. Q. Prinus—chestnut oak. Q. rubra—champion oak. Q. coccinea—scarlet oak. Q. falcata—downy oak. Q. tinctoria—black oak. Q. palustris—pine oak. Q. nigra—black jack oak. Q. Catesbæi—scrub oak. Q. Phellos—willow oak. Q. imbricata—shingle oak, very distinct. Q. Æsculus—Italian oak.

The Maples also comprise some very noble trees :—Acer macrophylla — the great Columbian maple. A. lobatum —Siberian maple. A. Lobelii—this kind grows erect, like the Lombardy poplar, and has violet shoots and striped bark. A. eriocarpum—Sir Charles Wager's maple, a fine silvery appearance. A. neapolitanum—Neapolitan maple, fine large foliage. A. obtusatum—Hungarian maple, very distinct from the Neapolitan maple, with which it is frequently confounded. A. colchicum rubrum—this kind has bright red twigs in winter, and is distinct and good. A. platanoides—Norway maple, very showy when in flower in the spring. A. Pseudo-Platanus purpurea—purple-leaved sycamore. Negundo fraxinifolia variegata—very showy.

Of other valuable subjects not running in such closely allied groups, the following will be found worthy of extensive use :—Catalpa syringæfolia, makes an ornamental tree near London. Laurus sassafras, Nyssa biflora and villosa—the autumn tints of these are fine ; they are natives of North America. Ulmus stricta, the Cornish elm ; Ulmus viminalis, distinct habit ; Ulmus vegeta, Chichester elm ; Ulmus nigra, Irish elm, a large timber tree of rapid growth ; Ulmus montana major, smooth elm ; and Ulmus americana, a very distinct tree, of large size. Planera Richardi, a fine tree, with a peculiarly distinct and striking mode of branching. Celtis occidentalis, the nettle tree. Juglans nigra, black walnut, fine foliage. Carya amara, C. alba, and sulcata,

N 2

fine foliage. Æsculus Hippocastanum flore-pleno — the flower lasts three times as long in perfection as that of the single horse-chestnut. Æ. rubicunda — the scarlet flowered. Tilia alba, or argentea—the Hungarian lime tree, the best. T. americana—Mississippi lime, very large leaves. Pavia rubra—native of the mountains of Virginia and Carolina, rather a small but ornamental tree. P. flava—this is a native of the same country, but attains a larger growth than the former species. P. indica—very distinct in foliage. P. californica—fragrant flowers in long spikes, a very handsome tree. Liriodendron tulipifera—the tulip-tree, and its variety obtusatum, the entire leaved form. Sophora japonica pendula—a highly ornamental weeping tree. It is somewhat tender in the north, but flourishes finely about London, and on dry soils generally. It can stand any amount of drought, and is therefore particularly well adapted for dry soils. When it flowers it is very ornamental. Virgilia lutea—a native of North America, has white pendulous racemes of flower, a little larger than those of the locust tree. It is very striking when in flower, and does best on a dryish soil. Robinia viscosa—a native of South Carolina and Georgia, comes in flower later than the common locust tree, and bears pink blossoms. Gleditschia sinensis (horrida) and ferox are very singular, from having their stems embellished with large and fierce spines. Gymnocladus canadensis, the Kentucky coffee tree, is remarkable for the beauty of its foliage during summer. Of the Cratæguses—Aronia, with large yellow fruit; Layi, with large red fruit used for preserving in China; Celseana (Leeana of the nurseries), pale red fruit; and Douglasi, black fruited, are among the most distinct.

Pyrus vestita is a large silvery species beautiful even among the many good things in its family. Ornus europæa, the flowering ash; Fraxinus lentiscifolia, Calabrian manna ash; Fraxinus americana, the broad-leaved American ash; F. pubescens, black American ash; F. epiptera (or lancea), the Canadian ash or lancewood. Platanus orientalis pyramidalis—fine variety. P. acerifolia—Spanish plane. P. umbellata—a fine variety. Liquidambar styraciflua—

fine tint in autumn.   Salisburia adiantifolia—var. macro-phylla.   Populus acerifolia, or nivea — the silver-leaved poplar, very fine.   P. angulata—Carolina poplar, leaves large, and sub-evergreen.   Alnus incana laciniata—cut-leaved alder.   A. cordata—fine large glossy leaves, very distinct from all the other kinds; sub-evergreen.   Betula angulata, or nigra—the black birch.   Elægnus argenteus, deciduous Magnolias, and Pterocarya caucasica — fine foliage and distinct tree.

CHAPTER X.

SUBTROPICAL PLANTS FOR THE FLOWER GARDEN.

THE term subtropical is popularly given to flower-
gardens embellished by plants having large and handsome leaves, noble habit or graceful outlines. It simply
means the introduction of a rich and varied vegetation, chiefly distinguished by beauty of form, to the
ordinarily flat and monotonous surface of the garden.
The system had its origin in Paris, where it was first
carried out on a small scale around the old Tour St. Jacques,
and is now adopted to a greater extent there than anywhere
else. Indeed, the presence of great numbers of fine-leaved
plants is one of the most marked features in the parks and
public gardens of that city. Mr. Gibson, the able and
energetic superintendent of Battersea Park, undaunted by
the popular nonsense about the great superiority of the
climate of Paris over that of London, boldly tried the
system, and with what a result all know who have seen his
charming " subtropical garden " in Battersea Park.

This system has taught us the value of grace and verdure amid masses of low, brilliant, and unrelieved flowers,
or rather has reminded us of how far we have diverged
from Nature's ways of displaying the beauty of vegetation.
Previous to the inauguration of this movement in England, our love for rude colour had led us to ignore the exquisite and inexhaustible way in which plants are naturally
arranged—fern, flower, grass, shrub, and tree, sheltering,
supporting, and relieving each other. We cannot attempt
to reproduce this literally, nor would it be wise or convenient to do so; but assuredly herein will be found the
source of true beauty in the plant world, and the more the
ornamental gardener keeps the fact before his eyes, the

nearer truth and success will be attained. Nature *in puris naturalibus* we cannot have in our gardens, but Nature's laws should not be violated, and few human beings have contravened them more than our flower gardeners during the past twenty years. We must compose from Nature, as the best landscape artists do, not imitate her basely. We may have all the shade, the relief, the grace, and the beauty, and nearly all the irregularity of Nature seen in every blade of grass, in every sea-wave, and in every human countenance, and which may be found too, in some way, in every garden that affords us lasting pleasure either from its contents or design. Subtropical gardening has taught us that one of the greatest mistakes ever made in the flower garden was the adoption of a few varieties of plants for culture on a vast scale, to the exclusion of interest and variety, and too often of beauty or taste. We have seen how well the pointed, tapering leaves of the Cannas carry the eye upwards; how refreshing it is to cool the eyes in the deep green of those thoroughly tropical Castor-oil plants with their gigantic leaves; how grand the Wigandia, with its wrought-iron texture and massive outline, looks after we have surveyed brilliant hues and richly painted leaves; how greatly the sweeping palm-leaves beautify the British flower garden;—and, in a word, the system has shown us the difference between gardening that interests and delights all the public, as well as the mere horticulturist, and that which is too often offensive to the eye of taste, and pernicious to every true interest of what Bacon calls the " Purest of Humane pleasures."

But are we to adopt this system in its purity? Certainly not. All practical men see that to accommodate it to private gardens an expense and a revolution of appliances would be necessary, which are in nearly all cases quite impossible, and if possible, hardly desirable. We can, however, introduce to our gardens most of its better features; we can vary their contents, and render them more interesting by a cheaper and a nobler system. The use of all plants without any particular and striking habit or foliage, or other distinct peculiarity, merely because they are " subtropical," should be tabooed at once, as tending to make much work, and to return—a lot of weeds; for "weediness" is all that I can write of many Solanums and stove plants of no real merit which have been employed under this name.   Selection of the most beautiful and useful from the great mass of plants known to science is one of the most important of the horticulturist's duties, and in no branch must he exercise it more thoroughly than in this.   Some plants used in it are indispensable—the different kinds of Ricinus, Cannas in great variety, Polymnia, Colocasia, Uhdea, Wigandia, Ferdinanda, Palms, Yuccas, Dracænas, and fine-leaved plants of coriaceous texture generally.   A few specimens of these may be accommodated in many large gardens; they will embellish the houses in winter, and, transferred to the open garden in summer, will lend interest to it when we are tired of the houses. Some Palms, like Seaforthia, may be used with the best effect for the winter decoration of the conservatory, and be placed out with an equal result, and without danger in summer.   The many fine kinds of Dracænas, Yuccas, Agaves, &c., which have been seen to some perfection at our shows of late, are eminently adapted for standing out in summer, and are in fact benefited by it.   Among the noblest ornaments of a good conservatory are the Norfolk Island and other tender Araucarias—these may be placed out for the summer much to their advantage, because the rains will thoroughly clean and freshen them for winter storing.   So with some Cycads and other plants of distinct habit—the very things best fitted to add to the attractions of the flower garden.   Thus we may enjoy all the benefits of what is called subtropical gardening without creating any special arrangements for them in all but the smallest gardens.

But what of those who have no conservatory, no hothouses, no means for preserving large tender plants in winter? They too may enjoy in effect the beauty which may have charmed them in a subtropical garden.   I have no doubt whatever that in many places as good an effect as any yet seen in an English garden from tender plants, may be obtained by planting hardy ones only! There is the Pampas Grass—which when well grown is unsurpassed by anything that requires protection. Let us in planting it take the trouble to plant and place it very well—and we can afford to do that, since one good planting is all that it requires of us, while tender things of one-tenth the value may demand daily attention. There are the hardy Yuccas, noble and graceful in outline, and thoroughly hardy, and which, if planted well, are not to be surpassed, if equalled, by anything of like habit we can preserve indoors.   There are the Arundos, conspicua and Donax, things that well repay for liberal planting; and there are fine hardy herbaceous plants like Crambe cordifolia, Rheum Emodi, Ferulas, and various fine umbelliferous plants that will furnish effects equal to those we can produce by using the tenderest. The Acanthuses too, when well grown, are very suitable to this style; one called latifolius, which is beginning to get known, being of a peculiarly firm, po- lished, and noble leafage.   Then we have a hardy Palm— very much hardier too than it is supposed to be, because it has preserved its health and greenness in sheltered positions, where its leaves could not be torn to shreds by storms through all our recent hard winters, including that of 1860.

And when we have obtained these we may associate them with not a few things of much beauty among trees and shrubs—with elegant tapering young pines, many of which, like Cupressus nutkaensis, have branchlets finely chiselled as a Selaginella; not of necessity bringing the larger things into close or awkward association with the humbler and dwarfer flowers, but sufficiently so to carry the eye from the minute and pretty to the higher and more dignified forms of vegetation.   By a judicious selection from the vast mass of hardy plants now obtainable in this country, and by asso- ciating with them where it is convenient, house plants that are stood out for the summer, we may arrange and enjoy charms in the flower garden to which we are as yet strangers, simply because we have not sufficiently selected from and utilized the vast amount of vegetable beauty at our disposal.

Let us next select the finer tender plants for this pur- pose, speak of the treatment they require, and the uses or

associations for which they are best adapted.   In selecting
tender plants of noble aspect or elegant foliage, suited for
placing in the open air in British gardens during the summer
months, we shall confine ourselves to first-class plants only.
It is necessary that they be such as will afford a distinct
and desirable effect if they *do* grow; and that is by no
means to be obtained from many subjects recommended for
subtropical gardening.   And above all we must choose
such as will make a healthy growth in sheltered places in
the warmer parts of England and Ireland at all events.
There is some reason to believe that not a few of the best
will be found to flourish much further north than is generally
supposed.   In all parts the kinds with permanent foliage,
such as the New Zealand flax and the hardier Dracænas,
will be found as effective as around Paris, and to such the
northern gardener
should turn his
attention as much
as possible. Even
if it were possible
to cultivate the
softer - growing
kinds like the
Ferdinandas to
the same perfec-
tion in all parts
as in the south
of England, it
would by no
means be every-
where desirable,
and especially
where means are
scarce, as these
kinds are not capa-
ble of being used
indoors in winter.
The many fine
permanent leaved

Fig. 61.

Variegated Agave.

subjects that stand out in summer without the least injury, and may be transferred to the conservatory in autumn, there to produce as fine an effect all through the cold months as they do in the flower garden in summer, are the best for those with limited means.

AGAVE AMERICANA and its variegated varieties are plants peculiarly suited for this kind of decoration, being useful for placing out of doors in summer in vases, tubs, or pots plunged in the ground, and also for the conservatory in winter. They are so well known and so long cultivated in this country that nothing need be said of their requirements or cultivation.

ARALIA PAPYRIFERA (the Chinese Rice-paper Plant).—This, though a native of the hot island of Formosa, flourishes vigorously around Paris in the summer months, and is one of the most valuable plants in its way. It is useful for the greenhouse in winter and the flower garden in summer. It is handsome in leaf and free in growth, though to do well it must be protected from cutting breezes, like all the large-leaved things. In some of the warmer parts of France

FIG. 62.

Aralia papyrifera.

the Peach does very well as a field tree—a low one, however. The wind is so strong that it would be destroyed if allowed to rise in the natural way, and so they train it as a dwarf bush, spreading wide. Tall subtropical plants have with us somewhat of the same disadvantage. If this Aralia be

planted in a dwarf and young state, it is likely to give more
satisfaction than if planted out when old and tall.   The
lower leaves spread widely out near the ground, and then
it is presentable throughout the summer.   Prefer therefore
dwarf stocky plants when planting it in early summer.   It
should have rich, deep soil and plenty of water during the
hot summer months.   The open air of our country suits it
better than the stove, and chiefly no doubt because it is
very liable to the mealy bug when kept indoors ; in the free
air this pest is washed away by the rain.   For the public
gardens of Paris it is kept underground in caves during the
winter ; but in private gardens with us it will doubtless be
worthy of a place in the greenhouse throughout that season.
It is easily increased by cuttings of the root.   It is usually
planted in masses, edged with a dwarfer plant ; but as a
small group in the centre of a bed of flowers, or even as an
isolated specimen in a like position, it is capital.   The stems
of this plant have a very fine pure white pith, which, when
cut into thin strips and otherwise prepared, forms the article
known as rice paper.   It is rare for a plant to be so useful
both in an ornamental and economic sense.

ACACIA LOPHANTHA.—This elegant plant, though not
hardy, is one of those which all may enjoy, from the freedom with which it grows in the open air in summer.   It
will prove more useful for the flower garden than it has
ever been for the houses, and, being easily raised, is
entitled to a place here among the very best.   The elegance
of its leaves and its quick growth in the open air make it
quite a boon to the flower gardener who wishes to establish
graceful verdure amongst the brighter ornaments of his
parterre.   It will furnish the grace of a fern, while close
and erect in habit, thus enabling us to closely associate it
with flowering plants without in the least shading them—
except from ugliness.   Of course I speak of it in the young
and single stemmed condition, the way in which it should
be used.   By confining it to a single stem and using it in
a young state, you get the fullest size and grace of which
the leaves are capable.   Allow it to become old and
branched and it may be useful, but by no means so much

so as when young and without side branches.   It may be raised from seed as easily as a common bedding plant.   By sowing it early in the year it may be had fit for use by the first of June ; but plants a year old or so, stiff, strong, and well hardened off for planting out at the end of May, are the best.   It would be desirable to raise an annual stock, as it is almost as useful for room decoration as for the garden.

ASPLENIUM NIDUS-AVIS.—This is a very remarkable fern, which has been placed out of doors in the garden in summer, but it is not vigorous or hardy enough to be generally

FIG. 63.

Asplenium nidus-avis.

recommended for this purpose.   However, as it may have been noticed in abundance at Battersea Park during the past season, I allude to it here.   It is a popular subject in places where large collections of tropical ferns are grown, and in such a plant may be tried in the open air in a very warm, shady, and perfectly sheltered position.

CALADIUM ESCULENTUM.—This species has proved the best for out-door work of a large genus with very fine foliage. It is only in the midland and southern counties of Great Britain that it can be advantageously grown, so far as I have observed ; but its grand outlines and aspect when well developed make it worthy of all attention, and of a prominent position wherever the climate is warm enough for its growth.   It does very well about London, and may have been noticed in considerable masses during the past year in the London parks, where it served to illustrate to some extent the disadvantages of that mode of planting. When seen in wide masses the effect is by no means so fine as when in a compact group or circle. The dead level line presented by the tops—which line, unlike that of the upper surface of the taller plants, is below the eye—neutralizes considerably the great lines of the leaves; but place the plant in a ring round a central object, or in some position where its fine leaves may contrast immediately with those of a different type of vegetation, and it is beautiful indeed. It may be used with great effect in association with many fine foliage plants; but Ferdinanda, Ricinus, and Wigandia usually grow too strong for it, and if

Fig. 64.

Caladium esculentum.

planted too close injure it. This may have been noticed particularly in cases where it was used as bordering to masses of the strong growing kinds above-named. With all kinds of stonework, vases, &c., it is peculiarly effective and beautiful. C. esculentum, though a stove perennial, is very easily kept over the winter in a dry spot under a stage or in boxes of sand in places where hothouse room is scarce. It is readily propagated by first starting the plants in heat, and when they have pushed forth eyes near the base, cutting them in pieces, an eye or bud in each. In spring the older plants should be potted and grown on in heat, so as to be fit to plant out about the middle of June. On the whole, although so fine and distinct, it is not suitable for any but mild and warm parts of the southern half of these islands. The nearly allied Colocasia odorata is sometimes employed, and effectively especially in the case of old specimens with stems.

FIG. 65.

Colocasia odorata.

THE CANNAS.—If there were no plants of handsome habit and graceful leaf available for the improvement of our flower gardens but these we need not despair, for they possess almost every quality the most fastidious could desire, and present a useful and charming variety. The larger kinds make grand masses, while all may be associated intimately with flowering plants—an advantage that does not belong to some free-growing things like the Castor-oil plant. The Canna ascends as boldly, and spreads forth as fine a mass of leaves as any; but may be closely grouped with much smaller subjects. The general tendency of most of our flower-garden plants is to assume a flatness and dead level, so to speak; and it is the very qualities possessed by the Cannas for counteracting this that makes them so valuable. Even the grandest of the other subjects preserve this tameness of upper surface outline when grown in great quantities: not so these, the leaves of which, even when grown in dense groups, always carry the eye up pleasantly from the humbler plants, and are grand aids in effecting that harmony between the important tree and shrub embellishments of our gardens and their surroundings, and the dwarf flower-bed vegetation, which is so much wanted. Another charm of these most useful subjects is their power of withstanding the cold and storms of autumn. They do so better than many of our hardy open-air plants, so that when the last leaves have been blown from the Lime, and the Dahlia and Heliotrope have been hurt by frost, you may see them waving as greenly and gracefully as the vegetation of a temperate stove. Many of the subtropical plants, used for the beauty of their leaves, are so tender that they go off in autumn, or

Fig. 66.

Canna nigricans.

require all sorts of awkward protection at that season; but the Cannas last in good trim till the borders must be cleared. All sheltered positions, places near warm walls, and nice snugly-warmed dells, are capital positions for them. They are generally used in great ugly masses, both about Paris and London; but their true beauty will never be seen till we

FIG. 67.

Canna atro-nigricans.

learn to place them tastefully here and there among the flowering plants—just as we place sprigs of graceful fern in a bouquet. A bed or two solely devoted to them will occasionally prove very effective; but enormous meaningless masses of them, containing perhaps several hundred plants of one variety, are things to avoid and not to imitate. As to culture and propagation, nothing can be

o

more simple: they may be stored in winter as readily as potatoes, under shelves in the houses, in the root-room, or, in fact, anywhere covered up from the influences of frost. And then in spring, when we desire to propagate them, nothing is easier than pulling the roots in pieces, and potting them separately. Afterwards it is usual to bring them on in heat, and finally harden them off previous to planting out; but a modification of this practice is desirable, as some kinds are of a remarkably hardy constitution, and make a beautiful growth if put out without so much as a leaf on them.

In rambling through an obscure part of Paris one evening, I encountered a tuft of Canna springing up strongly through and around a box-edging—pretty good evidence that it had remained there for some years. Upon inquiry of the proprietor of the garden I found this was the case, and that he had no doubt of the hardiness of several other kinds. They were planted not more than eight or ten inches deep. When we remember that the Cannas are amongst the most valuable plants we use for giving grace and verdure to the flower garden, this surely is a hint worthy of being acted upon. Considering their diversity of colour and size, their graceful pointed habit and facility of propagation, we must concede them the first place; but their capability of being used by anybody who grows ordinary bedding plants, and the fact that they may be preserved so very easily through the winter, enhance their value still more. The following are among the best of the hardiest kinds:—C. Annæi, musæfolia, gigantea, limbata, Warscewiczii, nigricans, maxima, and zebrina. Of course they will prove equally hardy with us. As it is desirable to change the arrangements as much as possible every year, it may not be any advantage to leave them in the ground, and in that case they may be taken up with the bedding plants, and stored as simply and easily as carrots, parsnips, or potatoes. A bed of Cannas, protected by a coating of litter, was left out in Battersea Park through the severe winter of 1866-7. During the unfavourable summer of 1867 they attained a height of nearly twelve feet.

THE DRACÆNAS.—Long as this noble family has been known in our gardens, we have yet to learn a great deal about its use and beauty. Hitherto only allowed to grace a stove or conservatory now and then, Dracænas in future will be among the most indispensable ornaments of every garden where grace or variety is sought. They are among the very best of those subjects which may be brought from the conservatory or greenhouses in early summer, and placed in the flower garden till it is time to take them in again to the houses, where we protect them through the winter. And if it were not necessary to protect them through the winter it would be almost worth our while to bring them indoors at that season, so graceful are they, and so useful for adding the highest character to our conservatories. One well filled with such plants presents a very different appearance to most English plant-houses in winter. The hardier and most coriaceous kinds, like indivisa and Draco, may be placed out with impunity very far north. The brightly coloured kinds, like terminalis, have been tried in the open air at Battersea, but not with success. It would be dangerous to try them in the open air much farther north, except in very favourable spots. The better kinds are indicated in the select list of subtropical plants. I have seen D. indivisa grow well in the open air in the south of England. It has been many years out at Bicton.

ECHEVERIA METALLICA.—This is scarcely elevated enough to be suitable for association with such plants as the foregoing, but it is so very distinct in aspect, and has been proved to grow so well in the open air during several unfavourable seasons, that we must not pass it by. I purposely exclude from this selection many things sometimes included in lists of " subtropical " plants, but which may be classed most properly with bedding subjects. But this, although not very large, forms an agreeable and distinct object, and is very well calculated for producing a striking effect among dwarf bedding and edging plants. It should, however, be placed singly, and among very dwarf things, such as Sedum, Sempervivum, and its dwarf relative E. secunda. It may be propagated by the leaves or by

o 2

cuttings, and requires a dry greenhouse shelf in the winter. Light sandy earth, not of necessity very poor, will suit it best in the open air. It is likely to become one of the most popular of all garden plants.

FERDINANDA EMINENS.—This is one of the tallest and noblest subtropical plants, growing well in the southern and midland counties : wherever it is supplied with rich soil and abundant moisture. It is also very much the better for being sheltered, and so are all large and soft-leaved plants. Where the soil is rich, deep, and humid, and the position warm, it attains large dimensions, sometimes growing over twelve feet, and suspending immense pairs of opposite leaves. It will in all cases form a capital companion to the Castor-oil plant, and, though it may not be grown with such ease in all parts, it should be in every collection, considering that it grows quite as well in the south of England as in the neighbourhood of Paris. It requires to be planted out in a young state, and grows freely from cuttings. Greenhouse treatment will do in winter. It is better to keep a stock in pots through the summer to afford cuttings, though the old ones may be used for that purpose.

FIG. 68.

Ficus elastica.

FICUS ELASTICA (India-rubber Plant).—Another fine old plant, for which we have lately found a new use. It is one of those valuable leathery-leaved things that are useful in hothouse, drawing-room, or flower garden. It not only exists in the open air in summer in good health, but makes a good growth under the influence of our weak northern sun. Never assuming the imposing proportions of other plants mentioned here, it is best adapted for select mixed groups, and in small gardens as isolated specimens amongst low bedding plants. It requires stove treatment, and is propagated from cuttings. In all cases it is better to use plants with single stems. It is especially valuable in consequence of doing perfectly well in the dry air of inhabited rooms, and this will enable many to enjoy a fine-leaved plant in the flower garden who have not a glass house of any kind on their premises.

Fig. 69.

Monstera deliciosa.

Monstera deliciosa.—This very remarkable-looking plant has been found to bear being placed in the open air with impunity in shady and sheltered spots. Its great perforated leaves look so singular that everybody should grow it who has a stove in which to do so, and it is so readily grown and propagated that a plant may soon be spared for placing in the open air during the warmer months.

Musa Ensete.—The noblest of all the plants yet used

in the flower garden is Musa Ensete—the great Abyssinian
Banana, discovered by Bruce.   The fruit of this kind is
not edible, like that of the Banana and Plantain (Musa
paradisiaca and sapientum), but the leaves are magnificent ;
and, strange to say, they stand the rain and storms of the
neighbourhood of Paris without laceration, while all the
other kinds of Musa become torn into shreds.   It is an
interesting and hitherto unknown fact, that the finest of all
the Banana or Musa tribe is also the hardiest and most
easily preserved.   When grown for the open air, it will of
course require to be kept in a house during winter, and
planted out the first week in June.   In any place where
there is a large conservatory or winter garden, it will be
found most valuable, either for planting therein, or for
keeping over the winter, as, if merely housed in such a
structure during the cold months, it will prove a great
ornament among the other plants, while it may be put out
in summer when the attraction is all out of doors.   Other
kinds of Musa have been tried in the open air in England,
but have barely existed, making it clear that they should
not be so cultivated in this country.   The Ensete is the
only species really worth growing in this way.   Where the
climate is too cold to put it out of doors in summer, it
should be grown in all conservatories in which it is
desired to establish the noblest type of vegetation.   It has
hitherto been generally grown in stoves.   It also stands the
drought and heat of a living room remarkably well, and
though, when well developed, it is much too big for any
but Brobdingnagian halls, the fact may nevertheless be
taken much advantage of by those interested in room decoration on a large scale.   The plant is difficult to obtain
as yet, but will, I trust, be sought out and made abundant
by our nurserymen.

Last September I saw a fine plant of this Musa that had
remained in the open ground in Baron Haussmann's garden
in the Bois de Boulogne during the preceding winter.   It
was left in the position in which it grew during the summer
of 1867, and in the month of November covered with a
little thatched shed, the space about the plant being filled

with dry leaves. All the leaves were cut off. In spring the protection was removed and the plant pushed vigorously. It had (on the 8th of September, 1868), sixteen leaves, not one of which was torn or lacerated, although it was in an exposed position. It was not more than five feet high, but more attractive than much larger individuals of the same species, from being so compact and untattered in its growth. As most people who grow it will have means of keeping it indoors in winter, and as it is so rare, this mode of keeping it is not likely to be taken advantage of with us at present; but that it can and has been so wintered is an interesting fact.

NICOTIANA WIGANDIOIDES.—This is a shrubby or rather tree-like species of Tobacco, which, when put out in a young and healthy state, makes a vigorous growth, and is an imposing subject both in the climate of Paris and London. The treatment given to such things as the Polymnia will suit it well. It is, however, scarcely so ornamental or generally useful as the large-leaved and bright-flowered variety of the common Tobacco spoken of further on.

FIG. 70.

Nicotiana wigandioides.

POLYMNIA GRANDIS AND PYRAMIDALIS.— These belong to the great composite order, and are distinguished by rich handsome foliage and rapid summer growth, which, moreover, never becomes objectionable from any trace of raggedness, the erect shoots growing away till the end of the season in our climate. Doubtless, there is a point at which in their native country seediness does arrive, but with us they, like the Ricinus of one summer, always look fresh and young, and are most appro-

FIG. 71.

Polymnia grandis.

priate for forming luxuriant masses of foliage in the flower garden, and for diversifying its aspect. P. grandis is best known in this country, and is second to no other plant for its dignified and finished effect in the flower garden ; but P. pyramidalis is also good and distinct. They are easily struck from cuttings taken from old plants and put in heat in spring. Like most large soft growing things in this way, they are best planted out in a young state, so as to insure a fresh and unstinted growth. P. pyramidalis is the newest of the group and that least known in cultivation. I saw it several times during the past season in Paris. The leaves are not so large as those of the other species, and differ in shape, being nearly cordate, but the growth is vigorous and the habit distinct. It pushes up a narrow pyramidal head of foliage to a height of nearly ten feet in Paris gardens, and will be found to do well in the south of England.

PHORMIUM TENAX (the New Zealand Flax).—This is tolerably well known among us as a greenhouse and conservatory subject, but not nearly so much grown as it ought to be. The French make a grand use of it, both indoors in winter, and in the conservatory and out-of-doors in summer. About Paris it is of course as tender as with us, and requires the same amount of attention, which, after all, is very little. They grow it by the thousand for the decoration of rooms, and in the great nursery of the city of Paris at Passy there are 10,000 plants of it, chiefly used for the embellishment of the Hôtel de Ville. I need hardly say that we are much worse off for graceful things for indoor decoration than the French, and should in consequence grow this plant abundantly, according to our space. When grown to a medium size its leaves begin to arch over, and when in that condition nothing makes a more graceful and distinct ornament for room or hall. It may be grown to presentable perfection in an eight-inch pot, or to a great mass of bold long leaves in a tub a yard in diameter. Generally with us it will be found to enjoy greenhouse temperature, though in genial places in the south and west of Ireland and England it does very well in the open air. Its best use is for the decoration of the garden in summer, a few specimens well grown and plunged in the grass or the centre of a bed giving a most distinct aspect to the scene. The larger such plants are, the better, of course, will be the effect. The small ones will prove equally useful and effective in vases, to which they will add a grace that vases rarely now possess. It is pre-eminently useful from its being alike good for the house, conservatory, and even the living rooms in winter. Wherever indoor decoration on a large scale is practised it is indispensable, and it should be remarked that, unless for vase decoration it requires to be grown into goodly specimens before affording much effect out of doors; but when grown large in tubs, it is equally grand for the large conservatory and for important positions in the flower garden.

RICINUS COMMUNIS (the Castor-oil Plant).—When well grown in the open air, there is not in the whole range of cultivated plants a more imposing subject than this. It may have been seen nearly twelve feet high in the London parks of late years, and with leaves nearly a yard wide. It is true we require a bed of very rich deep earth under it to make it attain such dimensions and beauty; but in all parts, and with ordinary attention, it grows well. In warm countries, in which the plant is very widely cultivated it becomes a small tree, but is much prettier in the state in which it is seen with us—*i.e.*, with an unbranched stem, clothed from top to bottom with noble leaves. Soon after it betrays a tendency to develope side-shoots, the cold autumn comes and puts an end to all further progress; and so much the better, because it is much handsomer in a simple-stemmed state than any other. The same is true of not a few other large-leaved plants—once they break into a number of side-shoots their leaf beauty is to a great extent lost. In the planting out of some other subjects, it has been considered well to raise the beds on lime-rubbish, &c., or in other words, to build them upon it, sloping up the edge with soil and turf. But to grow this to perfection, the best way is to deeply excavate the bed, and place some rich stuff in the bottom, making all the earth as rich as possible. It is as easily raised from seed as the common bean, requiring, however, to be raised in heat. The Ricinus is a grand plant for making bold and noble beds near those of the more brilliant flowers, and tends to vary the flower garden finely. It is not well to closely associate it with bedding plants, in consequence of the strong growth and shading power of the leaves, so to speak. It is a good plan to make a compact group of the plant in the centre of some wide circular bed and surround it with a band of a dwarfer subject, say the Aralia or Caladium, and then finish with whatever arrangement of the flowering plants may be most admired. A bold and striking centre may be obtained, while the effect of the flowers is much enhanced, especially if the planting be nicely graduated and tastefully done. It is a judicious combination of both the green and the gay that we are most in want of, and few things can do so much to effect it for us in the flower garden as the common Castor-oil plant. This combination may be effected in any way that taste may direct. A graceful handsome-leaved subject in the centre of a flower bed will help it out, and so will bold groups of fine-leaved plants towards the outer parts of the flower garden. These bold masses connect in some degree the larger ligneous vegetation that usually surrounds our flower gardens with the small and low-lying brilliant flowers. For such groups the varieties of the Castor-oil plant are not likely to be surpassed.

SEAFORTHIA ELEGANS.—This is perhaps the most elegant and useful of all palms which may be safely placed out in summer. It is too scarce as yet to be procurable by horticulturists generally, but should be looked for by all who take an interest in these matters, and have a house in which to grow it. It stands well in the conservatory during the winter, though generally kept in the stove, where of course it grows beautifully. There are hardier kinds—the dwarf fan palm for example, but on the whole none of them are so valuable as this. The following palms are suitable for like purposes :—Areca lutescens, Caryota urens, Caryota sobolifera, Chamærops humilis, Chamærops Fortunei, Chamærops Palmetto, Latania borbonica, Phœnix dactylifera, Phœnix sylvestris, Corypha australis.

THE SOLANUMS.—This family, so wonderfully varied, affords numerous species that look fine and imposing in leaf when in a young and free-growing state. In the nursery garden of the city of Paris there is a very large house entirely devoted to the family, in which are preserved over the winter months more than sixty species for the embellishment of

FIG. 72.

Solanum Warscewiczii.

Parisian gardens. But in selecting examples from this great genus we must be much more careful, as our climate is a shade too cold for them, and many of them are of too ragged an aspect to be tolerated in a tasteful garden.   Half a dozen species or so are indispensable, but quite a crowd of narrow-leaved and ignoble ones may well be dispensed with.   The better kinds—as seen both in

FIG. 73.

Solanum robustum.

London and Paris gardens—are marginatum, robustum, macranthum, macrophyllum, Warscewiczii, crinipes, callicarpum, jubatum, Quitoense, galianthum, hippoleucum, crinitum, and Fontainesianum, an annual with pretty leaves, crisped and distinct looking.

Most of these plants may be raised from seed, while they are also freely grown from cuttings.   As a rule hothouse treatment in winter is required, and in summer rich light soil, a warm position, and perfect shelter. S. marginatum, planted in a very dwarf and young state, furnishes a most distinct and charming effect : it should be planted rather thinly, so that the leaves of one plant may not brush against those of another. If some very dwarf plants are used as a groundwork so much the better, but the downy and silvery leaves of this plant are sure to please without this aid. It is very much better when thus grown than when permitted to assume the bush form. All the other kinds named are suitable for association with the larger leaved plants, though they do not attain such height and vigorous development as those of the first rank, like the Ricinus.

Fig. 74.

Uhdea bipinnatifida.

UHDEA BIPINNATIFIDA.—This is one of the most useful plants in its class, producing a rich mass of handsome leaves, with somewhat the aspect of those of the great cow-parsnips, but of a more refined type. The leaves are of a slightly silvery tone, and the plant continues to grow fresh and vigorously till the late autumn. It is freely propagated by cuttings taken from old plants kept in a very cool stove, greenhouse, or pit during the winter months, and placed in heat to afford cuttings the more readily in early spring. Under ordinary cutting treatment on hotbeds or in a moist warm propagating house, it grows as freely as could be desired, and may be planted out at the end of May or the beginning of June. It is well suited for forming rich masses of foliage, not so tall, however, as those formed by such things as Ricinus or Ferdinanda.

VERBESINA GIGANTEA.—To this and other members of the family, somewhat the same remarks will serve as have been applied to the preceding. They require about the same treatment, and are useful in the production of like effects. They, like their fellows, will be much the better for as warm and sheltered a position and as rich and light a soil as can be conveniently given them.

WIGANDIA MACROPHYLLA (caracasana).—This noble plant, a native of the mountainous regions of New Granada, is unquestionably, from the nobility of its port and the magnificence of its leaves, entitled to hold a place among the finest plants of our gardens. Under the climate of London it has made leaves which have surprised all beholders, as well by their size as by their strong and remarkable veining and texture. It will be found to succeed very well in the midland and southern counties of England, though too much care cannot be taken to secure for it a warm sheltered position, free good soil, and perfect drainage. It may be used with superb effect either in a mass or as a single plant. It is frequently propagated by cuttings of the roots, and grown on in a moist and genial temperature through the spring months, keeping it near the light so as to preserve it in a dwarf and well clothed condition; and, like all the other plants in this class, it should be very carefully hardened off. It is, however, much better raised from cuttings of the shoots, if these are to be had. W. macrophylla has the stems covered with short stinging hairs, and bearing brownish viscid drops, which like oil adheres to the hand when the stem is touched. Wigandia Vigieri is a plant of quick and vigorous growth, and remarkable habit. In the beginning of September, 1867, I measured a specimen with leaves three feet nine inches long, counting the leaf-stalk, and twenty-two inches across—the stem, nearly seven feet high and two inches in diameter, bearing a column of

Fig. 75.

Wigandia macrophylla (W. caracasana).

such leaves. It is known at a glance from the popular and older W. macrophylla—by the leaves and the stems being covered in a greater degree with glossy, slender, stinging bodies. These are so thickly produced as to give the stems a glistening appearance. W. urens is another species often planted, but decidedly inferior to either of the foregoing, except in power of stinging, in which way it is not likely to be surpassed. Seeds of the three species have been offered, and all may be raised in that way—W. Vigieri with unusual facility.

A select list of 100 *of the subtropical plants best suited for use in our climate. The most indispensable kinds are marked* *.

1. *Acacia lophantha.
2. *Agave americana.
3. Agave americana variegata.
4. Alsophila australis.
5. „ excelsa.
6. Anthurium Hookeri.
7. Aralia macrophylla.
8. „ *papyrifera.
9. „ reticulata.
10. Araucaria excelsa.
11. Areca lutescens.
12. Balantium culcitum.
13. Bocconia frutescens.
14. Brexia madagascariensis.
15. *Caladium esculentum.
16. C. bataviense.
17. Canna Annei superba.
18. „ *robusta.
19. „ *musæfolia hybrida.
20. „ *nigricans.
21. „ *grandiflora floribunda.
22. „ *Géant.
23. „ *discolor floribunda.
24. „ *metallica.
25. „ *rubra superbissima.
26. Carludovica palmata.
27. Caryota urens.
28. „ sobolifera.
29. Cassia corymbosa.
30. „ floribunda.
31. Chamæpeuce diacantha.
32. Chamærops humilis.
33. „ *excelsa.
34. „ *Palmetto.
35. Colea Commersoni.
36. *Colocasia odorata.
37. *Cordyline indivisa.
38. Corypha australis.
39. Cyathea dealbata.
40. *Cycas revoluta.
41. Dahlia imperialis.
42. Dicksonia antarctica.
43. *Dracæna australis.
44. „ *indivisa.
45. „ *Draco.
46. „ *braziliensis.
47. „ nutans.
48. „ Rumphi.
49. „ erithorachis.
50. „ *cannæfolia.
51. „ *lineata.
52. *Echeveria metallica.
53. Erythrina crista-galli, and its varieties.
54. *Ferdinanda eminens.
55. *Ficus elastica.
56. „ nympheæfolia.
57. „ Chauvieri.
58. Hedychium aurantiacum.
59. „ Gardnerianum.
60. Lomatia Bidwilli.
61. „ silaifolia.
62. Lomatophyllum borbonicum.
63. *Melianthus major.
64. Monstera deliciosa.
65. *Musa Ensete.
66. Neottopteris australasica.
67. Nicotiana wigandioides.
68. Papyrusantiquorum.
69. Philodendron Simsi.
70. „ macrophyllum.
71. Phœnix dactylifera.
72. „ sylvestris.
73. *Phormium tenax.
74. Polymnia grandis.
75. „ pyramidalis.
76. Pothos acaulis.
77. Rhopala corcovadense.
78. *Ricinus communis, in many varieties.
79. Sanseviera zeylanica.
80. *Seaforthia elegans.
81. Selinum decipiens.
82. Senecio Ghiesbreghti.
83. Senecio Petasites.
84. Solanum crinipes.
85. „ macranthum.
86. „ macrophyllum.
87. „ marginatum.
88. „ robustum.
89. „ Warscewiczii.
90. Sonchus laciniatus.
91. Sparmannia africana.
92. StadmanniaJonghii.
93. Tradescantia discolor.
94. „ zebrina.
95. Tupidanthus calyptratus.
96. *Uhdea bipinnatifida.
97. Verbesina gigantea.
98. „ *verbascifolia.
99. *Wigandia macrophylla.
100. „ Vigieri.

List of the best twenty-four subtropical plants.

1. Acacia lophantha.
2. Agave americana.
3. Aralia papyrifera.
4. Caladium esculentum.
5. Canna Annei superba.
6. Chamærops excelsa.
7. „ humilis.
8. Cordyline indivisa.
9. Cycas revoluta.
10. Dracæna Draco.
11. „ indivisa.
12. Ferdinandaeminens.
13. Ficus elastica.

14. Melianthus major.
15. Musa Ensete.
16. Phormium tenax.
17. Polymnia grandis.
18. Ricinus communis.

19. Seaforthia elegans.
20. Solanum marginatum.
21. „ Warscewiczii.

22. Uhdea bipinnatifida.
23. Verbesina gigantea.
24. Wigandia macrophylla.

Subtropical plants that may be raised from seed.

The best and readiest way to get up a stock of these plants is by raising them from seeds. Annuals, like the Castor-oil plant, must of course be raised from seeds in any case; but a number of the very finest perennial kinds may also be raised thus with great facility and pleasure to the raiser, in time quick enough to satisfy ordinary patience. And of those which cannot soon be grown to a presentable size from seeds like Palms, Dracænas, &c., it is yet very desirable to raise a batch, inasmuch as permanent dignified subjects like these are always of a greater value in any stage of their existence than the perishable rapid-growing subjects so usual with us. All the following have been offered in recent seed catalogues :—

Abutilon, in variety.
Acacia lophantha.
Andropogon bombycinus.
     „   formosus.
     „   Sorghum.
Aralia australis.
     „   elegans.
     „   papyrifera.
     „   Sieboldi.
     „   trifoliata.
Areca sapida.
Artemesia argentea.
Bambusa himalaica.
Bocconia cordata.
     „   formosa.
     „   frutescens.
     „   japonica.
     „   macrophylla.
Brugmansia, in variety.
Canna, in profuse variety.
Cassia corymbosa.
     „   floribunda.
Chamæpeuce Cassabonæ.
     „   diacantha.
Chamærops humilis.
     „   „ glauca.
     „   macrocarpa.
Cineraria platanifolia.

Cordyline indivisa vera.
     „   nutans   „
     „   superbicus.
     „   Veitchii.
Corypha australis.
Cyperus vegetus.
Dahlia imperialis.
Erianthus Ravennæ.
     „   violaceus.
Erythrina caffra.
     „   crista-galli.
     „   Hendersoni.
     „   laurifolia.
Eucalyptus globulus.
Ferdinanda eminens.
Grevillea robusta.
Hedychium Gardnerianum.
Humea elegans.
Latania borbonica.
Melianthus major.
     „   minor.
Musa Ensete.
Nicotiana grandiflora, a variety of N. tabacum.
     „   wigandioides.
Owenia cerasifera.
Paratropia tomentosa.

Paratropia venulosa.
Phormium tenax.
Phytolacca dioica.
Polymnia grandis.
Ricinus, in variety.
Seaforthia elegans.
Solanum acanthocarpum.
     „   auriculatum.
     „   giganteum.
     „   glaucophyllum.
     „   glutinosum.
     „   lanceolatum.
     „   macrocarpum.
     „   macrophyllum.
     „   marginatum.
     „   pyracanthum.
     „   robustum.
     „   verbascifolium.
Sonchus pinnatus.
Sparmannia africana.
Uhdea bipinnatifida.
Verbesina verbascifolia.
Wigandia macrophylla.
     „   urens.
     „   Vigieri.
Zea japonica variegata, and others.

P

CHAPTER XI.

This is a most important subject, and concerns every lover
of gardening in the British Isles; for, however few can in-
dulge in the luxury of rich displays of tender plants, or
however rare the spots in which they may be ventured out
with confidence, all may enjoy those that are hardy, and
that too with infinitely less trouble than is required by the
tender ones.   Those noble masses of fine foliage first displayed to us by tender plants have done much towards
correcting a false taste.   What I wish to impress upon the
reader is, that in whatever part of these islands he may live
he need not despair of producing sufficient similar effect to
vary his flower garden or pleasure ground beautifully by the
use of hardy plants alone; and that the noble lines of a
well-grown Yucca recurva, or the finely chiselled yet fernlike spray of a graceful young conifer, will aid him as much
in this direction as anything that grows either in tropical
or subtropical climes.   The herbaceous collections in the
Jardin des Plantes are very full, and correctly kept; and,
being much devoted to such plants, I rarely spent a week
without visiting them, chiefly to discover useful members of
this class; while, of course, such as are used in the various
public parks and gardens also came repeatedly under my
observation.   Of their relative importance and value I was,
perhaps, the better prepared to judge from having visited
for like purposes all the botanic gardens in the British Isles
within the past few years.   The following is the result of
my observations as to the finest subjects we can use :—

Acanthus latifolius.—This is a plant that anybody can
grow, and which is in all respects fine. The leaves are bold
and noble in outline, and the plant has a tendency, rare in

some hardy things with otherwise fine qualities, to retain its leaves till the end of the season without losing a particle of its freshness and polished verdure. In fact, the only thing we have to decide about this subject is, what is the best place for it? Now, it is one of those things that will not disgrace any position, and will prove equally at home in the centre of the mixed border, projected a little from the edge of a choice shrubbery in the grass, or in the flower garden; nobody need fear its displaying anything like the seediness which such things as the Heracleums show at the end of summer. In fact, few things turned out of the houses will furnish a more satisfactory effect. I should not like to advise its being planted in the centre of a flower bed, or in any other position where removal would be

Fig. 76.

Acanthus latifolius (lusitanicus).

necessary; but in case it were determined to plant permanent groups of fine-leaved hardy plants, then indeed it could be used with great success. Supposing we have an irregular kind of flower garden or pleasure ground to deal with (a common case everywhere), one of the best things to do with it is to plant it in the grass, at some little distance from the clumps,

and perhaps near a few other things of like character. It
is better than any kind of Acanthus hitherto commonly
cultivated in botanic gardens, though one or two of these
are fine. Give it deep good soil, and do not grudge it this
attention, because, unlike tender plants, it will not trouble
you again for a long time. A fine effect would be produced
by a ring of it around a strong clump of Tritomas (grandis
in the middle, and glaucescens surrounding it), the very
dark polished green Acanthus being in its turn surrounded
by the fine autumn-flowering Sedum spectabile. There
would be little difficulty in suggesting a dozen equally suitable uses for this fine plant. It is to be had now in some
London nurseries, and in nearly every Paris one. The plant
best known by this name is also sold under that of A.
lusitanicus. Both are garden names, the first the best. It
came into cultivation in the neighbourhood of Paris about
six or eight years ago, and has since spread about a good deal.
Nobody seems to know from whence it came. Probably it
is a variety of Acanthus mollis. The plant itself varies a
good deal; I have seen specimens of it about a foot high,
with leaves comparatively small and stiff and rigid, as if
cast in a mould, by the side of others of thrice that development, and of the usual texture.

ANDROPOGON SQUARROSUM is a hardy plant in the neighbourhood of Paris, or survives with but slight protection,
making luxuriant tufts seven feet high or more, when in
flower. It would probably make a beautiful object in the
warmer and milder parts of England and Ireland, and in
good soil, but, unlike the preceding, it is not a subject which
can with confidence be recommended for every garden. But
all who value fine grasses should try it.

ARALIA EDULIS.—This is a vigorous plant, well suited
for adding distinction to those positions in which we desire
a luxuriant type of vegetation. It is perfectly hardy, grows
six, seven, and even eight feet high in good soil, is of a fresh
and vigorous habit, and even so early as the end of June.
The leaves attain a length of nearly a yard when the plant
is strong, while the shoots droop a little with their weight,
and thus it acquires a slightly weeping character. It is a

little curious that plants so famous for their medicinal or other uses as the Castor-oil, the Chinese rice-paper, and the Indian-rubber plants, should have become so useful as ornaments in the garden. For this edible Aralia we may claim as high a position as a hardy plant, and for planting singly few things surpass it. It is very rare in this country now, but being easily propagated, may, it is to be hoped, not long prove so. I have seen it nine feet high; but as it dies down rather early in autumn it must not be put in important groups, but rather in a position where its disappearance may not be noticed. An isolated place, or one near the margin of an irregular shrubbery, fernery, or rough rockwork by the side of a wood walk, will best suit.

ARALIA JAPONICA.—A hardy woody species, and fine plant for varying the garden, bearing immense and graceful leaves, and delighting most in a warm and sheltered position—plenty of sun, but little exposure to wind. It is best when the stem is rather short and simple, and has an advantage that few things of the kind possess—it may be used with a stem of considerable height, or with a very dwarf one.

ARALIA SPINOSA, the angelica tree of North America, and resembling the preceding, is highly useful in this class, in consequence of its beauty of foliage and distinct aspect. Like many of the hardy things, it should be placed in positions where it would not be necessary to remove it, or closely associate it with tender plants requiring frequent disturbance of soil. Both this and the preceding kind may be had in our nurseries.

FIG. 77.

Aralia Sieboldi.

ARALIA SIEBOLDI is also a valuable species, usually treated as a greenhouse plant. It is perfectly hardy around Paris and London, at least on warm soils, and it not only remains healthful in living rooms during winter, but grows freely in them.

ARUNDO DONAX, the great reed of the south of Europe, is a very noble plant on good soils. In the south of England it forms canes ten feet high, and has a very distinct and striking aspect. It will do more than that if put in a rich deep soil in a favoured locality; and those who so plant clumps of it on the turf in their pleasure grounds will not be disappointed at the result. Nothing can be finer than the aspect of this plant when allowed to spread out into a mass on the turf of the flower garden or pleasure ground. It seems much to prefer dry sandy soils to moist ones; indeed, I have known it refuse to grow on heavy clay soil, and flourish most luxuriantly on a deep sandy loam in the same district. Like all large-leaved plants, it loves shelter. No garden or pleasure ground in the southern parts of England and Ireland should be without a tuft of it in a sheltered spot. But, fine as it is for effect and distinctness, its variegated variety is of more value to the flower garden proper.

ARUNDO DONAX VERSICOLOR.—This is a remarkably effective and beautiful plant, that is made little or no use of. We have already noticed several fine things for grouping together, or for standing alone on the turf and near the margin of a shrubbery border; and this is as well suited for close association with the choicest bedding flowers as an Adiantum frond is with a bouquet. It will be found hardy in the southern counties; and considerably north of London may be saved by a little mound of cocoa-fibre, sifted coal ashes, or any like material that may be to spare. In consequence of its effective variegation, it never assumes a large development, like the green or normal form of the species, but keeps tidy and low, and yet thoroughly graceful. It is of course suited best for warm, free, and good soils, and abhors clay, though it is quite possible to grow it even on that with a little attention to the preparation of the ground. But it is in all cases better to avoid things that will not

PLATE XXXIII.

BAMBUSA FALCATA.

grow freely and gracefully on whatever soil we may have to deal with; and it is to those having gardens on good sandy soils, and in the warmer parts of England, that I would specially recommend this grand variegated subject. For a centre to a circular bed, nothing can surpass it in the summer and autumn flower garden, while of course many other charming uses may be made of it. Not the least happy of these would be to plant a tuft of it on the green grass, in a warm spot, near a group of choice shrubs, to help, with many other things named, to fill up the gap between ordinary fleeting flowers, and the taller shrub and tree vegetation that is now nearly everywhere observed. It is better to leave the plant in the ground, in a permanent position, than to take it up annually. Protect the roots in the winter, whether it be planted in the middle of a flower bed or by itself in a little circle on the grass.

ARUNDO CONSPICUA is a worthy companion for the Pampas, though by no means equal to it, as has been stated by some writers. As a conservatory subject it is fine in flower, and it will be seen often in large conservatories after a few years. A large pot or tub will be necessary if grown indoors. The drooping leafage will always prove graceful, and then it sends up long silvery plumes, drooping also, and strikingly beautiful. Judging by its different appearance when freely grown in a tub indoors, and when planted out even on favourable spots, my impression is that it by no means takes so kindly to our northern climate as the Pampas grass. However, it is well worth growing, even in districts where it does not attain great development; it comes in flower before the Pampas, and may be considered as a sort of forerunner of that magnificent herb.

BAMBUSA VIRIDIS-GLAUCESCENS, and others.—I wish to call the attention of all horticulturists who live in the southern and more favoured parts of these islands to the fact that there are several bamboos and bamboo-like plants from rather cool countries that are well worth planting. Nothing can exceed the grace of a bamboo of any kind if freely grown; but if starved in a dirty hot-house, or grown in a cold dry place, where the graceful shoots cannot arch forth

in all their native beauty, nothing can be more miserable in
aspect.   On cold bad soils and exposed dry places in the
British Isles, these bamboos have little chance ; but, on the
other hand, they will be found to make most graceful
objects in many a sheltered nook in the south and south-

Fig. 78.

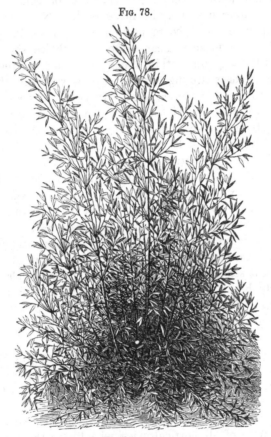

Bambusa aurea.

western parts of England and Ireland.   Nowadays there is
a growing taste for something else than mere colour in the
flower garden, and these will in many cases be found a
graceful help.   We have some knowledge of the capabilities
of one kind in this country.   In a well sheltered moist

spot at Bicton many have seen Bambusa falcata send up young shoots, long and graceful, like the slenderest of fishing-rods, while the older ones were branched into a beautiful mass of light foliage of a distinct type. The same plant has been grown in the county of Cork to a height of nearly twenty feet. This is the best known kind we have. At Paris I was fortunate enough to observe several other kinds doing very well indeed, although the climate is not so suitable as that of Cork or Devon. These are Bambusa edulis, aurea, nigra, Simmonsii, mitis, Metake, and viridis-glaucescens, the first and last of this group being very free and good. All the others will prove hardy in the south of England and Ireland, though, as some of them have not yet been tried there, it requires the test of actual experiment. Those who wish to begin cautiously had better take B. Simmonsii, viridis-glaucescens, nigra, and edulis to commence with, as they are the most certainly hardy, so far as I have observed. The best way to treat any of these plants, obtained in summer or autumn, would be to grow them in a cool frame or pit till the end of April, then harden them off for a fortnight or so, and plant out in a nice warm spot, sheltered also, with good free soil—taking care that the roots are carefully spread out, and giving a good free watering to settle the soil. There are no plants more worthy of attention than these where the climate is at all favourable, and there are numerous moist nooks around the British Isles where they will be found to grow most satisfactorily. The pretty little Bambusa Fortunei is also hardy.

Among the Centaureas there are a few subjects which might be used among hardy fine-leaved plants, but by far the most distinct and remarkable is the very silvery-leaved C. babylonica. This is quite hardy, and when planted in good ground shoots up strong spikes clad with yellow flowers to a height of ten or twelve feet. The bloom is not by any means so attractive as the leaves; but the plant is at all times picturesque.

BETA CICLA VAR. ("Chilian Beet").—Under the name "Chilian Beet" a very showy plant may have been seen in the neighbourhood of Paris during the past two years.

When well grown the leaves are often more than a yard long, and present a vivid and most striking coloration.

Fig. 79.

Centaurea babylonica.

Their midribs reach four inches or more across, and vary from a dark deep waxy orange to vivid polished crimson. The splendid hue of the lower part of the leaf stalks flows on towards the point, and spreads in smaller streams through the main veins and ramifications of the great soft blades of the leaf, often a foot and even fifteen inches in diameter, if the plant be in rich ground. The under sides of the leaves are the most richly coloured, and the habit such that these sides are well seen. It requires the treatment of an annual —to be raised in a gently heated frame, and afterwards planted out in very rich ground, though it may also be kept over the winter in pots. It varies a good deal from seed, and the most striking individuals should be selected before the plants are put out. Used sparingly, its effect would perhaps be more telling than if in quantity, and it is well suited for isolation—that is to say, placing singly on the grass near a clump of shrubs. Everybody who values a really distinct object in the flower garden should have it. During the past season it attained splendid dimensions and colouring in the nursery department of the Jardin des Plantes, and it doubtless will soon be seen everywhere with us.

CHAMÆROPS EXCELSA.—It may not be generally known that this palm is perfectly hardy in this country. A plant of it in her Majesty's gardens at Osborne has attained a considerable height. It is also out at Kew, though protected in winter. On the water side of the high mound in

Fig. 80

Chamærops excelsa.

the Royal Botanic Gardens, Regent's Park, it is in even better health than at Kew, though it has not had any protection for years, and stood the fearfully hard frost of 1860. If small plants of this are procured, it is better to grow them on freely for a year or two in the greenhouse, and then turn out in April, spreading the roots a little and giving deep loamy soil. Plant in a sheltered place, so that the leaves may not be injured by winds when they get large and grow up. A gentle hollow, or among shrubs on the sides of some sheltered glade, will prove the best places. The establishment of a palm among our somewhat monotonous shrubbery and garden vegetation is surely worthy of a little trouble, and the precautions indicated will prove quite sufficient.

CRAMBE CORDIFOLIA.—This is unquestionably one of the finest of perfectly hardy and large-leaved herbaceous plants. It is as easily grown as the common Seakale—easier, if anything; and in heavy rich ground makes a splendid head of leaves, surmounted in summer by a dense spray of very small flowers. Doubtless, if these be pinched off, a larger development of the fine glossy leaves may be expected, but as the shoots are so vigorously shot up and converted into a distinct and pretty inflorescence, many will prefer to " leave the plant to nature." In planting it, the deeper and richer the soil, the finer the result. It will prove a capital thing for every group of fine-leaved hardy plants, and may also be planted wherever a bold though low type of vegetation is desired. There is another species, C. juncea, which is also effective, but not so valuable as C. cordifolia.

CUCUMIS PERENNIS (Perennial Cucumber).—This has not the quality of leaf which we could desire, but it will prove interesting to many. It is perfectly hardy, and possesses, so to speak, great trailing power. Its leaves are strong, rough, and of a glaucous colour; and the shoots run about freely if the plant be in very rich soil. Where bold trailing plants for high trellis-work, or rough banks, or shaggy rockwork are desired, it will be found distinct; but withal we cannot give it a place in the front rank, and the small select garden without any of the above-mentioned appendages will certainly be better without it. For the botanical garden and curious collections it is indispensable. It is strong and lasting when well established, and may be allowed to fall over rough banks, stumps, or be trained up trellis-work, &c.

DATISCA CANNABINA. — The male plant of this has long been known as a very strong and effective herb—

graceful too; but I saw female plants associated with males
for the first time in the Jardin des Plantes, and since then
I have a higher opinion of the species.   The female plant
remains green much longer than the male, and being profusely laden with fruit, each shoot droops and the whole
plant improves in aspect.   It must not be forgotten in any
selection of hardy plants of free growth and imposing
aspect.   From seed will probably be found the best way to
raise it, and then one would be pretty sure of securing
plants of both sexes.

ELYMUS ARENARIUS.—This wild British grass—a strong-
rooting and most distinct-looking herb—is capable of adding
a striking feature to the garden here and there, and should
be quickly introduced to civilization.   Planted a short distance away from the margin of a shrubbery, or on a bank
on the grass, and allowed to have its own way in deep soil,
it makes a most striking object.   In short, it deserves to
rank fourth among really hardy big grasses, the Pampas
and the two Arundos alone preceding it.   I am not quite
certain that it is not more useful than the Arundo, being
hardy in all parts of these islands.   In very good soil
it will grow four feet high, and as it is for the leaves we
should cultivate it, if the flowers are removed they will be
no loss.   It is found frequently on our shores, but more
abundantly in the north than in the south.   The variety
called geniculatus, which has the spike pendulous, is also
worthy of culture, and in its case the flowers may prove worth
preserving.   It may possibly be useful for covert, and is
certainly so for rough spots in the pleasure ground and in
semi-wild places.

THE FERULAS.—I wish it were not necessary to write in
praise of such very fine plants as these, so noble in aspect
and beautiful in leaf.   If you grow 2000 kinds of herbaceous
plants, the first things that show clearly above the ground
in the very dawn of spring (even in January) are their deep
green and most elegant leaves.   In good garden soil they
look like masses of Leptopteris superba, that most exquisite
of ferns.   Their chief charm will probably be found to
consist in their furnishing masses of the freshest green and

highest grace in early spring. The leaf is apt to lose some of its beauty and fade away early in autumn, but this may to some extent be retarded by cutting out the flower-bearing shoots the moment they appear. Not that these are ugly; for, on the contrary, the plants are fine and striking when in flower. It is indispensable that the Ferulas, like some other hardy foliaged plants, be planted

Fig. 81.

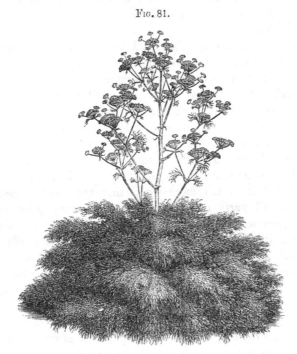

Ferula communis.

permanently and well at first, as it is only when they are thoroughly established that you get their full effect. At a first view, the best way to treat them would appear to be, so to arrange them that they would be succeeded by things that flower in autumn, and only begin their rich growth in early summer; but it will be equally wise to plant them near the margin of a shrubbery, where it is desired to have a diversified and bold type of vegetation. In the rougher and more solid ground, so to speak, near large rockwork or rootwork, they would of course prove grand. The Heracleums, so often recommended in garden literature for planting near water, &c., are mere coarse rags compared to the Ferulas, while the Ferulas may be used in the places recommended for Heracleums. We may look forward to the day when a far greater variety of form will be seen in English gardens than is at present observable, and these Ferulas are thoroughly well worth growing for their superb spring and early summer effect. The best species are F. communis, tingitana, and neapolitana. Probably a few others, including sulcata, ferulago, and glauca, may with advantage be added where variety is sought, but the effect of any of the first three cannot be surpassed. Among "aspects of vegetation" which we may enjoy in these cold climes, nothing equals that of their grand leaves, pushing up with the snowdrop. In semi-wild spots, where spring flowers abound, it will prove a most tasteful and satisfactory plan to drop a Ferula here and there in a sunny spot, and leave it to nature and its own good constitution ever afterwards.

GYNERIUM ARGENTEUM (the Pampas Grass).—This is so well known to the reader that there is no excuse for naming it here, except the opportunity to say a few words as to the splendid use we may make of it in the branch of gardening we are now discussing. It is to the Dublin Botanic Gardens we owe the introduction of this noble plant, now much grown in every country where ornamental gardening is pursued. It really deserves as much attention as any plant in cultivation, and yet how rarely is any thorough preparation made for its perfect development. A paltry class of tender plants may cost more labour and time in the course of a few months than would suffice to plant a field of the Pampas grass, yet such a glorious thing as this may be put in with a barrowful of mould to start on a bad soil, and then perhaps planted by the water or some other secondary spot called its "proper place." What is there growing in garden or in wild more nobly distinct and beautiful than the great silvery plumes of this plant waving in the autumnal gusts—the burial plumes as it were of our summer too early dead?　What tender plant so effective as this in giving a new aspect of vegetation to our gardens, if it be tastefully placed and well-grown?　Long before it flowers it possesses more merit for its foliage and habit than scores of things cultivated indoors for their effect—Dasyli-

Fig. 82.

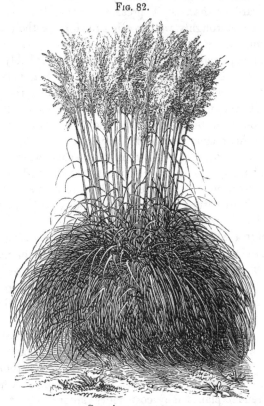

Gynerium argenteum.

rions, &c., for example, and it would be well worthy of being extensively used if one of its silken-crested wands never put forth in autumn.　It is not enough to place it in out-of-the-way spots—the general scene of every garden and pleasure ground should be influenced by it; it should be planted even far more extensively than it is at present, and given very deep and good soil either natural or made. The soils of very many gardens are insufficient to give it the highest degree of strength and vigour, and no plant better repays for a thorough preparation, which ought to be the more freely given when it is considered that the one preparation suffices for many years. If convenient, give it a somewhat sheltered position in the flower garden, so as to prevent as much as possible that ceaseless searing away of the foliage which occurs wherever the plant is much exposed to the breeze. We rarely see such fine specimens as in quiet nooks where it is pretty well sheltered by the surrounding vegetation. It is very striking to come upon noble specimens in such quiet green nooks ; but, as before hinted, to leave such a magnificent plant out of the flower garden proper is a decided mistake.

HELIANTHUS ORGYALIS.—They use this in some parts of the Continent as an ornamental-leaved plant in the pleasure ground, &c. It is as hardy as the common dandelion, grows to a considerable height, and is of a very distinct habit. Its distinction arises from the fact that the leaves are recurved in a peculiarly graceful manner. At the top of the shoots, indeed, their aspect is most striking, from springing up in great profusion and then bending gracefully down. It will form a capital subject for the group of fine-leaved, hardy plants, not running through the ground and requiring all the room for itself to spread about. As it is apt to come up rather thickly, the cultivator will act judiciously by thinning out the shoots when very young, so that those which remain may prove the stronger and the better furnished with leaves.

THE HERACLEUMS.—These are pretty well known for the rapid vigour and great size of their herbaceous vegetation ; but they are as a rule too coarse, and decay too early in summer, to be used in the flower garden or pleasure ground. They may, however, be employed with advantage where a robust and picturesque vegetation is desired in half wild spots on islands, and for furnishing distant effects.

HIBISCUS ROSEUS.—This is a very noble hardy perennial, growing from four to six feet high about Paris, and having

the upper part of each of its abundant shoots set thickly
with buds which produce flowers fully six inches across, of
a showy rose colour, with straight deeply coloured veins
running from the rich dark crimson base of the petals, and
gradually becoming lost towards the margin.   There is
reason to think it thoroughly hardy, and it is well worth a
trial in good soil in the southern and milder parts of England and Ireland.   The show it makes in autumn is really

FIG. 83.

Heracleum flavescens.

very fine, and it will probably be found a grand thing for
association with noble autumn flowers, like the Tritoma and
Pampas grass.   As regards leaf effect, it is scarcely subtropical—to use again that awkward term—and should
perhaps be classed with showy herbaceous plants; but as
it was used with pretty good effect in one of the Paris

parks, I name it here.   It should have a warm position, and deep, rich, and light soil.

MACLEAYA CORDATA.—This is a fine plant in free soil, but comparatively poor in that which is bad or very stiff.   It is quite distinct in habit and tone, and sometimes goes beyond six feet high.   The flowers are not in themselves pretty, but the inflorescence when the plant is well grown has a distinct and pleasing appearance.   It will prove a good thing for associating with other fine hardy plants suitable for making bold groups.   With some of the things before named, and with other perfectly hardy plants, there should be no difficulty in producing as bold and striking groups of vegetation as any ever seen either with us or in Paris, and afforded by costly and tender exotics requiring winter protection.

MELIANTHUS MAJOR.—This is usually treated as a green-house plant, and is sometimes put out of doors in summer. So grown, how-ever, the full beauty of the plant has not time to develope ; and much the better way is to treat it as a hardy subject, putting it out in some sunny and sheltered spot, where the roots will not suffer from wet in win-ter.   The shoots will be cut down with frost, but the root will live and push up strong ones in spring,

FIG. 84.

Melianthus major.

forming by midsummer a bush of very distinct and beautiful leaves.   I have grown it in this way to a much more pre-

Q 2

sentable condition than it ever assumes indoors, where it is usually drawn too much. I used to protect the roots in winter by placing leaves over them, and then covering all with a handlight, but have seen the plant survive without this precaution. It is, however, best to make quite sure by using protection, except where the soil and climate are particularly favourable.

MOLOPOSPERMUM CICUTARIUM.—There is a deep-green and fernlike beauty displayed profusely by some of the umbelliferous family, but I have rarely met with one so remarkably attractive as this species. Many of the class, while very elegant, perish quickly, get shabby indeed by the end of June, and are therefore out of place in the tasteful flower garden; but this is firm in character, of a dark rich green, stout yet spreading in habit, growing more than a yard high, and making altogether a most pleasing bush. It is perfectly hardy, a native of Carniola, easily increased by seed or division, but very rare just now. I doubt if it is even in our botanic gardens, but hope to see it in cultivation ere long.

FIG. 85.

Nicotiana macrophylla.

NICOTIANA MACROPHYLLA.—This is simply a garden name for a fine large variety of the common Tobacco. As it is so readily raised from seed, and grows luxuriantly in rich soil, I need not say it is a very desirable subject for association with the Castor-oil plant and the like, and especially suited for the many who desire plants of noble habit, but who cannot preserve the tender ones through the winter under glass. The flowers are very ornamental. It should be raised on a hotbed, and put out in May.

PANICUM BULBOSUM is a tall and strong grass, with a free and beautiful inflorescence. It grows about five feet high, and the flowers are very gracefully spread forth. It forms an elegant plant for the flower garden in which grace and variety are sought; for dotting about here and there, near the margins of shrubberies, &c.; and indeed for the sake of its flowers alone. P. virgatum is also a good bold grass. Both of these may be raised from seed, and are well worthy of cultivation.

PHYTOLACCA DECANDRA.—The true plant of this name forms a very free and vigorous mass of vegetation, and, though perhaps scarcely refined enough in leaf to justify its being recommended for flower garden use, no plant is more worthy of a place wherever a rich herbaceous vegetation is desired; whether near the rougher approaches of a hardy fernery, open glades near woodland walks, or any like positions.

POLYGONUM CUSPIDATUM.—This is an unusually large herbaceous species of a genus which, as cultivated in our botanical collections, does not appear likely to afford an elegant or a graceful subject for our gardens. But it is one of the best hardy things which can be recommended for their embellishment. The growth is rapid, the size unusual, perhaps eight or ten feet in very good soil, and the bearing of the plant not at any season shabby. It is covered with flowers in autumn. The same plant is often called P. Sieboldi, and frequently sold by that name. When planted singly, and away from other subjects, its head assumes a rather peculiar and pretty arching character, and therefore it is not quite fit for forming centres or using in groups, so much as for planting singly on the turf, there leaving it to take care of itself and come up year after year.

In this way it would be particularly useful in the pleasure
ground or diversified English flower garden. It is also
good for any position in which a bold and distinct type
of vegetation is desired, while of course, when we come
to have fine groups of hardy "foliage plants" in our
gardens, its use will be much extended. The deeper and

Fig. 86.

Rheum Emodi.

better the soil, the finer will its development prove. You
cannot make the soil too deep and good if you want the
plant to assume a striking character. As with tender
plants we have no end of attention to bestow, often daily,
the time and labour necessary to well prepare the ground
for a hardy subject should never be grudged. This plant
will probably be also found useful for game covert. It is

easily procured in our nurseries, and there is, or used to be, plenty of it at Kew.

RHEUM EMODI.—The Rhubarbs, from their vigour and picturesqueness, are well worthy of cultivation among hardy, fine-leaved plants. Some of the common kinds have recently been placed in our parks, but the most striking and distinct of the introduced kinds is the Himalayan Rheum Emodi, and it is the one that is seldomest seen.

RHUS GLABRA LACINIATA.—I have known this plant for about three years as a subject of much promise for garden decoration, and can confidently recommend it as one of the most useful and elegant dwarf shrubs we can employ to furnish an attractive effect. It is a small kind, with finely-cut and elegant leaves, the strongest being about a foot long when the plants have been established a year or two. When seen on a nicely established plant, these leaves combine the beauty of those of the finest Grevillea, with a fern frond, while the youngest and unfolding leaves remind one of the dainty ones of a finely cut umbelliferous plant in spring. The variety observable in the shape, size, and aspect of the foliage makes the plant charming to look upon, while the midribs of the fully grown leaves are red, and in autumn the whole glow off into bright colour after the fashion of American shrubs and trees. During the entire season it is presentable, and there is no fear of any vicissitude of weather injuring it. Its great merit is that, in addition to being so elegant in foliation, it has a very dwarf habit, and is thoroughly hardy. Plants at three years old and undisturbed for the past two years are not more than eighteen inches high. The heads are slightly branched, but are not a whit less elegant than when in a simple-stemmed and young state, so that here we have clearly a subject that will afford a charming fernlike effect in the full sun, and add graceful verdure and distinction to the flower garden. When the flowers show after the plant is a few years old they may be pinched off, and this need only be mentioned in the case of permanent groups or plantings of it. To produce the effect of a Grevillea or fern on a small scale, we should of course keep this graceful Rhus small, and propagate it like a bedding plant. The graceful mixtures and bouquet-like beds that might be made with the aid of such plants need not be suggested here, while of course an established plant, or groups of three, might well form the centre of a bed. Planting a very small bed or group separately in the flower garden, and many other uses which cannot be enumerated here, will occur to those who have once tried it. Some hardy plants of fine foliage are either so rampant or so top heavy that they cannot be wisely associated with bedding-plants. This is, on the contrary, as tidy and tractable a grower as the most fastidious could desire. It would be a pity to put such a pretty plant under or near rough trees and shrubs. Give it the full sun, and good free soil.

The Tritomas.—So hardy, so magnificent in colouring, and so fine and pointed in form are these plants, that we can no more dispense with their use in the garden where beauty of form as well as colour is to prevail, than we can with the noble Pampas grass. They are more conspicuously beautiful when other things begin to succumb before the gusts and heavy rains of autumn, than any plants which flower in the bright days of midsummer. It is not alone as component parts of large back ribbons and in such positions that these grand plants are useful, but in almost any part of the garden. Springing up as a bold close group on the green turf and away from brilliant surroundings, they are more effective than when associated with bedding plants ; and of course many such spots may be found for them near the margins of the shrubberies in the generality of pleasure grounds. It is as an isolated group flaming up amid the verdure of trees and shrubs and grass that their dignified aspect and brilliant colour are seen to best advantage. However, tastefully disposed in the flower garden, they will prove generally useful, and particularly for association with the finer autumn-flowering herbaceous plants. It seems we do not sufficiently appreciate the advantage of good hardy plants, however much we may grumble at the consumption of coals. Here are the finest of all autumnal flowers, never causing a farthing of expense for wintering, storing, or re-

planting, but merely asking for a little ordinary preparation of
the soil at first, and yet they are merely grown as adjuncts even
in good gardens, and in
many you can scarcely
find them. For every
quality that should
make a plant valuable
in the eyes of the
flower gardener, they
cannot be surpassed by
any subjects that re-
quire expensive care
all through the winter;
indeed we may say they
cannot be equalled by
any of such—a suffi-
cient proof that it is
not only those who
possess stoves, green-
houses, and glass-gar-
dens, so to speak, that
may enjoy the highest
beauty in their gardens.

Fig. 87.

Anemone japonica alba.

A most satisfactory result may be produced by associating these Tritomas with the Pampas and the two
Arundos, the large Statice latifolia, and the strong and
beautiful autumn-flowering Anemone japonica alba. This
is peculiarly suited for association with hardy herbaceous
plants of fine habit, and should be in every garden where a
hardy flower is valued.

VERBASCUM VERNALE.—Most of us know how very distinct and imposing are the larger Verbascums, and those
who have attempted their culture must soon have found out
what transient far-seeding things they are. Of a biennial
character, their culture is most unsatisfactory: they either
migrate into the adjoining shrubbery or disappear altogether.
The possession of a thoroughly noble perennial one must
therefore be a desideratum, and such a plant will be found
in the Hungarian Verbascum vernale. This is fine in leaf

and stature, and produces abundance of flowers. The lower
leaves grow eighteen or twenty inches long, and the plant
when in flower to a height of seven or eight feet, or even
more when in good soil. It is a truly distinct subject. for
helping us to variety, and may, it is to be hoped, ere long
be found in our gardens and nurseries. At present it is a
scarce plant in England, and perhaps not to be had in
many of our nurseries or botanic gardens, though it is certainly the best known plant of the genus to us in cultivation. I first saw it in the Jardin des Plantes.

THE YUCCAS.—Among all the hardy plants ever introduced
to this country, none surpass for our present purpose the
various kinds of Yucca, or " Adam's Needle," as it is sometimes called. There are several species hardy and well
suited for flower garden purposes, and, more advantageous
still, distinct from each other. The effect afforded by them,
when well developed, is equal to that of any hot-house plant
that we can venture in the open air for the summer, while
they are hardy and presentable at all seasons. They may
be used in any style of garden ; may be grouped together on
rustic mounds, or in any other way the taste of the planter
may direct. The best perhaps, considering its graceful and
noble habit, is Y. pendula, which is simply invaluable in
every garden. Old and well established plants of it standing
alone on the grass are pictures of grace and symmetry, from
the lower leaves which sweep the ground to the central ones
that point up as straight as a needle. It is amusing to
think of people putting tender plants in the open air, and
running with sheets to protect them from the cold and rain
of early summer and autumn, while perhaps not a good
specimen of this fine thing is to be seen in the place. There
is no plant more suited than this for planting between and
associating with flower beds. Next we have Y. gloriosa,
more pointed in habit and rigid in style, and also large
and imposing in proportions. Lacking the grace of pendula,
it makes up for it to some extent by boldness of effect,
while, like the preceding, it sometimes sends up a huge
mass of flower. Y. gloriosa varies very much when grown
from seed—a good recommendation, as the greater variety of

fine form we have the better. Then there is Y. glaucescens, with a sea-green foliage, and rather free to flower, the buds being of a pink tinge, which tends to give the whole inflorescence a peculiarly pleasing tone. This is a first-class

Fig. 88.

Yucca pendula.

plant. Y. filamentosa is smaller than these, but one which flowers with much vigour and beauty. It is well worth cultivating in every garden ; not only in the flower garden or pleasure ground, but also on the rough rockwork, or any

spot requiring a distinct type of hardy vegetation, and so is
its fine though delicate variegated variety. Yucca flaccida is
somewhat in the way of this, but smaller. It flowers even more abundantly and regularly than filamentosa, and is well worthy of cultivation. The preceding species, if not so much used in our gardens as they deserve, are at all events known in them. The following I met with for the first time in Parisian gardens :—

FIG. 89.

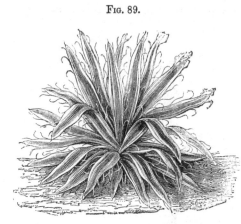

Yucca filamentosa variegata.

FIG. 90.

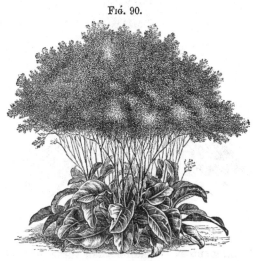

Statice latifolia.

Y. lutescens is a species of neat habit and slightly yellowish
tone of shining green, and very distinct. Y. flexilis is an

ornamental, though not large growing kind. Y. stricta is a rigid species scarcely so effective as the preceding; and Y. angustifolia has narrow pointed leaves and a distinct habit. Y. Treculeana is a very noble species, which will be found perfectly hardy on good soil and in warm situations. It has deeply furrowed and very large rigid leaves, and is well worthy of culture even in a cool house, in which it is sometimes kept in this country. If we had but this family alone, our efforts to produce an agreeable effect with hardy plants need not be fruitless. The freely flowering kinds, filamentosa and flaccida, may be associated with any of our nobler autumn flowering plants, from the Gladiolus to the great Statice latifolia. The species that do not flower so often, like pendula and gloriosa, are simply magnificent as regards their effect when grown in the full sun and planted in good soil; and I need not say bold and handsome groups may be formed by devoting isolated beds to this family alone. They are mostly easy to increase by division of the stem and rhizome; and should in all cases be planted well and singly, beginning with healthy young plants, so as to secure perfectly developed single-stemmed specimens.

List of Hardy Herbaceous and Annual Plants, &c., of fine habit, worthy of employment in the flower garden or pleasure ground.

Acanthus, several species.
Asclepias syriaca.
Statice latifolia.
Morina longifolia.
Polygonum cuspidatum.
Rheum Emodi, and several other species.
Euphorbia Cyparissias.
Datisca cannabina.
Veratrum album.
Tritomas, in variety.
Thalictrum fœtidum.
Crambe cordifolia.
Althæa taurinensis.
Geranium anemonæfolium.
Melianthus major.
Panicum, several species.

Spiræa Aruncus.
  „   venusta.
Astilbe rivularis.
  „   rubra.
Eryngium, several species.
Ferula, several species.
Seseli,   „
Chamærops excelsa.
Cucumis perennis.
Hibiscus roseus.
Rhus glabra laciniata.
Artemisia annua.
Phytolacca decandra.
Centaurea babylonica.
Lobelia Tupa.
Peucedanum ruthenicum
Heracleum, several species.

Dipsacus laciniatus.
Alfredia cernua.
Cynara horrida.
  „   Scolymus.
Carlina acanthifolia.
Telekia cordifolia.
Echinops exaltatus.
  „   ruthenicus.
Helianthus argyrophyllus.
  „   orgyalis, and others.
Gunnera scabra.
Funkia subcordata.
  „   japonica.
Tritoma, in varieties.
Arundo Donax.
  „   conspicua.
Gynerium argenteum.

Elymus arenarius.
Bambusa, several species.
Arundinaria falcata.
Yucca, several species.
Verbascum vernale.
Aralia spinosa.

Aralia japonica.
„ edulis.
Macleaya cordata.
Panicum bulbosum.
„ virgatum.
Kochia scoparia.

Datura ceratocaula.
Silybum eburneum.
„ marianum.
Onopordon Acanthium.
„ arabicum.

*List of Hardy Plants of fine habit, that may be raised
from Seed.*

Among suitable hardy plants that may be raised from
seed, the following are offered in recent seed catalogues:—

Acanthus latifolius.
„ mollis.
„ spinosus.
Artemisia annua.
Astilbe rivularis.
Campanula pyramidalis.
Cannabis gigantea.
Carlina acanthifolia.
Datura ceratocaula.
Echinops, several species.
Eryngium bromeliæfo-
lium.
„ campestre.
„ cœlestinum.

Eryngium giganteum.
Ferula communis.
„ tingitana.
Geranium anemonæfo-
lium.
Gunnera scabra.
Gynerium argenteum.
Helianthus argyrophyl-
lus.
„ orgyalis.
Heracleum eminens.
„ giganteum.
„ platytænium.
Kochia scoparia.

Lobelia Tupa.
Morina longifolia.
Onopordon arabicum.
„ tauricum.
Centaurea babylonica.
Panicum, several species.
Phytolacca decandra.
Salvia argentea.
Silybum marianum.
„ eburneum.
Statice latifolia.
Tritomas, in variety.
Yucca, several species.

PLATE XXXIV.

THE PALACE AND GARDENS OF VERSAILLES.

CHAPTER XII.

This being one of the most celebrated gardens in the world it behoves us to examine it somewhat in detail—were we, however, to treat of it in proportion to its real merits as a garden, a very small amount of space would suffice. Let us pass through the vast stone courtyard and take up our position near the garden front of the palace. Standing near the walls, looking over the gardens, and following the vista of the canal into the low country beyond, the eye first rests on a vast spread of gravel, some marble margins of great water basins, sundry protuberances from the level of the water, and away in the distance an effect like that afforded by a suburban canal in a highly practical and unlovely country. A few Lombardy Poplars help this remote vista, but the effect of the whole is from this point of view lamentable. To the right of the palace there is a rather pleasing garden, with big box-edgings, clipped conical Yews and other trees, and numerous statues well shown against dense woods of Horse-chestnut trees. To the left there is one of those spreads of gravel, grass, a few stumpy clipped Yews, &c., generally known as geometrical gardens, the Horse-chestnut groves starting up rather abruptly and relieving the whole so as to render it tolerable. Advancing from the palace, the lower terrace and its surroundings come into view, and the effect improves. The faces of the terrace walls are hedged with green ; the flower borders are somewhat after the fashion of those at the Tuileries, and surrounded by a line of well-grown Orange trees. Above the terrace walls Yew trees are planted and clipped very regularly ; in the centre there is a fine and costly fountain, and the dense groves of trees near at hand again save the scene from bald formality,

not to say hideousness.   Versailles is a vast garden, much
of its interest being hidden behind these kindly groves of
trees, but we have about here the broadest effects of this
far-famed place, and may judge in how far they are worthy
of the praise bestowed on them and of our admiration or
imitation.

Versailles is held up by the French and others as the
queen of geometrical gardens, and however this position
may be dissented from, it cannot be denied that it is a vast
illustration of the formal school of gardening.

There are in books many dissertations on the several
styles of laying out gardens; indeed some have taken us
to China and Japan, and others gone into Mexico for illustration; but when all is read and examined, what is the
result to anybody who looks from words to things?   That
there are really two styles: one straitlaced, mechanical,
fond of walls or bricks, or it may be gravel; fond also of
such geometry as the designer of wall papers excels in, often
indeed of a much poorer and less graceful kind than that;
fond too of squirting water in an immoderate degree, with
trees in tubs as an accompaniment, and perhaps griffins and
endless plaster and stone work.   The other, with true
humility and right desire, though often awkwardly and blun-
deringly, accepting nature as a guide, and endeavouring
to multiply, so far as convenience and poor man-power will
permit, her most charming features.

Mr. Ruskin tells us that " we are forced, for the sake of
accumulating our power and knowledge, to live in cities:
but such advantage as we have in association with each
other is in great part counterbalanced by our loss of fellowship with nature.   We cannot all have our gardens now,
nor our pleasant fields to meditate in at eventide.   Then
the function of our architecture is, as far as may be, to
replace these; to tell us about nature; to possess us with
memories of her quietness; to be solemn and full of tenderness, like her, and rich in portraitures of her; full of
delicate imagery of the flowers we can no more gather, and
of the living creatures now far away from us in their own
solitude."

PLATE XXXV.

VIEW FROM THE TAPIS VERT.

What, then, are we to think of those who carry the dead lines and changeless triumphs of the building and the studio into the garden, which, above any other artificial creation, should give us the sweetest and most wholesome "fellowship with nature"?

Simply that it is presumption and bad taste, founded upon ignorance of what a true garden ought to be, and of knowledge that the deadliest thing you can do with it is to introduce any feature which, unlike the materials of our world-designer, never changes. Away, then, with the wretched affectation of pretending to enjoy—away with the ignorance which asserts or blindly believes that there is some mysterious and occult beauty in, or necessity for, such gardens as this!

It is perfectly true that there are some positions where an intrusion of architecture and embankments into the garden is justifiable—nay, even now and then necessary; but the misfortune is that they are often said to be so when such is not the case. It would be a waste of space to quote the nonsense that is printed and urged about things being "in keeping,"—the necessity of making an architectural garden associate with some particular style of building, and so on. The best terrace gardens in continental countries are those built where the nature of the ground most calls for them and usually in positions where the ground is steep and rugged; and it is in positions most like these that they best succeed, and are most wanted in this country. Why, then, talk of "congruity" in the matter, when it is considered right to place the most geometrical kind of garden in the spots where the ground is most picturesque and irregular? There is no code of taste resting on any real foundation which proves that garden or park should have any extensive stonework or geometrical arrangement. Many instances could be given to prove that the natural or nearly natural disposition of the ground is far preferable to the great majority of expensive mathematical gardens.

Among other not often urged objections to great expenditure on architectural embellishments, costly fountains, and statues, instead of on the development of the real life and interest of a garden, is the fact that outdoor artistic embellishment, good or bad, is by no means so appropriate in these cool and gusty climes as in warmer countries. Where people can live out of doors the greater portion of the year, and where the winter is merely a pleasant dry kind of atmospheric tonic; where the native can dine for more than half the year in a bower of vines, and breathe the spices of Orange trees and Magnolias—in the south of France, Spain, and Italy, and all along the shores of the Mediterranean—it is more desirable to have the nude form in marble in the open air, independently of the fact that the lichens and moss do not so soon begin to embellish the carving, or grass to grow out of its interstices, in countries near the sun. Leave art indoors—where, unfortunately, we must content ourselves for the most part—use as few wallpaper patterns and as little stonework as possible in our gardens, and arrange them so that when our sunny season does come they may be full of life and change, and that all our efforts therein may tend to their improvement in the right direction.

In discussing this phase of gardening we have a capital example in the case of the Crystal Palace, in the region of the great fountain basins, where a more horrid impression is received than in any part of Versailles, though the upper terrace at the Palace illustrates the best features of the system, and shows as well as anything I know of in how far it may be safely adopted near a great building. But both at the Palace and Versailles the vast expense for a poor theatrical effect is not the most regretful of present features; that, perhaps—not to look deeply into the blemishes of such positions—is the dirty, wide, changeless water basins, with their squirting pipes and perhaps crumbling margins; for the purse that creates such delights frequently fails, if it does not get tired of expenditure that never produces the changeful beauty for which the heart of man yearns. To me there is nothing more appalling than the walls, fountain basins, clipped trees, and long canals, &c., of such a place as Versailles, not only because they utterly fail to satisfy in themselves, but inasmuch

PLATE XXXVI.

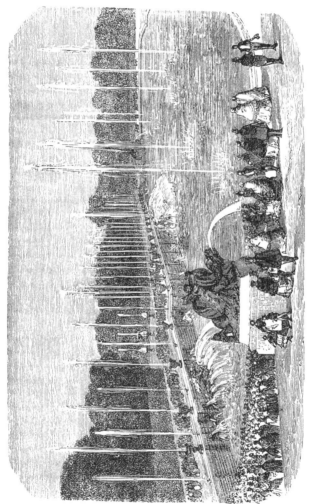

FOUNTAINS OF THE BASIN OF NEPTUNE.

as they are ever accompanied by a day-ghost of wasted effort—of riches worse than lost.

In connexion with the Crystal Palace one thinks of ruined shareholders; and with Versailles, of the enormous sums wrung from an oppressed people, and put to such a miserable use, that one can scarcely regret a wild blood-dance of revolution came and put an end to it all. And this was the kind of good effected with the money so hardly wrung from starving millions! It was merely burying wealth—indeed, it might have been better to have buried it, for many would prefer the naked earth to these gyra-tions, which must be kept in repair at great cost or they become intolerable even to their builders and designers.

When a private individual indulges in expensive fancies, he has small influence to injure any one but himself; but when the place is a public one, and set up as an example of all that is admirable, then, in addition to the first wasteful expenditure, we have an object hurtful to the public taste, and sowing the seed of its ugliness all over the country.

It may be said that our taste in England is sufficiently assured against this; but it is not so. I have known those whose lawns were or might readily be made the most beautiful of gardens, ruin them, and for the mere sake of having a terraced garden. There is a modern castle in Scotland where the embankments are piled one above another till the thing looks as if the Chinese who carve the ivory balls had been invited to make a corresponding ar-rangement in the fortification style. Were it a matter of trifling cost, or which could be easily abolished or even avoided, it would not be worthy our attention; but being so expensive that it may curtail for years the legitimate outlay for a garden, and prevent expenditure in live interest rather than in slow crumbling monotony, too much cannot be urged against it. The style was in doubtful taste in climates and positions more suited to it than those of northern France and England; but he who would now adopt it in an age when civilization has set its formal brand upon everything, and in the presence of the inexhaustible and magnificent collections of trees and plants which we

now possess, is an enemy to every true interest of the
garden.

We will next visit a few of the more interesting of such
features as are hidden from the general scene, first, however, glancing at the Tapis Vert—the grassy avenue which
leads from the parterre to the Bassin d'Apollon and the
Grand Canal.　From it the effect is much better than from
the terrace above, and it, like many parts of the place, is
bordered or hedged with numbers of costly statues and
vases.　They seem as profuse as if the gold and marble had
been dug up on the spot, and as if this had been the reason
why a great garden had been made in such a very bad
position.

The Orangery here, in a sunk garden to the south of
the Palace and the Parterre du Midi, is probably the most
remarkable known.　It is most permanently and massively
built in the face of a terrace, and is more than thirteen
hundred feet long by thirty-six wide.　It is in fact an
immense archway, lighted at one side.　The height from
the balustrade of the terrace above to the walk in front of the
Orangery is about forty-six feet, and once on the occasion of
a night fête a poor English visitor, thinking this balustrade
was merely a dividing line between two parterres, jumped
over, and was found nearly killed below.　The collection of
Orange trees here is immense; but as we have already discussed this unhappy phase of horticulture in the chapter on
the Tuileries gardens, little need be said here.　One of the
trees, however, is deserving of especial remark, and, indeed,
I hoped to give an exact portrait of it, and should have
done so were it not for an unpunctual photographer.　This
tree was produced from seeds sown in 1421, by Leonora of
Castille, wife of Charles III., King of Navarre, and after
enduring between 400 and 500 years, is still healthy and
verdant in its leafage, though a little tottering, and requiring
to be carefully propped up.　That it should have lived so
long under the circumstances is indeed very remarkable,
for of course a tree put into a half-lighted building in
winter, and placed in the open air in summer, and at all
times liable to vicissitudes at the roots, runs great danger

PLATE XXXVII.

LA TOILETTE D'APOLLON.

compared to one in the open air. The large collection of Orange-trees is usually placed in the open air about the 15th of May, and under cover not later than the 15th of October, so that the trees only enjoy the free air and sun for five months out of the twelve. In addition to the Orange-trees, a few other exotics were kept in this structure in winter, and submitted to the same treatment as the Orange-trees at all other seasons. These are Justicia Adhatoda, Olea angustifolia, Jasminum azoricum, and Edwardsia grandiflora. They seemed to do remarkably well under the treatment usually given to Orange-trees on the Continent, and the Justicia and Jasminum, and perhaps the others, are in my opinion more worthy of being thus grown than the Orange, inasmuch as they display their fine flowers in the open air in summer, and they are less costly than when grown in stoves or conservatories. The specimens of the Madeira Jasmine are the finest I ever saw; the rich green shoots drooping gracefully and bearing abundance of flowers. The Justicia and others were said to flower abundantly in their seasons. This, considered in connexion with the success which attends the culture of the Oleander and the sweet-scented Pittosporum under like circumstances, and even when preserved in cellars during the winter, would seem to point to the desirability of adopting the system, or a modification of it. It has not spread among us, but it certainly is as practicable in England as in many parts of the Continent where it is seen. With us the nearest thing to it is the practice of putting handsome evergreens in tubs for placing in terrace gardens, &c. But surely it is scarcely worth while doing this with things that we see in every shrubbery! If we do go to the expense of growing plants thus, let us select those that will not bear the open air of our winters, but which succeed well out of doors in summer.

It is needless to describe the numberless gardens, fountains, &c., of Versailles, and indeed impossible, unless one could devote a book to the subject. Very few of the spots indicated on the plan will please the visitor more than the garden or Bosquet du Roi, near the Orangery; and simply because the artificialness, the stonework, and want of repose which are characteristics of the greater part of Versailles, are here absent. It is simply a sweep of grass, surrounded by handsome trees, with a few flower beds and fine-leaved plants here and there. It is but one of a thousand types of scene which pure taste and a knowledge of hardy trees and plants may produce, and yet it is sufficient to show the vain, unsatisfactory, and trumpery character of the various far more costly gardens in the immediate neighbourhood. All visitors should see it after surveying the general dreariness of the rest.

Fig. 91.

The Tapis Vert.

To the south of the Tapis Vert, and near the Jardin du Roi, the Colonnade is very well worth seeing, and perhaps the happiest feature of the architectural gardening. The grove encloses a peristyle in marble about one hundred feet in diameter. It is composed of thirty-two columns of marble in different colours, with the capitals in white marble, and all most richly ornamented. Under each arch is placed a vase-like basin in marble from which springs a jet d'eau, and in the centre of the arena is a group in marble representing the Rape of Proserpine. The effect of

PLATE XXXVIII.

THE COLONNADE AT VERSAILLES.

the whole, closely surrounded as it is by a dense grove, is very fine.

To the right of the château, the most interesting spot, is the Bosquet des Bains d'Apollon. This is simply a large and picturesque surface of rock, well-backed by trees and with a pillared grotto or recess about its centre, containing the group in white marble shown in the accompanying plate. To the right and left of this magnificent group, and also exposed on the rocky face, are two groups in marble, representing the horses of the sun watered by tritons. These three groups form the most imposing ensemble of sculpture at Versailles. When the waters play, a cascade tumbles from the rocks into the piece of water at their base. The banks in front of them are wildly clothed with trailing shrubs, the Polypody densely mantles the rocks; the vegetation around is tastefully arranged to suit the scene, which is on the whole the most striking and satisfactory in the gardens at Versailles.

On the fountains and waterworks of Versailles skill and gold were lavished by their creators. The Bassin de Neptune is the most important. As the waters only play on special occasions, and as they cost about 10,000 francs every time they do play, one is justified in considering the basins in their usually dormant aspect. Nothing can look more wretched than any garden exhibiting large fountain basins. Early in the morning of the 24th of September, 1868, I strolled round this fountain basin, endeavouring to discover some beauty in it. Had it not been for the fruits falling abundantly from the Horse-chestnut groves close at hand and a poor woman gathering them for fuel, I should have imagined myself in a dead world. The formality of the surroundings, the mouldering, faded margins and indescribable emptiness and ugliness of the scene, seemed only worthy of some sphere of geometrical craters and pools.

The figure showing this basin in full play obviates any necessity of describing it. The upper margin of the basin has twenty-two large vases in lead, ornamented with bas-reliefs, while on the face of the wall are three immense groups in the same material—the central one representing Neptune and

Amphitrite seated in a vast shell surrounded by nymphs, tritons, and sea monsters. The western group represents Ocean resting upon a sea unicorn—that to the east Proteus, while at the angles are two colossal dragons bearing Cupids. This fountain is the last to play of all those at Versailles, the grand final scene of the day, usually beginning about five o'clock, when all the others have ceased. The effect is very good of its kind, and attracts great crowds of people.

FIG. 92.

One of the statues on the upper terrace.
Vase from the basin of Neptune.   Vase Borghèse.

It is hopeless to attempt an enumeration of the riches possessed by almost every part of the garden in statues and vases. The accompanying figure may serve to give an idea of their execution.

The Bassin de Latone has five circular basins, rising one above the other in a pyramidal form, surmounted by a group representing Latona with Apollo and Diana, the goddess imploring the vengeance of Jupiter against the peasants of

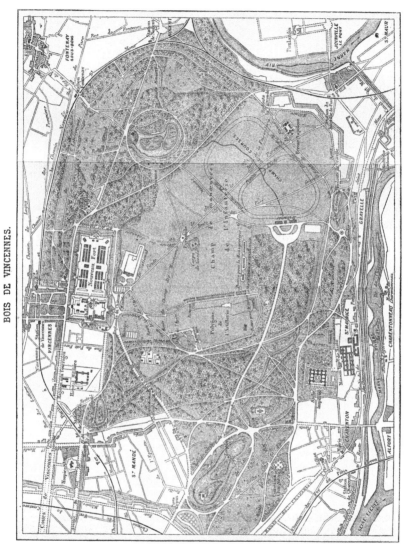

BOIS DE VINCENNES.

The material originally positioned here is too large for reproduction in this reissue. A PDF can be downloaded from the web address given on page iv of this book, by clicking on 'Resources Available'.

Libya, who refused her water. The peasants, changed, or in the process of being changed, into frogs, tortoises, and lizards, are placed on the different levels, and shoot water upon Latona from every direction. The tablets are of red marble, the group white marble, and the frogs, &c., in lead, and the effect of the whole is very striking when the waters play.

There are many other fountains, basins, &c., of minor importance which the visitor will be delighted to escape from into the garden of the Petit Trianon, at the north end of the park.

The Grand and the Petit Trianons are simply two villas

Fig. 93.

Temple de l'Amour in the gardens of the Petit Trianon.

at the extremity of the park, each with extensive gardens. Those of the first mentioned are among the most angular, ugly, and cheerless it has ever been my fortune to see—those of the Petit Trianon by far the best gardens at or near Versailles. To pass into them from these interminable gardens, where the "genius of Le Notre" has been so successful in stealing from nature every grace, is as refreshing as being suddenly transferred from some gigantic cotton-opolis to a green and sunny Piedmontese valley. It was the favourite residence of Marie Antoinette, and the gardens were in great part laid out by her in what the French call the "English" or natural style. Most of the expense has been devoted to the planting of choice kinds of hardy trees, of which there are many fine specimens in the grounds, many planted before the Revolution, and a great many newer species since. Quiet and refreshingly verdant glades, a tiny streamlet picturesquely meandering through them, a well designed piece of water, a little Swiss village, dairy, &c., erected by Marie Antoinette, constitute the chief charms of the place apart from its associations. But trees and grass, and shrubs and flowers, a streamlet and rocks, a lake and

Fig. 94.

View in the garden of the Petit Trianon.

a few picturesque buildings, how much you can do with them! We travel far and wide to see natural combinations of a few of these, and find them lovely and different in every clime. Even so may we make our gardens vary, and infinitely more so, because we may combine the vegetable beauties of many climes, while in nature we only get a comparatively limited view of them in any one spot.

No visitor to the Petit Trianon should omit to see the Jardin des Fleurs in front of the house of M. Charpentier,

PLATE XXXIX.

VIEW IN THE GARDENS OF THE PETIT TRIANON.

chief gardener of the Trianons. It is a garden in the best sense of the word, not large, but containing a rich variety of plants tastefully arranged. There are many handsome hardy trees, groups of Arundo and Pampas dotted on the grass, the Tamarix used in like manner, and very effective with masses of Cannas, Salvias, New Zealand Flax, and numerous other tender plants put out for the summer. These are usually very gracefully arranged, the boxes containing the larger specimens carefully concealed by dwarfer subjects and plenty of verdant grass as well as brilliant flowers.

In the immediate neighbourhood of this garden are very extensive and well managed nurseries for the supply of the imperial gardens of Paris. To the professional horticulturist they will prove worthy of a visit.

Fontainebleau, and the Gladiolus Grounds of M. Souchet.

Fontainebleau is one of many places in France not likely to be remembered with much pleasure for their gardens.

Fig. 95.

Canal in the gardens of Fontainebleau.

The formality of the water and the avenues and the lines of fusty clipped Lime trees render it impossible for the eye to find in such a place any of the solace or charms of a true garden. The portion planted as an " English garden " has indeed some peace about it, but unhappily the strictest

formality governs every line of the vegetation on the garden
front of the palace.   It is that type of garden which has
not a curious corner in it nor a ray of novelty, and consequently to describe it in detail would be to waste space.   The
illustrations show it fully, but with a touch of grace which
it does not really possess, for an artist has too much feeling
to draw it in all its angularity and baldness, and weaves in
over the grateful surface a little of the freedom and grace
of nature.   The only feature of any practical horticultural interest in the place is the famous Treille du Roi, an

Fig. 96.

View in the forest of Fontainebleau.

enormous wall of vines, which, bordering the park, is said
to annually produce 8000lbs. or 9000lbs. of grapes.   It is
planted with the Chasselas de Fontainebleau, and the wall
is well managed, as indeed one would expect it to be, so
near Thomery.

Of the things to be seen at Fontainebleau, those best
worth remembering are far away from the château and
even from the garden.   It is tedious work getting away
from the interminable long straight roads that lead from
the château in every direction; but, once in the midst of

PLATE XL.

THE PARTERRE AT FONTAINEBLEAU: VIEW FROM THE CHATEAU.

one of those wilds where huge rocks and indigenous trees
are scattered in about equal profusion, the visitor will
hardly ask himself why Rosa Bonheur resides in the neighbourhood.

But the most beautiful of those glorious wilds in the
forest does not present a greater charm to the landscape
artist or the lover of the picturesque than M. Souchet's
collection of Gladioli affords to the horticulturist.   It
is by far the finest collection in the world, and a remarkable example of high cultivation.   M. Souchet is the Emperor's gardener, and has been so for many years.   He has
been cultivating the Gladiolus for more than thirty years ;
and it was cultivated also here by his father.   This is the
most noble of our autumnal flower garden ornaments, and
one comparatively neglected by us.   There is no flowering
plant so well calculated to improve the aspect of the
autumnal garden, of no matter what style.   M. Souchet
grows it in fields, surrounded by white stone walls.   His
ground was for the most part formerly occupied by market
garden cultivators, and these usually surround their
gardens by such walls.   He altogether occupies, with the
culture of his favourite, from eight to nine hectares of land,
or say about twenty acres English.

The first thing noticeable in this ground is that about
half of the land is unoccupied for the current year.   That
bare portion is ploughed, and manured, and cultivated
throughout the summer as well as in winter, and thus he
has fresh land in capital condition for his bulbs every year.
Besides, the fact that the ground is bare for a year helps to
counteract to some extent the particularly vicious insect
enemies with which he has to contend, as, having no food on
the ground for the summer, they are not attracted ; and when
the ground is rolled between the ploughing and manuring
the tracks of the courtelière are easily seen, and it may be
readily destroyed.   This idle ground is thoroughly tilled,
ploughed, or in some way disturbed six or seven times
during the season, and he would like to do it a dozen
times if time or labour would permit.   The ground planted
this year will of course be empty next, and so on.   Over

the whole of the extensive grounds planted with Gladioli
you could not notice a decayed leaf, and all the plants were
in the rudest health, some of the varieties growing as much
as six feet high.   It was a fine sight at any time of the
day to see the magnificent stretch of varied bloom; but
the days about the time of my last visit were very hot, and
one was obliged to get up very early in the morning to see
it at its best.   Although very showy at noon, yet the hot
sun had caused the most open flowers to flag a little.   But
in the early morning, when the dew hung upon the bloom,
and every petal was braced with its freshness, the flowers
were magnificent.

The insect enemies of M. Souchet would prove enough
to deter and defeat most men.   He makes ceaseless war
against them, and if they do succeed in destroying a bulb

Fig. 97.

Courtelière (Mole Cricket).

now and then, it generally forms the guide to their detection
and destruction.   If the courtelière or mole cricket were
allowed his own way for a fortnight in these grounds, I
fear some of the great bulb houses would suffer from their
want of Gladioli in autumn.   When this strong and well
armed little fellow gets into a bed of choice Gladioli, you
cannot well dig him out as you could if he happened to be
in an open spot. The way he is killed here is so interesting
and effective that I must relate it.   M. Souchet explained
it to me; but so many receipts for exterminating vermin
are not worth the trouble of trying the second time, that
probably I should not have noticed it had he not called a
workman and given me an illustration on the spot.   When
the mole cricket goes about, he leaves a little loose ridge,
like the animal after which he is named: and when his

PLATE XLI.

VIEW FROM THE PARTERRE AT FONTAINEBLEAU, SHOWING THE BAD EFFECT OF THE LINES OF CLIPPED
LIME TREES SO COMMON IN FRANCE.

presence is detected in a closely planted bed of Gladiolus
at Fontainebleau, they generally press the ground quite
smooth with the foot, so that his track and halting-place
may be more distinctly seen the next time he moves about.
This had been done in the present instance in the case of
a young bed of seedlings.   We saw his track, and a workman, who brought with him a jar of water and one of
common oil, opened a little hole with his finger above the
spot where the enemy lay.   Then he filled it with water
twice, and on the top of the water poured a little oil.   The
water gradually descended, and with it the oil, which,
closing up the breathing pores of the mischievous little
brute, caused it to begin to suffer from asphyxia, and in
about twenty seconds we had the pleasure of seeing it put
forth its horns from the water, go back a little when it saw
us, but again come forth to die on the surface, hindered
for ever from destroying valuable bulbs.   Being very strong
and well armed, a single mole cricket can do a deal of damage
in a bed of Gladiolus, and therefore the moment the workmen of M. Souchet see a trace of the pest they take means
to catch it as described, jars of water and oil being always
kept at hand.

The mole cricket is only one enemy—the ver blanc is
worse.   Of what a vile opponent this is,
some idea may be formed when I relate the
precautions M. Souchet is obliged to take
against it, even for the sake of enjoying a
few Rhododendrons.   He has built a private house near his Gladiolus grounds, and
wishing to have a couple of beds of these
shrubs within view of the windows, he has
had to build strong cemented walls deep
into the earth around each bed, and fill
in the bottom with a deep bed of fine

Ver blanc (grub of
the cockchafer).

sand, so as to guard against the entrance of this dreaded
worm into the bed.   But it is among his bulbs that most is
to be feared.   He employs a great number of people to gather
the parent insect at the egg-depositing season, has the larvæ
picked up after the plough, and one way or another avoids

their ravages, though at great cost of time and money. The ver blanc is simply the larva of the cockchafer.

The damage done by these enemies to horticulture and agriculture in France is almost incredible. They are productive of far greater injuries than any that we are visited with through insects; and though their ravages are not so noticeable in this country, it is very likely that they occasionally do a good deal of damage perhaps without being suspected. Where they happen to be plentiful in or near gardens they are sure to be at mischievous work, and should be watched accordingly. Gardens and fields and whole districts are sometimes ruined by the ver blanc in France; and there are even places where it is impossible to cultivate any kind of vegetable in consequence of the ravages of the mole cricket.

The soil is a very sandy, not a fluffy one, observe, but one with some holding power, and yet when you get a dry bit of a clod of it, and crumble it fine on a silk glove, you find that most of it sinks through to the palm of your hand, in the form of nearly impalpable sand. It is well manured, and pretty rich and deep, from having been long used as kitchen-garden ground. Horse manure is preferred, and that as well rotted as possible. The time of planting is, perhaps, one of the most important things to be acquainted with, and they do it here from April till the early part of June. The late planting is not often resorted to however. They prefer the beginning of May for the general and the safest planting. The medium-sized bulbs give the best flowers as a rule, the biggest often breaking into several stems instead of giving one good one. To plant at various times of course will lead to a succession of bloom. The seedlings flower in their third year. The time of taking up is October, and, from the great quantity to be stored, this process sometimes goes on to the beginning of November. The plants are mostly in beds, about four feet wide, placed in rows across the bed, from fifteen to eighteen inches apart. The beds are all covered with short litter to keep the soil moist. In very hot weather they are well watered. Each kind is numbered, the scraps of lead on which the numbers are stamped being wrapped round bits of Vine prunings, stuck in the earth. The beds are also carefully examined during the blooming season, so as to destroy all those not true to name, or what are termed "rogues." Such are the chief points as to cultivation—next for a selection of the varieties.

There are altogether in cultivation here between 250 and 300 varieties. Of these, we first selected the undermentioned as best, and then went over them again, marking the very best of all. This second or choicest selection is indicated by an asterisk to all those so chosen :—

Achille.	*Lord Byron.	Prince of Wales.
Anaïs.	*Madame Furtado.	Princess of Wales.
Belle Gabrielle.	Madame Leséble.	*Princesse Clothilde.
Charles Dickens.	Madame de Sévigné.	*Princess Mary of Cam-
Cherubini.	*Madame Vilmorin.	bridge.
*Dr. Lindley.	Maréchal Vaillant.	*Queen Victoria.
*Duc de Malakoff.	*Marie Dumortier.	Reverend Mr. Berkeley.
El Dorado.	Mazeppa.	Roi Léopold.
Fulton.	Météore.	Rubens.
Galilée.	*Meyerbeer.	*Shakspeare.
*Impératrice Eugénie.	*Milton.	Sir William Hooker.
*James Veitch.	*M. Ad. Brongniart.	Stephenson.
*John Waterer.	*Napoleon III.	Stuart Low.
Lady Franklin.	Newton.	Thomas Moore.
Laquintinie.	Ophir.	*Sir Joseph Paxton.
*Le Poussin.	Oracle.	Vesta.
La Titiens.	*Pénélope.	*Sir Walter Scott.
Linné.		

It is evident there is an ample field from which to select, and a sufficient variety to please the most fastidious. M. Souchet grows exclusively for wholesale houses, and a large proportion of the bulbs of these attractive autumnal flowers, which are met with in the stores of the Paris and London nurserymen or seedsmen, are derived from the grounds of this most successful of cultivators. I cannot close this without acknowledging the great kindness of M. and Madame Souchet, both as amiable and excellent in private life as M. Souchet is distinguished in horticulture; and some of the pleasantest of the many agreeable visits I have made to great gardens were those paid to M. and Madame Souchet and the forest and gardens of Fontainebleau.

In France the Gladiolus is cultivated much more abun-dantly than with us—a state of things which should not

long continue, as nothing can be more worthy of general
cultivation, or more calculated to improve the general aspect
of our ornamental gardens.   Perhaps one of the best recommendations of this fine bulb is that its flowers continue
to open long after the spike is cut, and bloom in a vase of
water as freely as in the open garden.   I have never seen
anything more beautiful or effective than large Sevres vases
filled with the spikes of the finer kinds in M. Souchet's
house.   Many of his varieties grow five feet or more high;
when cut, a yard or more of the spike is preserved, no
other arrangement being needed except to insert their bases
in the mouth of the vase, and allow their heads to spread
widely forth, placing a few branches of evergreens, or any
verdure at hand, among the stems, just to give them a little
relief.   There is no one kind of flower that could make
such a noble combination, and the effect within the cool,
thick-walled French house, on hot days, was of the highest
character.   It may safely be said that the Gladiolus is the
finest of all our flowers for indoor decoration in autumn, its
tall and noble spike entirely preventing it from being used
to produce the dumpling-like effects given by Dahlias and
other popular flowers; and in the open air its uses are even
more valuable.

It should be premised, however, that in all cases either a
naturally sandy, rich, and deep light soil should be given
to it, or one made so artificially.   There are many stiff and
sticky soils on which it would be much better to avoid its
culture, and turn one's attention to things more tolerant of
the soil.   But the question of soil once settled, let us take
the case of a bed of choice Roses in some position near the
house.   Most probably this bed will present a somewhat
disappointing aspect after the Roses are past their best; and
even if they continue to flower well, the peeping forth of
some splendid spikes of Gladioli here and there will surely
not detract from their beauty.   To secure this, all we have
to do is to insert some bulbs of the various kinds of Gladioli
in the spaces between the Roses in the early part of May, or
thereabouts, planting them singly here and there, and at about
three or four inches deep, and taking up the roots in the

PLATE XLII.

THE GARDENS AND PALACE OF ST. CLOUD.

month of October. Is it necessary to suggest a score of other analogous uses? Need it be said how tastefully they may be introduced just within the edge of the low choice plantation, or in beds of valuable shrubs on the lawn? Groups of them in the centre of flower-beds would be splendid; and planted thinly here and there among beds of low-growing subjects—say Saponaria, Mignonette, &c., they would rise above these, and their effect above the surfacing flower would prove very fine indeed. They may be placed in groups or rings around Standard Roses; they will make the most valuable groups in the mixed border; and finally, we may make grand beds of them by themselves, or associated with Lilies or Irises.

St. Cloud, popular as it is, is perhaps one of the most uninteresting gardens known. It is, however, worth seeing, if only to get an idea of how much " the genius of a Le Notre" may do to spoil a place naturally beautiful. The canals, the lines of ugly clipped trees, and every base feature of geometrical gardening are there, but nothing worth remembering as an example. The situation is one of the most beautiful that gardening man could desire, and would be ravishing if tastefully and simply laid out in the natural style. The lamentable

FIG. 99.

effect of clipping the trees is well shown in the plate; it is very evident the poor trees do not like it. It would be difficult to find a more striking example of labour worse than thrown away than that bestowed on clipping trees in many French gardens. Not only are the trees themselves robbed of all individual beauty or character, but many noble places are spoiled by their presence. Frequently the trees become hideous from disease consequent upon mutilation; and what they are in perfection may be seen by the accompanying model tree figured by a professor in one of the best French books on arboriculture.

A French ideal of tree-beauty.

Any real necessity for this clipping does not exist.

s 2

When trees are planted in close lines to form a shady
avenue, their natural tendency is to form a beautiful
and formal, though picturesque arch, so that clipping them
to obtain this is a futile barbarism. Do we want to prevent
them spreading forth and filling the streets with their great
wide heads ? If so we may select trees almost pillar-like in
their habit, as the Lombardy Poplar, the fastigiate Acacia,
and various trees of similar aspect. Do we require them
flat-headed and low, so that while shading the hot street

FIG. 100.

Meudon.

they may not darken all the windows ? If so we have the
Paulownia, of great shading power, and fine as a street tree
on dry soil, without a disposition to mount much higher
than the headed-down Limes we notice in so many London
street gardens.

Meudon is much less known and visited by English
people than St. Cloud or Versailles, though it could hardly
fail to please as much as the former. For a charming view
of Paris and the intervening country it is superior to St.
Cloud, and equal to any spot near Paris. This view may

be best enjoyed from the terrace, which is tastefully planted and not disfigured with Orange trees in tubs, as many continental gardens are. As to its gardening proper, the most novel feature is an ornamental orchard—a good idea. This is simply formed by planting shrubberies, groups, and single specimens of well trained fruit trees in an undulating piece of pleasure ground instead of the ordinary subjects generally used. The plan is well worthy of adoption, nearly all of our best fruit trees being highly valuable as ornaments alone.

CHAPTER XIII.

The plant decoration of apartments.

THE graceful custom of growing plants in living rooms is
very much more prevalent on the Continent than with us.
It is true that we often see a display of flowering plants in
rooms, though we very rarely rise to the use of subjects
distinguished by beauty of form, or select those that are
peculiarly adapted for culture indoors. But the day is approaching when the value of graceful plants as house ornaments will be very fully recognised ; and that the substitution
of life and changeful interest for much that, however costly
or well executed, is without these qualities, will prove a gain
few will doubt. Apart altogether from their effect as ornaments, what can more agreeably introduce us to the study
of natural history ? The influence of the graceful form of
a young Palm in the hall, the fascinating verdure of Ferns
and fine-leaved plants from many countries in the drawing-
room, and flowers, from the Orchids of the uplands of
Mexico to the tiny bulbs of Europe, in your Lilliputian
room-conservatory, is surely more eloquent in that direction
than any book teachings. You cannot deny, as Kingsley
says, that your daughters " find an enjoyment in it, and are
more active, more cheerful, more self-forgetful over it, than
they would have been over novels and gossip, crochet and
Berlin wool. At least you will confess that the abomination
of ' fancy work'—that standing cloak for dreamy idleness
(not to mention the injury it does to poor starving needle-
women)—has all but vanished from your drawing-room

since the 'Lady Ferns' and 'Venus's Hair' appeared; and that you could not help yourself looking now and then at the said 'Venus's Hair,' and agreeing that nature's real beauties were somewhat superior to the ghastly woollen caricatures which they have superseded." Ferns, to be sure, have been a great help and a great attraction; but they are by no means so readily grown in rooms as some things to be presently mentioned; nor are they supreme as regards verdure and elegance. By a combination of all the plants suitable for this purpose, we may not only find very agreeable indoor employment, but create the highest kind of ornament and interest in the house at all seasons.

FIG. 101.

Maranta fasciata.

Not only are we deficient as regards the better kinds of plant ornament in houses, but also in large gardens—with far greater means of readily developing it than occur on the Continent. Merely displaying a few popular or showy subjects is not plant decoration in any high sense! Rooms are often overcrowded with ornaments, many of them exact representations of natural objects; but in the case of the plants we may, without inconvenience, enjoy and preserve the living natural objects themselves. Those we employ for this purpose now are mostly of a fleeting character, and such as cannot be preserved in health for any length of time in living rooms. But if in addition to the best of these we select handsome-leaved plants of a leathery texture, accustomed to withstand the fierce heats of hot countries, we shall find that the dry and dusty air of a living room is not at all injurious to them, and that it is quite easy to keep them in health for months and even for years in the same apartments.

FIG. 102.

Dracæna terminalis.

They would speak to us of many distant lands; interest us by their growth under our care; teach us the wonderful variety and riches of the vegetable kingdom, and prove themselves quiet, unobtrusive friends. The variety of form and grace of outline which many of these plants possess, may to some extent be judged of by the illustrations scattered through

this chapter. Many of them are exotics that in this country
are rarely seen out of stoves, while about Paris they are sold in
abundance for the decoration of apartments.   The demand
for use in private houses gives rise to a large and special
branch of trade in many of the nurseries, and I know one
Versailles cultivator who annually raises and sells 5000 or 6000

Fig. 103.

Gymnostachyum Verschaffelti.

plants of the bright-leaved Dracæna terminalis (Fig. 102)
alone, and by far the greater part for room decoration.

As compared with the plant decorations of one of the balls
at the Hôtel de Ville, anything seen in the British Isles is
poor indeed ; while the way plants are arranged at the
Linnean and Royal Societies and other important places, on

special occasions, is almost sufficient to prevent people tolerating them indoors at all, and yet the plants are much
better grown in England than they are in France. The difference is caused by exceedingly tasteful and frequently peculiar arrangement, and by employing effective and graceful
kinds.   What the Parisians do as regards arrangement may
perhaps be best gleaned if, before selecting the kinds most

Fig. 104.

Dieffenbachia seguina maculata.

deserving of indoor culture, I describe the decorations for
one of the balls at the Hôtel de Ville.

Entering the Salle St. Jean, the eye was immediately
attracted by a luxuriant mass of vegetation at one end;
while on the right and immediately round a mirrored recess
was a very tasteful and telling display made as follows :—In
front of the large and high mirror stretched a bank of moss,

common moss underneath, and the surface nicely formed of
fresh green Lycopodium denticulatum, the whole being
dotted over with the variously-tinted Chinese Primulas—a
bank of these plants, in fact, high enough in its back parts
to be reflected in the mirror with the taller plants which
surrounded it, gradually falling to the floor, and merging
into the groups of larger plants on either side of the bank,
the whole being enclosed by a low gilt wooden trellis-work
margin. The groups at each side contrasted most beautifully
with this. Green
predominated, but
there was a suffi-
ciency of flower,
while beauty of
form was fully
developed. In the
centre and back
parts of these
groups were tall
specimens of the
common Sugar-
cane (Saccharum
officinarum) which
held their long
and boldly arching
leaves well over
the group. These
were supported
by Palms, which

FIG. 105.

Alocasia metallica.

threw their graceful lines over the specimen Camellias, which
were, in their turn, graced here and there by the presence of
a Dracæna or dwarf Palm; and so down to the front edge,
where Cinerarias, forced bulbs, Primulas, and Ferns, finished
off the groups, all very closely placed, so that neither the
lower part of the stems, nor a particle of any of the pots,
could be seen. Any interstices that happened to remain
between the bases of the plants were compactly filled with
fresh green moss, which was also pressed against the little
gilt trellis-work which enclosed the whole, so that from the

uppermost point of the Cane leaves to the floor nothing was
seen but fresh green foliage and graceful forms, enshrouding
the ordinary flowers of our greenhouses, that are infinitely
more attractive when thus set in the verdure of which
nature is so profuse, and which is always so abundant where
her charms of vegetation are at their highest.

A scene such as this explains the prevalence of these
graceful and noble-leaved plants in Paris gardens and in

FIG. 106.

Æchmea fulgens.

Parisian flower-shops and windows; for you may frequently
see elegant little Dracænas ornamenting windows there, and
as they look as well at Christmas as at midsummer, I need
hardly suggest how highly suited they are for purposes of
this kind.   The number of Dracænas cultivated in and
around Paris is something enormous, and among the newer
species of these—not alluding to the coloured-leaved kinds
—are some that combine grace with dignity as no other

plants combine them. They are useful for the centres of noble groups of plants in their larger forms, while the smaller species may be advantageously associated with the Maidenhair Fern and the Cinerarias of the conservatory bench. They are of the greatest utility in these decorations, and are largely used in all parts. So are most kinds of fine-leaved plants, from Phormium to Ficus. Young Palms are also cultivated to an enormous extent about Paris ; and so is every green and gracefully-leaved plant, from the Cycads to the common trailing Ivy,—used a good deal to make living screens of. With such plants they have but little trouble to find materials for this kind of embellishment.

The wide staircase ascending from the entrance hall had also a charming array of plants so placed that the visitors seemed to pass through a sort of floral grove — fine-leaved plants arching over, but not rising very high, and having

Fig. 107.

Caladium argyrites.

a profusion of flowering things among and beneath them. As the bank of Primulas and the groups of tall plants were placed opposite this staircase, and reflected in the great mirror behind, the effect when descending the staircase was fascinating indeed. A still finer effect was produced in a room near the great dancing saloon, and through which the guests passed to the magnificent ball-room. Against each pillar in this saloon was placed a tall palm with high and arching leaves like those of Seaforthia elegans, and others with longer leaves and pendulous leaflets. These meeting, or almost meeting across, produced a very graceful and imposing effect, while round them were arranged other plants distinguished either by beauty of leaf or flower, and the groups at each pillar connected by single rows of dwarf plants, closely placed, however, and well mossed in, as in the

Fig. 108.

Caladium mirabile.

case of the more important groups. The very close placing of the plants is a peculiar part of the arrangement—you cannot notice any dividing marks or gaps, yet there is no awkward crowding. The fact is that with an abundance of plants distinguished by beauty of form, it is almost impossible to make a mistake in arranging them.

These arrangements are infinitely varied at the great balls, both public and private; rocks, water grottoes, and similar decorations, are occasionally introduced, both indoors and in the open air, and in the gardens behind private houses. The Tuileries Gardens at the time of the great fêtes were largely decorated in this way, each of the numerous lamp-posts having a bed of flowers around it, and the whole scene being turned into a kind of conservatory in a few days. The number of flowers required to do this was

FIG. 109.

Pteris cretica albo-lineata.

something enormous; and when it is considered that at the same time great quantities of plants were arranged both indoors and out, in other great public and private buildings, some faint idea may be formed of the enormous extent to which plant decoration is carried out in Paris.

To go more fully into details would be useless—very few words serve to explain the difference between their system and ours. It simply consists in the use of a far greater number of fine-leaved subjects on their part. This, of

course, has a greater effect in popularizing the use of plants
in houses; for how can you make beautiful arrangements in
this way if you ignore the higher beauties of plant form?
The fashion as carried out in such instances as the above
carries its influences through every grade of society. Thus
you see people with
a graceful Yucca
or young Palm, or
New Zealand Flax,
in their windows
and rooms, who, if
in England, would
not in all probabi-
lity have had an
idea of the exist-
ence of such things.
The extent to which
the floral embellish-
ment of the Hôtel
de Ville is carried,
may be judged
from the great
numbers of plants
grown at Passy for
that purpose—the
New Zealand Flax
which is so very
useful for indoor or
outdoor decoration
being grown to the
extent of upwards

FIG. 110.

Begonia dædalea.

of 10,000 plants, and Palms and all plants with fine leaves
in great quantity.

The following few notes on the principal plants which
serve for window and room decoration in Paris are by
M. A. Chantin, a cultivator of plants for these purposes
on a large scale, and the possessor of a very rich collection
of Palms and other exotics distinguished by beauty of leaf
or habit. Among these, the Palms, without doubt, occupy

the most important position, are most generally used, because of their hardy character and moderate price, and among the very best are the fan-palms — Chamærops humilis and excelsa.   Corypha australis, although now but little known as a house plant, is destined in a short time to occupy a foremost place in the decoration of apartments.   It is conspicuous for its peculiar beauty, and the number of its leaves, and is, I believe, the most hardy and

enduring of all the Palms for indoor culture.   Cocos coronata and flexuosa are very elegant, and produce a charming effect.   Latania borbonica is certainly one of the most valuable plants of this family, and is valued as much for the deep yet fresh green of its leaves as for its hardiness and elegant appearance.   Phœnix dactylifera,

Fig. 111.

Maranta rosea-picta.

leonensis, and reclinata are very much sought after, and are highly esteemed, also Areca alba, lutescens, and rubra.

The following Palms could be used with great advantage in the decoration of apartments; but their high price and great rarity cause them to be not much known, although they accommodate themselves to the atmosphere of rooms as well as any of those previously mentioned.   Areca

T

sapida, most of the species of the genus Caryota, Chamædorea amazonica and elatior, Chamærops Palmetto, Elæis
guineensis, Euterpe edulis, with its finely-serrated and very
graceful foliage; Oreodoxa regia, young plants of which
are very frequently used; Phœnix pumila, Phœnix tenuis,

FIG. 112.

Dieffenbachia Baraquiniana.

Rhapis flabelliformis, Thrinax argentea, T. elegans, and
Leopoldina pulchra.

Next in importance to the Palms we must place the
Dracænas. Those which are the most frequently noticed
are Dracæna australis, cannæfolia, congesta, indivisa, in-

divisa lineata, rubra, stricta, terminalis, and umbraculifera. Those most easily managed, and therefore the most popular for window ornaments, are Dracæna congesta, rubra, and terminalis. Pandanus utilis, amaryllifolius Vandermeerschi, and javanicus variegatus; Cycas revoluta, and varieties of Aspidistra, occupy also a very important place in the decoration of apartments.

The plants compos-
ing the following list,
although suitable and
distinct in appearance,
require somewhat more
care   and   attention
than   the   preceding.
Several   species   of
Aralia, more especially
Aralia Sieboldi; Bam-
busa japonica variegata
and B. Fortunei varie-
vata;   the   different
varieties of Begonia;
most of the Bromelias;
Billbergias and allied
families are very use-
ful,   including   the
variegated Pine-apple,
which forms a splen-
did object for placing
in large warm rooms
on special occasions.

Fig. 113.

Gesnera cinnabarina.

Caladium odorum, for winter decoration, and the species with the beautifully-spotted and mottled leaves, for the summer; Carludovica palmata and plicata; Croton pictum, pictum variegatum, and discolor; Curculigo recurvata, and several species of the genus Dieffenbachia. The Ficus elastica is a capital plant for window ornament, and some years ago was very much employed for that purpose; but since it has become somewhat common Ficus Chauvieri has been substituted for it in many places. There are many

T 2

other Ficuses which are suitable for this purpose, and will be found most useful when they become plentiful enough. Maranta zebrina is the only species of Maranta suitable for cultivation in apartments, as all the other species should be grown and kept in the houses, and only introduced to the house when extra attractions are desired for special occasions. Several species of Musa are favourites,

FIG. 114.

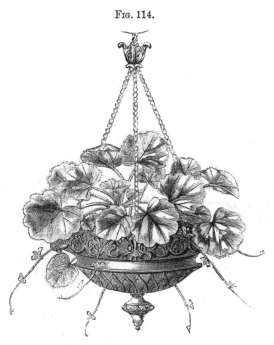

Saxifraga Fortunei tricolor.

but principally M. discolor and M. rosea; Musa Ensete is particularly suitable for room culture, but it is still so scarce, and of such a high price, that it is but seldom met with. Monstera deliciosa was much sought after during recent winters, and has in most places thriven so well that it has given general satisfaction. Several varieties of Beaucarnea are suitable for rooms, and produce a very beautiful and graceful effect when grown in suspended vases or baskets. Rhopala corcovadense is a plant that exhales a somewhat disagreeable odour, but it is nevertheless much sought after, on account of its very elegant and graceful appearance during the development of its young leaves. Hecktia pitcairnifolia is capital for suspending in baskets. Tradescantia discolor, Phormium tenax, Rhododendrons, Camellias, Grevillea robusta, Euonymus, Aucubas, Bonapartea, Agaves, variegated Yuccas, &c., are also frequent. In addition to the common Saxifraga sarmentosa, which is frequently seen with its slender runners pendant from window baskets in England, several other allied species would prove equally useful in the same way —Saxifraga Fortunei variegata, and S. cuscutæformis, for example.

FIG. 115.

Maranta vittata.

The family of Ferns, although classed among plants with delicate tissues, and having a great dislike to dry hot atmospheres, nevertheless furnishes numerous examples which, with careful management, add very much to the beauty of apartments. Thus I have very frequently remarked several species of Adiantum, which, wherever they can be preserved in good health, produce without doubt a most pleasing effect. Pteris argyræa, P. cretica albo lineata, and P. serrulata variegata likewise produce a good effect with their prettily marked fronds. Alsophila australis and Dicksonia antarctica are also sometimes employed for decorative purposes in rooms of large dimensions, where their magnificent appearance never fails to produce a pleasing impression. Nephrolepis exaltata is universally useful, and stands the air of rooms without the slightest injury.

Until recently, I had little belief in the utility of Orchids for this purpose, but experience has shown me that they may be introduced into a drawing-room with perfect success, the plants not having suffered in the least by the change of atmosphere. The most suitable Orchids are the various species of Cattleya, Vanda, Aerides, and Cypripedium. Doubtless the time is not far distant when we may venture to try many more kinds than we can now afford to do ; but even from what we have already done in that way, I entertain no doubt that the Orchid family will eventually furnish the most valuable of all plants for room decoration. True they may not live throughout the year in rooms as Ficuses and such plants do, but that is not desirable— their appearance, as a rule, not being prepossessing when out of flower. The quality that they do possess, and that which makes them so valuable, is, the thick succulent texture of the flowers generally. This enables them to continue a long time in bloom in a room, and a like kind of texture enables the leaves to stand during the blooming time without injury.

FIG. 116.

Tillandsia splendens.

We ourselves are foremost so far as flowering plants are concerned, ours being as a rule better grown. One plant, however, cultivated in great abundance around Paris for winter blooming, is well worthy of increased attention—

Epiphyllum truncatum. There are several varieties, and they certainly form most beautiful objects on dull December

FIG. 117.

Maranta zebrina.

FIG. 118.

Pandanus javanicus variegatus.

days. The employment of simple materials is also deserving of commendation. Thus the variegated form of the common

Roast-beef plant—Iris fœtidissima—may be seen occasionally used with good effect. We mostly use hot-country plants if we want those that live long in our dwelling rooms, but this is a true hardy native which well deserves culture indoors, though in the open air it never presents a very striking variegation—looks rather undecided, in fact. It forms a very pretty plant for room decoration, requires none but the most ordinary attention, and is easily obtained. In France the plant is rather commonly used as an edging. The Acanthuses too, and particularly A. lusitanicus used so effectively out-of-doors, are also grown abundantly in rooms, where they do very well. Everything proved to do well indoors without the protection of a case is a gain to the very large class who, from choice or necessity, like to grow plants in rooms.

Reform in the Conservatory.

There are few things more worthy of the attention of the numbers interested in indoor gardening in this country than the superior mode of embellishing conservatories and winter gardens which is the rule in France and on the Continent generally. Conservatories and similar structures are, it is true, scarcer abroad than at home, but whenever they are erected they are gracefully verdant at all times, being filled with handsome exotic evergreens, planted and arranged so as to present the appearance of a mass of luxuriant vegetation, and not that of a glass shed filled with pots and prettiness with which we are all so familiar.

We build more glass houses than any other nation, but have as yet nearly everything to learn as to the arrangement of the most important of them, or what is usually called the conservatory. This in some form is an adjunct to a large class of country and suburban houses; sometimes it is well placed and an ornament to the house, but more frequently a thing which to a tasteful person would seem better placed among the out-offices. When unsatisfactory, the cause is in the structure of the building or in its contents—often in both, generally in the latter. As regards the form and style of building little need be said, as the improvement required seems so obvious. When conservatories are built near the house they should always present a somewhat permanent and architectural character, and be removed as far as possible in stability and appearance from glass sheds. This is desirable for several reasons—chiefly the propriety of having a presentable and lasting structure in such an im-

FIG. 119.

Cordyline indivisa.

portant position, and the fact that plants and flowers show to greater advantage in a subdued light than in that of the glass shed. For growing plants you cannot as a rule have too much light; but when in flower their effect is much heightened by being placed in a subdued light. Those who consider for a moment the charming effect of the flowers under the thick canvas of the great flower-show tent in the

Regent's Park, as compared with the aspect of the same
plants in a well-lighted conservatory or placed in the open
air, will have no difficulty in appreciating the truth of this.
It should also be borne in mind that things that are worthy
of culture for their leaf-beauty alone always associate well
with substantial surroundings.

FIG. 120.

Tree Fern for Conservatory.

But the grand improvement to be effected is in the contents of conservatories. They will never be truly enjoyable
until we display in them beauty of form. Numerous reasons urge us to endeavour to make a change in this respect.
The aspect of the greater number of conservatories through-

out the country is paltry in the extreme, except perhaps
when the flush of flower in early summer diverts our eyes
from the faults of a structure so little conservative of the
elegant forms and bewitching grace which make the vegetable world so attractive.   Having these structures staring
point-blank at our drawing-rooms in numerous instances, it
is clearly desirable to make them presentable.   We build
hothouses to enable us to enjoy the vegetation of warmer
and more favoured countries.   Let us enjoy it, then, and
not delude ourselves by cramming our conservatories with

FIG. 121.

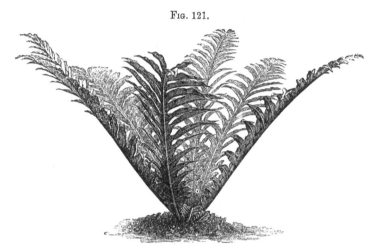

Polypodium morbillosum.

all the popular small fry, from the Cineraria to the Azalea.
Such things may please the enthusiastic amateur of these
and like plants; but plants are capable of higher work
than that, and nothing can be hoped for the conservatory
until a nobler type of vegetation is not only represented,
but predominates.

Flowers of a similar, if not of a nearly identical type to
the popular ones mentioned, abound in our gardens during
summer, and there is consequently no necessity for letting
them predominate indoors; while on the other hand those
wonderful aspects of vegetation which we can never produce

out of doors in this country may be obtained under glass
without difficulty. The temperature of conservatories generally is sufficient to develope as noble a type of vegetation
as the hottest stove, so objectionable from its heat and
moisture. The grandest of all the Banana tribe (Musa
Ensete) thrives healthfully in a cool house, while the Palmetto Palm of the Southern States of America, the Fan
Palm of Europe, Chamærops excelsa, and the graceful
Seaforthia elegans, and many other Palms, do the same.

Fig. 122.

Blechnum brasiliense.

Nothing even among Palms can surpass the effective
grace of such Dracænas as lineata, Rumphi, umbraculifera,
Draco (before it gets very old and branched), brasiliensis,
cannæfolia, and australis; and they all grow well in the cool
and agreeable temperature of the conservatory. Numerous
Ferns, from those great Dicksonias which at the Antipodes
rival or surpass the Palms in grace, to the Woodwardias,

which spread about such great fronds covered with swarms of little plants on their surface, grow under such conditions without trouble, compared to what the commoner and smaller stuff, as the gardeners call it, requires. For instance, a Dracæna, growing in a pot or planted out, never requires any attention beyond watering, whereas a Pelargonium must be cut down and regrown every year,

FIG. 123.

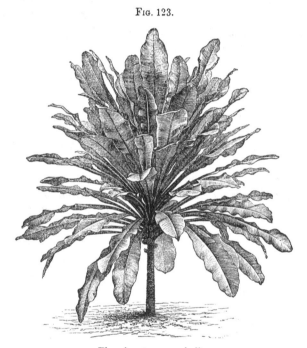

Theophrasta macrophylla.

causing much labour for staking, &c. And it is not only the Palms, Cycads, Tree Ferns, Dracænas, and fine-leaved plants generally which thrive throughout the year in a cool temperature that we may enjoy therein; nearly all similar plants that flourish in stoves would well bear being introduced to the cool conservatory or winter garden after their vigorous spring and early summer growth had been matured. Left there during the hottest months they would be more appreciated than if in a hot stove, and they could be taken back to their winter quarters in early autumn.

But perhaps the best plea in favour of the fine-leaved gracefully-built plants that can be urged to the generality of cultivators, is that they enhance the beauty of the ordinary flowering subjects in a remarkable degree, and that by their aid one-sixth the amount of flowers will suffice to produce a more beautiful effect than was ever obtained by the use of the blooming plants alone. This is a great point at all times, and particularly in winter, when flowers are scarce. In winter too the aspect of houses arranged on the system I advocate is quite as good as in summer, and more grateful from its contrast with the surrounding dreariness; and in summer, when abundance of flowering plants may demand more space in the " show house," many of the fine-leaved ones may be placed in the open air, much to their benefit and the improvement of the flower garden.

The greater part of the foregoing having appeared in *The Field*, the following response was drawn from one of its correspondents :—" This subject has long been engaging my attention. We do build more glass houses than any other nation, for every suburban villa boasts nowadays of the so-called conservatory ; but whether these adjuncts are ornaments or not is most questionable. In nine cases out of ten, I affirm, they are far from ornamental, whether viewed from the inside or the outside, and it is a wonder to me that people consent to have these ill-shapen, ill-adapted greenhouses stuck on to their residences. Any one visiting the villas built within twelve or fifteen miles of London must have noticed the conservatories, so named, attached to the houses. I ask, are they even sightly ?

" But there is a point I wish to insist upon much more than upon the external ; it is the arrangement of the plants inside. What do we find as a general rule ? Long lines of white stages with sickly, leggy plants in pots all round the house ! If people could all hire efficient gardeners, the thing would be different ; the conservatory might then be filled with show plants and specimen shrubs creditable alike to the owner and to his gardener. It is needless to say that the gardener could do but little with only one house; what I want to point out is the advantage to be derived from a totally different arrangement of the house.

FIG. 124.

Cycas circinalis.

As you say, ' Let us enjoy it, then, and not delude ourselves by cramming our conservatories with all the popular small fry, from the Cineraria to the Azalea.' Just so. For goodness sake get rid of all those weakly, insect-infected

Cinerarias, Primulas, Geraniums, and others, and plant in borders round the house plants and shrubs alike easy of cultivation and beautiful. You put forward a plea for the fine-foliaged plants which it would be needless for me to insist upon. Your readers must see that what I am aiming at is a graceful and novel kind of shrubbery adjoining the drawing-room, rather than a house full of pots. Why not make round the house rich borders of the same width you would have devoted to these unartistic stages, and plant Camellias, Ficus, and other such things ? You mention the names of many suitable plants of the Palm and Fern tribes, and the list could be added to a hundredfold if it were necessary. Let us only see the attention of the owners of conservatories directed towards this point, and lists of plants will soon be published by the horticultural firms.

" As you say, the aspect of these houses is equally beautiful in summer and in winter. This is the most thorough praise that can be given to the system. To pursue the subject yet further I will illustrate. In St. Petersburg, where the climate is intensely rigorous, conservatories are even more appreciated than here at home. When people cannot afford them, you will find their rooms crowded with plants of the Palm tribe and numerous creepers, which thrive well all the winter ; and it must be remembered that the windows are not opened from October till April. In the conservatories of the wealthy what do we see ? A shrubbery—a maze of luxuriant foliage. It matters not whether there be 50 degrees or 60 degrees of frost : the promenade round the greenhouse—truly a *greenhouse*—is always agreeable, always charming. No words of mine could give your readers a true idea of the beauty of these places, nor of their utility to those deprived of plants and trees for six months in the year. One requires to see these plant houses thoroughly to appreciate them.

" Your readers may object that they are more suited to Russia than to our country. Not so. Is it not a melancholy exhibition to see our conservatories naked, nearly destitute of bloom, during December and January, and equally disheartening to see them full of flowers only when the gardens are becoming gay ?   Depend upon it, what we want, and what will some day be the cry, is an agreeable promenade attached to the house—not a swarm of little plants in pots, which none but the gardener can name or appreciate.   And then, again, look at the simplicity of the cultivation of the plants whose cause I advocate.   Plant them fairly in the border, and they will always thrive. Azaleas, Geraniums, &c., are constantly requiring to be

FIG. 125.

Alsophila.

smoked or watered with manure water.   Are you to take them outside, or into another house, each time they require such attention ?   If not, and the conservatory adjoins the drawing-room, there will be a decidedly unpleasant aroma there when either of the above-mentioned processes takes place.   I could go on to show other advantages connected with the system I am endeavouring to put forward ; I could attempt a description of the plant house of the wealthy

U

banker, Outchine, at St. Petersburg; but I feel I have already said too much. To my thinking, it is, however, a truly important topic, and I hope to live to see more interest taken in the beauty of the conservatory, of its tout ensemble, and less of the rarity of the plants and flowers."

To any person with a knowledge of what the beauty of vegetation really is there can be no doubt of the correctness of these views. The rule therefore in every conservatory in the land should be to use plants of handsome foliage or noble habit. Plant them in beds or borders; grow them in pots or tubs; the means, size, and requirements of the place must determine on what scale the thing may be carried out. In some degree the effect desired may be produced in the smallest greenhouse; where the space is large enough to develope the effect of the finer plants named, the aspect that may be wrought by their tasteful use will prove ravishing, compared to that of the old display of small-leaved, ordinary-looking vegetation.

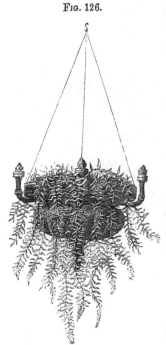

Fig. 126.

Goniophlebium in suspension basket.

In planting out, select things that are graceful and ornamental during the whole course of their natural lives. Do not plant subjects which, like Acacias, run up to the roof in no time, giving you a mass of bloom for a week or two in spring, and a great mop-head of ugliness for the remainder of the year. A great many greenhouse plants grow like these; but if you plant out a Palm like Chamærops, or a thing like the New Zealand Flax or the superb Musa Ensete, they are presentable and satisfactory at all seasons; and besides, do not run up against the roof in a few years, like many New Holland and other greenhouse plants.

Every conservatory should possess, in proportion to its size, a certain number of green and graceful plants, or those distinguished by some peculiar beauty of habit, which are ready at all times for fresh combinations, and look as well in mid-winter as in June. These are not sought for by horticulturists generally, but certain it is that without them we cannot succeed in the successful arrangement of a conservatory at all seasons without great expense, or even with it. What are flowers unless set in the graceful green among which we find them nestle in a wild state?

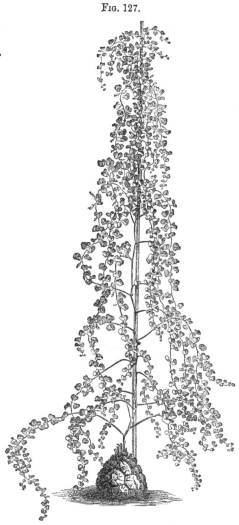

FIG. 127.

Testudinaria elephantipes (Elephant's-foot plant).

By the selection of a great number of things which flower profusely—so profusely as to hide the leaves in many instances,

u 2

your modern cultivator has contrived almost to annihi-
late leaf beauty. Nature is strongly vigorous in the
production of leaves, and in the widest spread of Heath over
a mountain, in the densest mass of Bluebells in a wood, or
in any natural display of bloom whatever, you find the
mass toned down by pointed leaves, and in the case of the
spreading Heather by fringe of Polypody and cushion of
moist mountain moss, if you go near to it and examine it.

Fig. 128.

Maranta micans.

Everywhere Na-
ture sets her
flowers in clouds of
refreshing green,
and therefore those
who merely culti-
vate dense flower-
ing things, and do
not take care to
relieve them with
others possessed
of sweet grace and
verdure, outrage
nature, and offer
nothing worthy
of admiration to
the educated or
tasteful eye. To
have all the
flowers dished up
without a bit of
green, is like eat-
ing your dinner in the form of a pill—a great saving of
time no doubt, but still utterly destructive of the joys of
the table.

A not unimportant merit of the subjects I so strongly
recommend for general culture is the great ease with which
they are cultivated; no neat staking, delicate attentions,
or repeated pottings, being required. They may be
grown with nearly equal facility in pots or tubs or planted
out. The continental plan of divesting the interior of the

conservatory of all formality is well worthy of imitation with us. Usually an attempt to create a picturesque scene in some small spot with formal surroundings has a ridiculous ending; but in consequence of the luxuriant growth of many plants that flourish in the temperate greenhouse, it is possible to sufficiently hide almost every trace of the building

FIG. 129.

Caladium.

in a few years. With little lawns made of Lycopodium denticulatum, tiny winding streamlets bordered with New Zealand Flax and graceful Grasses, Ferns, and the like; groups of Tree-ferns, Camellias, and Palms, and a plant of the noble Musa Ensete, I have seen some winter gardens made really worthy of the name, and quite as charming as veritable bits of nature in climes the most favourable to

vegetation. Whether the natural system of arrangement be
adopted or not, every attempt should be made to soften the
lines of the building and to shroud the spot with graceful
verdure.   The use of hanging baskets with creeping plants
cannot be too much recommended where this end is to be
attained, while climbing or twining plants with a pendulous
rather than an erect habit in the branchlets, flowering-spray,
or leaves should always be preferred.   A claim too deserves to be made in favour of singular and curious plants—
subjects like the Elephant's-foot plant for example.   The
Monstera with the perforated leaves, figured in the chapter
on Subtropical Gardening, one of the most curious as well
as one of the handsomest of fine-leaved plants, thrives
tolerably well in the conservatory in summer; indeed I
have known it grown well where it had to pass all the
winter in a cool house.

   In large places where stove plants are grown, much improvement might be effected by introducing some of them
into the conservatory during the summer months.   Stoves
are so warm during that period that they are seldom visited,
and rarely enjoyable, and it could hardly be otherwise than
a gain to see some of their best ornaments in the cooler
and shaded conservatory during three or four of the warmer
months.   Considering the number of exotics that have been
placed in the open garden at Battersea during the past few
years, there is little need to say that the temperature of a
conservatory will be enjoyed by many stove plants during
summer.   The host of handsomely marked Caladiums, and
other fine-foliaged plants that are now in cultivation, should
not be kept at all times in a steamy house, but when at
their best exposed where people may see and enjoy them.
Take, for example, that noble plant the variegated Pine-
apple—a subject never seen in our conservatories.   Usually
treated as a stove plant, and growing best in a warm temperature, it may, when fully grown, be employed in the conservatory, and will therein attract the attention of everybody
who sees it.   That it will not suffer there, is evident from
the fact of its having been employed in the open air in
summer in the gardens at Cliveden, and with telling effect.

It also bears well the dry air of a living room ; but there are many stove plants with leaves having a soft open texture which, while they will not bear the air of rooms, will not suffer in that of a conservatory in summer.

As for Orchids, hothouse Ferns, and other stove plants, which do not bear without injury the temperature of the conservatory, an arrangement might be readily made by which they also could be enjoyed in this structure. A conservatory heated to stove temperature would be intole-

Fig. 130.

Ananassa sativa variegata.

rable near the house, and is not desirable elsewhere, the heat of the temperate house being so agreeable to our senses. The best way to secure means for the display in the conservatory of the very tender subjects alluded to, is by making a closely-glazed case in some convenient spot therein, and fitting it up with rustic shelves. In this might be placed any Orchids, choice Ferns, or not over-large stove plants that come in flower at any time, particularly in winter, spring, and autumn, and by interspersing them with the pretty, and in many cases beautifully marked plants so common of late in our stoves, a charming feature may be added to the conservatory. As the plants would only remain in this case during their period of flowering, and the "foliaged plants" perhaps a few weeks longer, the position of the case as to light matters little. Against the back or some other wall of the house is of course the best position; and if there be an arched recess, or anything in that way, it would be the very place in which to put the case. The best example of this that I am acquainted with is heated with a few small pipes from the kitchen, which is nearly under it; the little apparatus being distinct from that required for heating the conservatory in cold weather. Of course it could be readily heated in that way, but it is found more convenient and economical to heat it distinctly. To heat a little boiler sufficiently to keep any desired temperature in such a case would be of very easy accomplishment, and to do it with gas would be very convenient indeed to many persons. The boilers attached to some of the propagating houses in the Jardin Fleuriste at Passy are thus heated most effectually, and the propagator informed me that he could regulate the temperature to a degree with this mode of heating. To make the wall and the shelves in this case of a rustic character is a good and tasteful plan; they should be studded with Moss, which if kept moist will give off the vapour so congenial to stove plants, and particularly Orchids and Ferns, and the windows or folding-doors should be fitted with large glass, kept clear at all times. It would be easy to induce the common Lycopodium and other stove mosses to crowd over the back wall, or even to grow on turves placed along the front shelves; and if the rustic-work were well done, to stud every spot not used as a standing-place for a plant with seedling-ferns, trailing plants, &c.

Palms.

In conversing one day with M. Barillet, the superintendent of the parks and gardens of Paris, he informed me that he was more surprised at the marked absence of Palms in

English gardens than by any other want, and he thought this the more remarkable from the fact that the superb collections of exotics grown in many parts of this country are quite unrivalled. That the plants which combine the qualities of dignity and grace as no others do, should be so neglected in a country where vast sums are spent upon Orchids and almost every other tribe of exotics, and where these are culti-vated better than anywhere else, is indeed somewhat singular.

The Palms are plants that we know very little about as a rule; but this is not at all surprising, for practically they belong to a dif-ferent world to ours. The oppo-sites in every vein of their structure of our wiry twig-ged and tortuous Oaks and Elms, they are as far removed from them geographically as structurally. Avoid-ing the cold grim North, they luxuriate in the hottest and moistest regions of the earth, spread for thousands of miles along the banks of the Amazon and Orinoco and their tributaries, running north all the way through the Isthmus and Mexico, crossing the Mississippi, and

Fig. 131.

Chamædorea latifolia.

fringing the Gulf. They appear again in abundance in the Eastern Archipelago, they form impenetrable forests in tropical Africa, they occur frequently in North Australia and the Pacific Isles, and flourish, in fact, in almost every torrid country, gradually dying out towards the Poles, but going a little further north than south, and ascending nearly up to the snow line in Asia. We have in northern and temperate regions our gay dwarf meadow flowers set in the sweetest grass; our Oaks and Ashes and graceful Birches; and our Firs, which are among the finest and most majestic subjects of the vegetable kingdom. We have our exquisite alpine vegetation, confessedly inferior to none; but we have not a trace of the noblest of all plants as regards form, the Palms. They are therefore more worthy of being grown artificially than numerous other exotics, which though requiring as much or more heat than Palms are by no means so distinct from all northern types of vegetation. There are few of us who have not read of their grace and magnificence in the Indian isles or Amazonian forests; but the rather humiliating fact remains that in our practical horticulture they are almost unknown. From an ornamental point of view, it is not easy to over-estimate the loss this is to high gardening. A perfect idea of what they are capable of doing for us can hardly be obtained until small, well-grown specimens of the most elegant kinds are seen in abundance at our flower shows and in our plant houses.

In this particular respect we are behind the Belgians and the French who, long ago, recognised the superiority of Palms, now cultivate them by thousands, and employ them for every purpose of plant decoration in rooms either permanently or for special occasions, in greenhouses, stoves, and for the open garden in summer. Nurseries like Chantin's at Paris, and Verschaffelt's at Ghent, have house after house filled with Palms, in great variety, some very rare and dear, many cheap enough for the purse of the poor window gardener. It should be also noted that they cannot be propagated in quantity and with rapidity like many popular plants, so that the formation of such collections has taken up much time and pains. It should be observed that while a new Verbena or Pelargonium may be fashionable for a season, or attractive

Fig. 132.

Seaforthia elegans.

for a few months, they are soon lost sight of or perish; whereas Palms, under ordinary treatment, go on prospering from year to year, and increasing in value. Some kinds

are costly in the beginning; but there is a great difference
between growing subjects which at the end of several years
will be more valuable than when you obtained them, and
propagating those which multiply so fast with yourself and
your neighbours that they soon become of only nominal
value.   In consequence of the value to which Palms are
sure to attain in the future for the decoration of large conservatories, stoves, and any plants that became too big for
small inexpensive greenhouses or stoves, could be sold or
exchanged to those wanting large subjects.   This may meet
the objection of those who regard them as only suited for
houses like the great Palm stove at Kew.   They may be
grown by everybody in possession of a snug pit, greenhouse,
stove, conservatory, or fernery, and it will be found eventually that not a few of them—thanks to their leathery texture
—will flourish in the dwelling-house without protection.
Everybody possessing such structures and in the habit of buying plants, should secure some few examples, as few others will
furnish such lasting satisfaction to the buyer; and there are
certainly no plants in existence more worthy of becoming
the fashion.   To make them abundant in a country abounding with things grown for their colour alone, will be to
ennoble its gardening.

It is tempting to trace them through the warmer zones—
to speak of their almost innumerable uses, one species yielding Palm oil, another Cabbages; of their striking diversity of
size, from a little Oreodoxa with a stem no thicker than one
of our grasses, to Jubæa, whose stem is nearly four feet in
diameter; of the species that spread their leaves on the
ground, and there rest stemless and content, to those that
shoot up as straight as the columns of a cathedral, to a
height of between two and three hundred feet, waving their
plumes far above forest vegetation as vast as our own woods.
Apart from its beauty, the family is perhaps the most useful
of all to man; but we, deriving our food directly or
indirectly from the grasses, frequently forget their great
interest in this respect.

This is far from being the case with the owner of a cottage on the banks of the Rio Negro.   The rafters of his

dwelling are formed by the straight cylindrical stems of the
Jara Palm; the roof is thatched with large triangular
leaves neatly arranged in regular alternate rows, and bound
to the roof with forest creepers: the leaves are those of the
Carana Palm. The door of his house is a framework of
thin hard strips of wood neatly thatched over with the split
stems of a species of Palm. In one corner stands a heavy
harpoon for catching fish, made of the black wood of the
Pashiuba variegata. By its side is a blowpipe ten or twelve
feet long, and hanging near it a little quiverful of small
poisoned arrows, with which the Indian brings down birds
for his food, or for the sake of their gay feathers, or even
slays the wild Hog or Tapir; it is from the fierce spines of
two species of Palm that they are made. His great bassoon-
like musical instruments are made of Palm stems; the cloth
in which he wraps his most valued feather ornaments is a
fibrous Palm spathe; and the rude chest in which he keeps
his treasures is woven from Palm leaves. His hammock,
his bow-string, and his fishing-line are from the fibres of
leaves which he obtains from different Palm trees, according
to the qualities he requires in them. The comb which he
wears on his head is ingeniously constructed of the hard
bark of a Palm; and he makes fish-hooks of the spines, or
uses them to puncture on his skin the peculiar markings of
his tribe. His children eat the red and golden fruit of the
Peach Palm, and from another species he prepares a
favourite drink which he offers you to taste. The carefully
suspended gourd contains oil which he has extracted from
the fruit of another species. The plaited cylinder used for
squeezing dry the pulp that makes his bread is made of the
bark of one of the singular climbing Palms.

What veneration this man must have for the noble
family of Palms, which not only furnishes him with many
comforts and conveniences, but affords him a choice, so
that he nicely selects the kinds that best suit his wants.
Should we wonder if Palm worship were a common creed on
the Rio Negro? At least let us hope that they never kneel
down to a carved idol while such living benefactors as
those generous Palms are to be found! These manifold

uses came within the observation of only one gentleman,
Mr. Wallace, and in an almost unexplored country : how
marvellous would the uses of the tribe appear to us if we

Fig. 133.

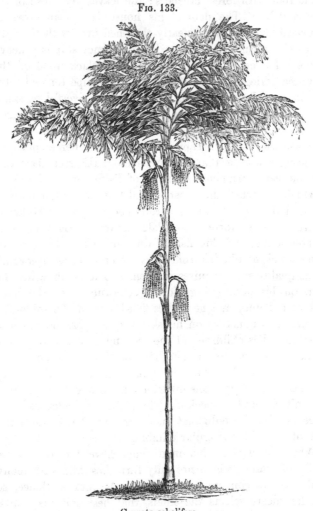

Caryota sobolifera.

could but glance at the various races of men who manage
to exist upon Palms alone! The cocoa-nut, so plentifully
grown on the coasts of all tropical countries, is alone said

to have as many uses as there are days in the year. It yields everything, from cordage to candles, to say nothing of arrack, door-mats, and Fern fronds. Upwards of 170,000 cwt. of a valuable oil, afforded by its kernel and used for soap-making, were imported in the year 1862. Of the oil afforded by Elæis guineensis of west tropical Africa, near 1,000,000 cwt. are imported annually! The uses of Palms are as infinite as their grace is inimitable. Sago, upon which whole races solely depend for food; dates, which feed dusky hosts in Arabia and North Africa; toddy, which affords one of those strong drinks the human race always manage to squeeze out of something or other in every known country, to the peril of their souls and destruction of their stomachs; resins, wax, brooms, books (the old Sanscrit was written on Palm leaves), sugar, and the bottoms of chairs.

Clearly they are of the highest interest from an economical as well as ornamental point of view; and we may confidently look forward to seeing them abundantly grown in the gardens of this country before many years have elapsed. The object of amateur growers of these plants should be to secure a suitable selection, preferring such as are hardy and small in their proportions. Hitherto Palms have been for the most part confined to our botanic gardens, and in them all sorts were welcome; but for the purposes of private collections we must be more select and choose them for their ornamental qualities rather than their botanical interest, particularly favouring all the dwarf kinds. There can be no doubt that when once the taste for these plants is established among us, plant-hunters will search for all that is diminutive in this vast family, so that the owner of a town greenhouse may enjoy his Palms in pots as well as the owner of a conservatory big enough to hold an old Date Palm. The collections found in the nurseries of this country are by no means so rich as those of the Continent, and particularly Belgium; but many of our larger nurserymen keep them in stock, and no difficulty need be experienced by the purchaser in getting them. The following selection has been made from the collections in our botanic gardens, and from those in continental and British nurseries.

As there are a greater number of persons who can grow greenhouse than stove Palms, and from the fact that the former class will as a rule prove doubly useful by adorning the open garden in summer as well as the houses in winter, we will begin with a list of Palms that may be grown in cool houses—*i.e.*, winter gardens, conservatories, greenhouses, and even in orchard-houses.

Chamærops excelsa
　„　Fortunei (sinensis).
　„　Ghiesbreghtii.
　„　humilis.
　„　Palmetto.
　„　tomentosa.
Cocus australis.
　„　Bonnetii.
　„　campestris.
Corypha australis.
Latania borbonica.

Molinia chilensis (syn. Jubæa spectabilis).
Phœnix dactylifera.
　„　farinifera.
　„　humilis.
　„　leonensis.
　„　pumila.
　„　reclinata.
　„　sylvestris.
　„　tenuis.
Rhapis flabelliformis.

Sabal Adansonii.
　„　Mocini.
Seaforthia elegans.
　„　robusta.
Thrinax parviflora.
　„　tunicata.
Areca lutescens.
Brahea calcarea.
　„　dulcis.
Diplothemium maritimum.

Of Stove Palms there is abundance. Some of those that grow well in the cooler houses flourish healthfully in the warmer ones, simply growing faster therein; and the constitution of hot-country Palms is generally such that they flourish without much special care in hothouses. The following are among the more desirable kinds:—

Areca aurea.
　„　monostachya.
　„　nobilis.
　„　rubra.
　„　sapida.
　„　speciosa.
Astrocaryum mexicanum.
Bactris, all obtainable species.
Calamus elegans.
　„　dealbata.
　„　Verschaffeltii.
Caryota sobolifera.
　„　elegans.
Cocos coronata.
　„　flexuosa.
Copernicia cirrifera.
Elæis guineensis.
Latania glaucophylla.

Latania aurea.
Pritchardia pacifica.
Euterpe edulis.
　„　oleracea.
Geonoma fenestralis.
　„　magnifica.
　„　paniculata.
　„　pumila.
　„　speciosa.
　„　Verschaffeltii.
Hyphæne thebaica.
Maximiliana elegans.
Oreodoxa regia.
　„　Ghiesbreghtii.
Oncospermum fasciculatum.
Phœnicophorium sechellarum.
Phytelephas macrocarpa.

Arenga obtusifolia.
Calamus adspersus.
Rhaphia Hookerii.
　„　tædigera.
Sabal princeps.
Thrinax argentea.
　„　elegans.
　„　radiata.
Verschaffeltia melanochætes.
　„　splendida.
Ceroxylon audicola.
Chamædorea excelsa.
　„　paniculata.
　„　atro-virens.
Dæmonorops plumosus.
Leopoldina pulchra (expensive and rare).
Calamus Getah.
　„　oblongus.

Most of the above range in price from two shillings,

and even less, to half a guinea each, while some of the rarer kinds go much higher, and strong well-grown specimens of all are of course much more expensive than the small and young plants to be bought for the prices above given.

The Ivy, and its Uses in Parisian Gardens.

The Irish Ivy is a very old friend that is often seen beautifying old walls and like positions, and one, as we may have thought, sufficiently appreciated and employed. Gaiety and grace I was led to expect in Parisian gardens, but that they should take up our Hibernian friend, so partial to showers and our mossy old ruins, and bring him out to such advantage in the neighbourhood of new boulevards and sumptuous architecture, was not to be expected. That " a rare old plant is the Ivy green when it creepeth o'er ruins old," we Britons all know, but that it is no less admirable when mantling objectionable surfaces with its dark polished green in winter, would not appear to have yet sufficiently dawned upon us. Apart from the fact that the Ivy is the best of all evergreen climbers, it is the best of all plants for softening the aspect of town and suburban gardens in winter, not to say all gardens. The Parisian gardeners know this fully, and they, taking it out of the catalogue of things that receive chance culture, or no culture at all, bring it from obscurity and make of it a thing of beauty.

To rob the monotonous garden railings of their nakedness and openness, they use it most extensively, and there are parts about Passy where the Ivy, densely covering the railings, makes a beautiful wall of polished green along the fine wide asphalte footways, so that even in the dead of winter it is refreshing to walk along them. And if it does so much for the street, how much more for the garden? Instead of the inmates of the house gazing from the windows into the street swarming with dust, or splashing with mud, a wall of verdure encloses the garden; privacy is perfectly secured; the effect of any flowers contained in the garden is much heightened; and lastly, the heavier rushes of dust are kept out in summer, for so admirably are the railings

x

covered by planting the Ivy rather thickly, and giving it some rich light soil to grow in, that a perfectly dense screen is formed. Railings that spring from a wall of some height around the larger houses are covered as well as those that almost start from the ground. Frequently the tops of the rails are exposed, and often these are gilt, while wire netting on the inner side supports the Ivy firmly.

One day, as I was passing near the Hôtel de Ville, and looking at its traceries, my eye was caught by something more attractive than these: a gilt-topped railing densely covered with Ivy, and between the mass of dark green and the bared spikes at the top a seam of light green foliage, here and there besprinkled with long beautiful racemes of pale purplish flowers. That was the Wistaria, one of the most beautiful of China's daughters, here gracefully

FIG. 134.

Railings densely covered with Ivy. This figure also shows the wide asphalte pathway, the grating over the ground at the base of one of the trees, and the cage used to protect its stem.

throwing her arms round our Hibernian friend, and forming a living picture more pleasing to the eyes of a lover of nature than any carving in stone. If there are tall naked walls near a Parisian house, they are quickly covered with a close carpet of Ivy. Does the margin of the grass around some clump of shrubs or flower beds look a little angular or blotchy? If so, the Parisian town gardener will get a quantity of nice young plants of Ivy, and make a wide margin with them, which margin he will manage to make look well at all times of the year—in the middle of winter when of a dark hue, or in early summer when shining with the young green leaves.

When the Ivy is planted pretty thickly and kept neatly to a breadth of, say, from twelve to twenty inches, it forms a dense mass of the freshest verdure, especially in early summer, and of course all through the winter, in a darker state. The best examples of this description of edging that I know of anywhere are to be seen around the gardens of the Louvre, and in the private garden of the Emperor at the Tuileries. In the latter the Ivy bands are placed on the gravel walks, or seem to be so; for a belt of gravel a foot or so in width separates them from the border proper. The effect of these Ivy bands outside the masses of gay flowers is excellent. They are the freshest things to look upon in Paris during the months of May, June, and July. They form a capital setting, so to speak, for the flower borders—the best, indeed, that could be obtained; while in themselves they

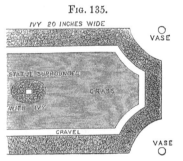

Fig. 135.

Ivy edgings in geometrical garden.

possess qualities sufficient to make it worth one's while to grow them for their own sakes. In some geometrical gardens we have panels edged with white stone—an artificial stone very often. These Ivy edgings associate beautifully with them, while they may be used with advantage in any style of garden. A garden pleases in direct proportion to the variety and the life that are in it; and all bands and circles of stone, all unchangeable geometrical patterns, are as much improved by being fringed here and there with Ivy and the like, as are the rocks of a river's bank.

It should be observed that an Ivy edging of the breadth of an ordinary edging is not at all so desirable as when its sheet of green is allowed to spread out to a breadth of from fifteen to eighteen inches. Then its rich verdure may be seen to full advantage. It must of course be kept within straight lines if the garden be symmetrical : if it be a natural kind of garden, you may let it have its own wild way to some extent. In nearly every courtyard in Paris the Ivy is tastefully used. I do not think I ever saw the scarlet Pelargonium to so great advantage as in deep long boxes placed against a wall densely covered with it, and with Ivy planted also along their front edge, so as to hang down and cover the face of the boxes. One of the best known of the floating baths on the Seine has a sort of open air waiting-room immediately outside its entrance—a space

FIG. 136.

Section of circular bower formed of a single plant of the Irish Ivy grown in a tub.

made by planks, and communicating with the quay by a gangway. On this space there are seats placed around, on which in summer people may sit and wait for their turn if so disposed, while the whole is elegantly embowered with Ivy, which looks as much at home as if the river was not gurgling rapidly beneath. This is secured by placing deep boxes filled with very rich light soil here and there on the bare space ; then planting the Ivy at the ends of each box and devoting the remainder of the space to flowers, keeping the soil well watered, and training the shoots of the Ivy to a neat light trellis overhead.

In the garden of the Exposition a pretty circular bower was shown perfectly covered with it, the whole springing from a tub. Imagine an immense green umbrella with the handle inserted in a tub of good soil, boards placed over this tub, so as to make a circular seat of it, and you will understand it in a moment. That and the like could of course be readily made on a roof, wide balcony, or any such position. One sunny early summer day, when the Ivy was in its youthful green, I met with a shallow bower made of it that pleased me very much. It was simply a great erect shell of green not more than five or six feet deep, so that the sun could freshen the inside into as deep a verdure as the outer surface.

The Ivy may be readily grown and tastefully used in a dwelling-house. I once saw it growing inside the window of a wine-shop in an obscure part of Paris, and on going in found it planted in a rough box against the wall, up which it had crept, and was going about apparently as carelessly as if in a wood. If you happen to be in the great court at Versailles, and, requiring guidance, chance to ask a question at a porter's little lodge seen to the left as you go to the gardens, you will be much interested to see what a deep interest the fat porter and his wife take in Cactuses and

Fig. 137.

Variegated Ivy in suspension basket.

such plants, and what a nice collection of them they have gathered together, but more so at the sumptuous sheet of Ivy which hangs over from high above the mantelpiece. It is planted in a box in a deep recess, and tumbles out its abundant tresses almost as richly as if depending from a Kerry rock.

The Ivy is also used to a great extent to make living screens for drawing-rooms and saloons, and often with a very tasteful result. This is usually done by planting it in narrow boxes and training it up wirework trellises, so that with a few of such a living screen may be formed in any desired part of a room in a few minutes. Sometimes it is permanently planted; and in one instance I saw it beautifully used to embellish crystal partitions between large apartments.

To make the Ivy edgings which are so abundantly employed in and around Paris, plants are easily procured in pots, and at a very cheap rate, at the markets on the quays, or of the nurserymen at Fontenay aux Roses, who every year grow it in large quantities. It is planted thickly in borders, and trailed along in strips from twelve to sixteen inches in width, according to the size of the beds. It is laid down with wooden pegs, a layer of earth being placed over the stems. When once planted, it only needs to be kept clear of weeds, and to be moderately watered. Under this

Fig. 138.

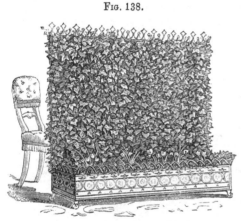

Ivy screen for the drawing-room with flowers at its base.

treatment, it forms healthy borders the year after it is planted. In preparing the Ivy for growing against railings and trelliswork that encloses the various parks and gardens, it is trained carefully during the first one or two years, so that all empty spaces may be filled up. At the end of the second year, the railings will be completely covered, and for the future it is only necessary to keep it properly pruned.

The Ivy used by the City of Paris for ornamenting the flower beds in the squares, the trunks of trees, &c., is grown and propagated at the nurseries in the Bois de Boulogne. Towards the end of the summer the propagation of the Ivy by means of cuttings is carried on. Three or four leaves are left on each cutting, and they are planted very thickly in lines in a half-shady position. When they have taken root sufficiently, which generally takes place in the following spring, they are transplanted into pots of four or five inches in diameter. Afterwards stakes are fixed along the lines of pots, from which are stretched lines of thin galvanized wire, and to this slender but firm trellis from three to five feet high the plants are trained several times during the growing season. At the end of the second or third year the plants are strong enough to be employed to cover railings, and for many similar purposes. The nurserymen in the suburbs of Paris generally propagate them by layers. For this purpose old plants are placed at a certain distance from each other, and are allowed to grow long. Pots from four to six inches in diameter are then plunged in the ground around, the Ivy being fixed in them by means of small pegs, one shoot in each pot. Afterwards stakes are placed in the pots, and the Ivy trained against them as it grows. When the layers are sufficiently rooted, they are separated from the old plants, and towards the end of the second or third year it is ready for use. If a wide belt of Ivy is desired, the young plants may be put in in two or three rows, as the French do when making such excellent Ivy edgings as are here described. In any case, after the plants are inserted the shoots must be neatly pegged down all in one direction.

The reason why Ivy edgings when seen in England look so poor compared with those in Paris, is that we allow them to grow as they like, and they get overgrown, wild, and entangled, whereas the French keep them the desired size by pinching or cutting the little shoots well in, two or even three times every summer, after the edging has once attained size and health. The abundant supply of established plants in small pots enables the French to lay down these edgings so as to look well almost from the first day.

CHAPTER XIV.

FRUIT CULTURE : HOW ARE WE TO IMPROVE?

THE discussion on French and English fruit growing which
emanated from my letters to the *Times* in August, 1867,
and afterwards spread through all the gardening papers,
was too desultory to leave any impression on the public
mind as to the best course to pursue. For this reason
therefore, and to prevent misrepresentation, I entered on
the question of the general improvement of our hardy fruit
culture in the *Times* in May, 1868 ; and in this chapter I
propose to enter more fully into this very important question. Some have so little understood me as to suppose that I
had recommended the cordon for orchard culture, which
would be a stupendous blunder only worthy of some nurseryman very anxious to sell his trees. Therefore, although
the present subject may seem wide of the aim of the book,
it is necessary to enable the reader to estimate the value—
be it small or great—of what we may learn from continental
fruit growers, and how we may improve our supplies. The
fruit question is not one that merely concerns those who
can afford to keep gardeners, or even the much larger class
who can devote some time and money to the pleasant and
healthful amusement that amateur gardening affords ; it is
a question for the public in its widest sense, and of especial
importance when considered in relation to the enormous
and badly supplied masses in our ever-growing great cities
and towns.

I shall first deal with the Pear, for several reasons :—
1st. Considering its hardiness, keeping qualities, and rich
variety, it is the most delicious and valuable fruit that can
be grown in northern latitudes. A perfect Peach may be preferred to a first-rate Pear, but by properly selecting varieties

of Pears we may have them in perfection during eight or
nine months of the year—or even longer—and the variety
in flavour is perhaps greater than in the case of any other
fruit. 2nd. We are quite behind the French growers in
its production. Our stocks of Apples are usually good and
abundant; our stocks of Pears are frequently scarce and
very poor in quality. I have seen many large gardens in
the British Isles where a really good Pear was almost as rare
as a Mangosteen. 3rd. I believe we can increase the quantity and quality of our Pears in a tenfold degree over the
greater part of England and Ireland, and even in time to
come export the fruit that we now import so largely.

It is indisputable that the brighter sun of France is more
favourable to the culture of the Pear than our own climate;
but it is equally as true that by the aid of walls for some
sorts, by judicious selection of ground, locality, and kinds,
we may grow it to perfection. The quantity of pears the
French send to our markets is surprising. Messrs. Draper,
the salesmen of Covent-garden, showed me by their books
that from one importer alone they sell from 60*l.* to 100*l.*
worth of French garden produce (chiefly Pears) each market
day; and a fruit merchant has told me of one dealer in
pears who annually collects in France and sells in our
markets 10,000*l.* worth of that fruit. Are not these signi-
ficant facts for the British cultivator?

It is quite a mistake to suppose that the climate does all
this for the French—the winter and spring in many parts of
northern and north central France being quite as difficult
for the fruit grower as those of England. The pear loves a
moist, genial climate, and in many parts of England and
Ireland our advantage in this respect will be found to
compensate in some degree for the difference in sunlight.
Some pears are grown better in England than in France,
and it is a curious fact that some that ripen and go off
quickly in the neighbourhood of London remain in an
eatable state much longer and acquire a more delicious
flavour in the cooler climate of Yorkshire. Let it be borne
in mind that we are talking of the culture of a fruit which
grows in a wild state as far north as southern Sweden, and

not of the Pomegranate or any really tender subject; and
then the objections of those who say that our climate
prevents any improvement, and perhaps immediately afterwards assert the superior quality of British-grown fruit,
will pass for no more than they are worth. If one
individual can grow a first-class Pear, why not a score or
more persons in the same neighbourhood? Nature is our
willing handmaid in this matter, and I firmly believe that
we have it in our power to place this fine fruit within
the reach of all, and render ourselves quite independent
of the French. I do not say we could grow such big Belle
Angevine Pears as they sell at Covent-garden for a guinea
and a half apiece; but that is of no consequence, as these
are at best only fit for show or kitchen use, and are, in fact,
little better for eating than a raw turnip.

There are various ways in which we may improve the
culture of the Pear, and the first and best is by paying
more attention to it as a naturally developed standard
tree—in a word, by an improved system of orcharding.
This also applies to other hardy fruit trees, and is treated
of at greater length further on. Upon orchards we must
chiefly depend for the supply of our large cities and towns.
This subject, in its commercial aspect, may be left to the
growers of fruit for the market, but the country gentleman
and large farmer—in fact, everybody possessing a hedgerow,
field, or shrubbery—cannot be too strongly urged to use the
great opportunities they have for growing Pears. They
grow useless shrubs and weedy trees in many places where
the finest fruit might be grown without any attention,
expense, or trouble beyond gathering it. There are
plenty of landed proprietors who at present know not what
it is to have the luxury of a stock of good Pears, who might
gather them from spots now utterly useless; there are multitudes of farmers who hardly ever see a good fruit of this
kind, in possession of lines of hedgerow where the tree would
stand as healthfully from among the lower brushwood as
any subject that now embellishes them; and there are
thousands of owners of villas and suburban gardens who
now go to market for their fruit who might gather it from

places in their little shrubberies, at present entirely devoted
to that miserable shrub the Privet, and some of its most
worthless allies.   I know well the kind of objection that is
made to some of these suggestions—the boys would gather
the fruit, &c.   Small blame to the poor boys for making an
occasional attempt on the little fruit that comes in their
way, and for exercising a little ingenuity in getting at what
is for them such wholesome and delicious food; but if the
fruit were as plentiful as it ought to be they would not be
so tempted.

It should be remembered that some of our hardy fruits
are capable of affording quantities of wholesome food to the
people; but before they do so efficiently we must take
them out of the class of things that are carefully walled
in gardens, overdone with kindness, or perhaps mutilated
to death by excessive and unnatural pruning, and recognise
and take full advantage of the fact that many excellent
kinds are as hardy and easily grown as the Blackberries
and Sloes of the hedges.   For the purposes herein suggested thoroughly hardy and free-growing sorts should alone
be selected; but it must not be supposed that first-class
fruit, even of the continental varieties, cannot be produced
in this way.   The other day in visiting the gardens at Oak
Lodge, Kensington, my attention was attracted by a very
large and handsome Pear tree growing among the Rhododendrons and other choice shrubs which adorn the margin
of a piece of rock-bound water.   Upon further inquiry I
found it was a fine old tree of the Beurré Diel, which,
without pruning or attention of any kind, produced abundantly such good fruit, that of twelve samples of the same
fine variety recently laid before the Fruit Committee of the
Royal Horticultural Society the fruit of this tree was pronounced the best.   I by no means mention this as a
remarkable instance, but merely to prove that the finest
Pears may be grown by the simplest means, and that the
tree is worth cultivating for its beauty alone.   The garden
of Oak Lodge is the best designed town garden I have yet
seen, and Mr. Marnock, who arranged it, left several
of these old Pears in conspicuous positions when laying

out the place solely for their beauty as trees, apart from
their fruiting qualities. Therefore it is clear that we may
effect considerable improvement by planting this tree in
shrubberies, pleasure grounds, and like positions, and in
many wild and semi-wild places, both in enclosed private
grounds and in the open country. There can be no doubt
that enormous quantities of good fruit could be grown upon
railway banks now useless, and from which fruit could be
so readily conveyed to market. The French are nearly as
backward in these matters as ourselves, but they have at all
events taken the initiative, as described in another part.
It is really astonishing that such beautiful objects as most
of our fruit trees are when in flower do not more frequently occur outside the garden walls in this country.

The second way in which we may improve the cultivation
of the Pear is by planting it to a greater extent as a pyramidal tree, and grafted on the Quince where the soils are
rich, moist, or deep. On many dry and sandy lands the
Pear must be grown on its own stock, and for orcharding
purposes generally that may safely be pronounced the best.
Indeed, one writer suggested this as the remedy for all our
wants in this way ; but it is not so. We shall never have a
cheap supply for market till we pay more attention to the
Pear as a freely developed standard tree ; we shall never have
a first-rate supply of winter pears till we pay better attention to walls than we do at present. The French, from
whom we have adopted the pyramidal form, employ it to an
enormous extent, but do not stop there. It is in planting
the pyramid that most of our improvement in this direction
has taken place for a good many years back. Almost every
nurseryman has now a stock of the tree in this form, and
we cannot employ it too much, provided sorts that ripen
well in ordinary seasons are selected ; but there are other
ways of equal importance. The pyramid is so pleasing in
outline, and indeed in all other respects, that, although so
highly suited for the kitchen or fruit garden, it should by
no means be confined to either. Handsome specimens may
well be introduced in favourable spots in the pleasure ground
and shrubberies, and thus the owners of those numerous

small ornamental gardens near towns may gather fine fruit. However, this form is so well known, and has been so much recommended for many years, that I shall now turn to the third way of improving the culture of the Pear, and one that has been comparatively neglected for some years past.

I mean the Pear on walls. Here we are certainly behind-hand, and do not appear to have made much progress for a very long time. Perhaps it may be thought that the French might dispense with walls; but no such thing. They find them indispensable for the perfect culture of the finer winter Pears; and were it not for their use, they could never obtain such a stock of them as they have. Yet we have for a long time past been paying attention to almost every kind of garden improvement but this very important one. It is true that walls are expensive, but once up it is a great pity to neglect them; and, apart even from garden walls, there are numerous places with as much wall surface naked and use-less as, if properly covered, would yield a good supply of fruit to the family. Few things combine beauty and utility more effectively than a well-covered wall of Pear trees; and the creation of such is not a matter of mystery or difficulty, but what anybody can perform. With walls it may be safely said that our climate is as good as that of northern France. Indeed, there can be no doubt about it, as I and many others have eaten as good fruit off well-managed English wall-trees as ever grew; but unfortunately there is but little attention paid to them compared to what they deserve. Most large gardens would be benefited by having a much greater proportion of wall-surface than they have at pre-sent; to many small ones they would prove a great ad-dition. Fortunately, a recently-invented, or revived, process offers an opportunity of building them very much cheaper than before, and as good as could be desired.

I allude to Tall's plan for making concrete walls, which has not as yet been utilized by horticulturists, but which is certain to prove of the greatest use to them, and to have a marked influence on our horticulture for the future. I have seen it employed with much success in the building of the Emperor's model houses for workmen near the Bois de

Vincennes, and from the day of visiting them I have had no doubt whatever that it will prove a great gain in our fruit growing. The building of houses by its means is simple and as easy as could be desired, although none but the roughest labourers are employed; that of walls may be effected even with greater ease. The Paris houses were built with very rough gravel dug up on the spot. The same or any like material may be used for like purposes, as may burnt clay, stony rubbish of any kind, or even such material as clinkers, abundant and hitherto useless in many districts. I need not and cannot here go into the plan, but it consists in little more than mixing a small portion of cement and sand with the rougher material, and throwing the mass between boards firmly adjusted to the size of wall required. The mass hardens in twenty-four hours or so, then the boards are elevated, another layer of concrete thrown in, and so the work goes on. It will be clearly seen that nothing can be better suited for garden purposes. In addition to this mode, I know no reason why walls of adhesive earth on a brick or stone foundation should not be used with us as well as on the Continent. I have seen many of these garden walls and houses perfectly sound and strong many years after their erection, and looking no worse, indeed better, than ordinary brick walls.

No matter of what material the wall be made, it will be desirable to whiten its surface and keep it white. Black and dark coloured surfaces absorb heat in the daytime, and give it out again during the night in the form of radiant heat; from which facts we might draw the conclusion that walls for training fruit trees against should be black, or at any rate of a dark colour. Direct experiment was, however, necessary to settle this question, and M. Vuitry, who employs his leisure in arboriculture, has communicated the results of his experiments in this direction to M. du Breuil, which leave no doubt as to the proper colour to be chosen for walls against which fruit trees are to be trained. He has proved—1st. That a thermometer hung during the day with its face turned towards a white wall, at a distance from it equal to that of a fruit tree trained against it—*i.e.*, about

an inch and a quarter—always showed a mean temperature of
nearly 6 deg. Fahr. higher than one hung against a black wall
under precisely similar circumstances. 2nd. That during
the night the difference of temperature shown by these two
thermometers was inappreciable. Contrary therefore to
the opinions entertained by many persons, it seems to be
evident that the walls must be whitened when we wish to
give the trees trained against them the maximum amount of
heat to be obtained from the particular climate and aspect.
Indeed, it is precisely the plan that has already been pursued by the fruit growers of Montreuil for Peach trees, and
of Thomery for their Vines, it having been frequently remarked that trees trained against white walls were healthier
than those nailed to more or less dark-coloured ones. This
result is easily explained, for not only does the lighter colour
reflect more heat back to the trees, but by this means they
receive a greater quantity of light ; and it is well known
how greatly vegetation is stimulated by these agencies.
Walls of a light tint are advantageous in another way, for
they not only reflect light and heat on the particular trees
trained against them, but also on the others in their immediate neighbourhood. By abundantly planting the finer
winter Pears against walls with a warm exposure and whitewashed surface, we may within half a dozen years gather
such crops of the really valuable winter Pears as have never
before been seen in this country.

Another improvement must of necessity accompany this,
and that is the French method of wiring garden walls.
We cannot use nails with concrete and earth walls, and
if we could the deliverance from nails would be a great
point gained. To me the most lamentable of all garden
sights is that of men handling those miserable shreds and
nails during winter time, and blowing heat at their fingers
and patting their toes to keep up the circulation. Our way
of wiring a wall is so expensive and cumbrous that many
still prefer the nails, but the French mode of employing a
little raidisseur or tightener on each wire, and using very
slender galvanized wire, is quite perfect in its way. When
adopted with us it will be found to save much time

and greatly improve the appearance of garden walls. We must also adopt the improved kind of espalier which the French are beginning to employ so extensively, and which is elsewhere described and figured.

Of all our wants in connexion with the Pear, that of the spread of good varieties is perhaps the greatest. Naturally, or rather I should say in a wild state, the Pear is a poor fruit about an inch and a half long; and from this in the course of thousands of years the splendid race we now possess has sprung. Scattered through our gardens and orchards in all parts of this kingdom, there are scores of kinds which are practically of little more use than the wild fruit trees of the woods and hedgerows. But apart from all these worthless varieties, named and unnamed, that occupy valuable ground, there are numbers which are regularly sold in our nurseries, possessing fine names and pedigrees, and yet which are practically useless to the cultivator, and it may be mischievous to the amateur. Let us suppose the case of a person wishing to commence Pear culture—he has some slight knowledge of other branches of horticulture, and expects that the long list of the varieties of Pears which he finds in his nurseryman's catalogue will resemble each other pretty much as his Verbenas or Pansies do. Taken by the different names and descriptions, he goes in for collection instead of selection, seeks variety and finds disappointment. The truth is that a wide selection of varieties is an evil in every way. It requires much sagacity on the part of men who have studied gardening all their lives to know what to avoid in these lists; how very dangerous, then, for the amateur, or for those who have neither amateur nor professional knowledge of the matter, to make a selection! Let us glance for a moment into some of the fruit catalogues. It is needless for us to state how much the Pear varies. Here is a catalogue naming, describing, and numbering nearly 400 kinds. What a danger for those who suspect not how few are the really good varieties of Pears suited to this climate! People suppose that giving long lists of this kind is for the sake of selling a great number of varieties; but that course would be so clearly a mistaken one, that one cannot suppose an intelligent person persisting in it. The presence of bad and unsuitable Pears everywhere throughout the country simply tends to retard the culture of this noble fruit; whereas the distribution of the really good kinds in abundance would create such a demand for them as would cause the trade in young trees to increase tenfold.

The compilers of the above catalogues do not follow the example of the famous M. de la Quintinye, chief gardener to Louis XIV. at Versailles, whose list was lengthy, although published so very long ago, but who conscientiously divided it into several sections—viz., " good pears," " indifferent pears," and " bad pears !" This was honest in De la Quintinye, and would be admirable in a British nurseryman. The spirit of expurgation was strong in this famous old gardener, and he follows the bad with another list—a long one—heading it—" Besides the pears which I know not, here is a particular list of those which I know to be so bad that I counsel nobody to plant any of them." And that is followed by another :—" A list of those which I esteem not highly enough to counsel any gentleman to plant them, nor yet so much despise as to banish them out of the gardens of them that like them." Here was an instance of a most praise-worthy desire to weed out the bad, followed by others to exterminate the middling and the not very good. This, observe, was in France, where a greater number of kinds arrive at perfection than is the case with us, and where a greater number of varieties are grown. Although our nurseryman friend, with his long list, is somewhat of an exception, the lists of others of our fruit-tree raisers are much too long to be of any real guidance to the amateur.

The following list comprises the cream—the best Pears of the many hundred kinds known :—Doyenné d'Eté, Jar-gonelle, Williams's Bon Chrétien, Louise Bonne of Jersey, Jersey Gratioli, Urbaniste, Fondante d'Automne, Beurré d'Amanlis, Suffolk Thorn, Seckel, Comte de Lamy, Flemish Beauty, Désiré Cornelis, Marie Louise, Baronne de Mello, Thompson's, Beurré Bosc, Duchesse d'Angoulême, Beurré Diel, Beurré Hardy, Maréchal de la Cour, B. Superfin,

Y

Doyenné du Comice, Glou Morceau, Winter Nelis, Beurré Rance, B. Sterckmans, Joséphine de Malines, Bergamotte Esperen, Easter Beurré.

Of the above, Marie Louise, Beurré Bosc, Duchesse d'Angoulême, Beurré Diel, Doyenné du Comice, Glou Morceau, Beurré Rance, Joséphine de Malines, Bergamotte Esperen, Easter Beurré, Beurré Sterckmans, Désiré Cornelis, and Winter Nelis should be grown against walls. In some cases they may afford a satisfactory result away from them, but if grown against white walls they will in all cases be highly improved, and some of the very best of them are only to be had in perfection when thus grown. As wall space is often limited, and as it is necessary to have the warmest walls to perfect the finest winter pears, it is desirable to be very particular indeed when selecting pears for wall culture; and I should advise Easter Beurré, Doyenné du Comice, Glou Morceau, Beurré Rance, Joséphine de Malines, and Bergamotte Esperen to be abundantly planted against walls wherever a prime supply of first rate winter pears is a want—and of course it is a very general one.

Some of our authorities on fruit growing give the Easter Beurré as one which should be planted as a bush or pyramid, and say it is " mealy and insipid from walls." To show how worthless is this opinion, I have merely to point to the fact that the splendid Easter Beurrés which adorn our tables in winter and spring are grown on walls in France. All of the same variety for the imperial table are grown in like manner at Versailles; and as soon as a wall is cleared of other varieties of Pear trees there it is immediately planted with the Easter Beurré—so much is this fine variety esteemed. The quantity of its fruits sold in the markets of Europe during the winter season is something incredible. It is perhaps the most valuable of all winter Pears; and the chief, I may say nearly the whole supply comes from France. The climate does it, some will say, but such is not the case; for if left to the climate unaided, we should have few fine Easter Beurré pears in Coventgarden in winter. And the same remark applies to other varieties of winter Pears. The flavour is said to be inferior when grown against walls. Let us try them against white walls as the French do, and see if we cannot nearly or quite equal their pears in size, and quite equal them in flavour. We have been for years planting them as bushes and pyramids, and paying little or no attention to their culture against walls; hence our deficiency of good winter pears— those which are by far the most valuable of all.

Having taken sufficient care to select the very best varieties, and to place them in positions where they are likely to succeed, there is more to be done in getting rid of the bad ones. They abound in every part of the country, and take up space in which the most delicious kinds may be grown. This prevalence of bad kinds not only results from the greater scarcity of the good varieties in bygone days, but also from the large number of inferior kinds that are still offered for sale. In very many cases the tree is worthless, because it has not been planted in a position to insure success. It frequently happens, for example, that the very finer kinds of winter Pears, and those which the French grow against walls around Paris, are in Britain sent out as suitable for pyramids. All worthless Pear trees should be destroyed, and good kinds planted or (happily there is an alternative) regrafted with good sorts. Instead of sacrificing a plantation we may cut the trees close in, regraft the branches with the best kinds, and thus in a short time have established trees of the finest sorts that arrive at perfection in any given locality. This may be performed with either standard, pyramid, wall, or espalier trees, " crown " grafting being the best for this purpose. Another great point would be gained if the custom of growing inferior kinds from pips—which is common among farmers in some fruit growing districts, with Pears as well as other hardy fruits—were abandoned, and only first-rate and hardy kinds planted or grafted.

It was touching the utility of the low cordon for the production of superb dessert Apples that we have had most discussion; and a much wider experience with French fruit gardens enables me to say that it is worthy of all the praise that I have given it, and certain, when well managed, to give the highest satisfaction. The reader will kindly ob-

serve that I specially recommend only one species of cordon for the Apple. There are many kinds, with various merits.
The grower for market will also oblige me by bearing in
mind that I only recommend it for the garden and for a
special purpose. Emphatically I say that a good hardy
kind on a well-managed standard or naturally developed
tree is the best for the supply of the markets with all but
the best fruits, and for all ordinary purposes; and that the
system of orcharding in the London market gardens is on
the whole a good and safe one. Generally speaking our
apple-culture is not to be complained of, though it may
certainly be improved. There is in this country a large
demand for fruit of the finest quality that can be obtained,
both in the case of those who buy all they use and those
who grow their own. In these islands it is also generally
admitted that to keep the sun from the general contents of
our gardens by shading them with Apple trees is anything
but desirable, and therefore I recommend the cordon trained
as an edging, and on wire, tightened as before described,
at one foot from the ground. I have ascertained beyond
all question that where well managed these will, if placed
alongside the walks in the kitchen and fruit gardens, furnish abundance of fruit without planting any others.

Now, in many places the positions in which this plantation
may be made are quite unoccupied, and therefore the system
will prove a decided gain. It will have to encounter prejudice and bad management; but once well managed specimens are seen in our gardens it will spread rapidly through
these islands, and prove a great boon wherever perfect fruit
is desired. I have passed through many parts of northern
England and Scotland during the past year or two—districts
in which every ray of sun is required; and yet in these you
see in all directions the gardens shaded and half destroyed
by old standard Apple trees. This is especially the case with
the smaller class of gardens, in which you may frequently
see gouty old trees shading and souring the aspect of the
very house itself. The objections urged against the simple
cordon are dealt with in the chapter devoted to this system
of culture. Let no person think he has fairly tried

the cordon system if he employs what is called the " English Paradise " stock.

And now a few words about the Peach.   This fruit attains the finest possible condition when well grown against walls in England.   In other countries it may be grown freely as a standard tree ; in none can they produce finer or better fruit than may be gathered from walls in England and Ireland.   France has very diverse climates—some in which the Peach grows well as a standard—but the best Peaches grown in France are gathered from walls in those parts where the climate is most like our own.   In the middle of September, 1867, I ate capital specimens of Crawford's Early Peach, gathered from pyramid trees standing in the open quarters of the Rev. Mr. Benyon's garden in Suffolk.   I by no means mention this as an example to be followed, but simply to prove that in the midland and southern parts of the British Isles the Peach may be grown against walls to the highest degree of perfection ; and in favourable parts of the south, the Early York Peach may be grown with success as a standard or bush tree, away from all protection.

There can be no doubt whatever about the fact, that if we pay as much attention to the Peach as the cultivators of Montreuil do, we can attain quite as good a result.   The fact cannot be too widely known that no fruit tree nailed against walls furnishes a more certain and regular crop than the Peach tree when well treated ; and yet it is hardly possible to buy a good Peach in London.   In Covent-garden, it is true, excellent Peaches may be bought at 8d. and 1s. each, but those sold by most fruiterers at 3d. and 4d. are worse than those procurable in Paris for a sou, and are only fit for pig-feeding.   And in numbers of private gardens the fruit is by no means common.   Our good gardeners understand its culture well enough ; but of late years public attention has, by various means which I will not detail here, been called away from the fact that, with walls, we can produce the finest fruit in the world, and without them do little or nothing with the choicer fruits.   The " power of the climate " in Paris may be very wonderful to some people, but there is one thing it cannot do better than our own—it cannot produce a better Peach than I have often gathered from walls both in England and Ireland. It would be thought, perhaps, that with their fine climate, the French would be able to dispense with protection to the trees in spring, and altogether leave their trees more to nature than the British gardener; but the fact is exactly the reverse. The French peach-grower takes care to have a good protecting coping to his wall. With us it is not uncommon to see the culture of the Peach and Nectarine attempted, and even with success, without any coping at all. The French cultivator frequently places iron rods eighteen inches or two feet long, and furnished with a catch at the end, just under the permanent coping of his wall, which rods enable him to slip on a most efficient protection in the shape of a temporary coping just under the permanent one. I know one grower who has 4000 yards of this temporary coping, made of tarpaulin, stretched on cheap light frames.

This is, I trust, a sufficient commentary upon the climatic advantages possessed by the two sets of gardeners! Of course we want this protection as badly as the French, if not worse. Over the greater part of the country, without question, the Peach may be grown to the highest degree of perfection, and yet, though few Englishmen could manage, as Johnson did, "seven or eight large peaches of a morning before breakfast began," they may well say with him that getting "enough" of them was indeed a rarity. It is stated in a recently published book on fruits that for the majority of the population to partake plentifully of this fruit, "the only hope that can be held out involves nothing less than an emigration across the Atlantic!" The present state of matters justifies the writer in the remark. The quality of the Peaches sold at the lowest, but by no means a low price, is such as to prevent anybody making a second investment in them, and therefore the fruit is, as the writer remarks in describing it, "a luxury confined to the wealthy." Before it is otherwise, good fruit must be sold at a price that will put it within tasting reach of others than those provided with a powdered footman to convey it from the fashionable fruiterers to the carriage waiting at the end of the " Row." To market gardeners I may with respect to this fruit offer a word of advice, though I have not ventured to do so with regard to other matters. To succeed with the Peach you must remove it altogether from the chance culture now bestowed upon it; you must employ men to give it full attention in spring and early summer; you must select suitable soil in the first instance, and thus avoid expense for what is called made ground. You must take care to protect the trees in spring, as the careful French cultivators do; and you must take advantage of the very cheap and excellent way of erecting walls that I have alluded to. No chance culture on any walls that may happen to surround the place will alter matters much.

The same remarks apply to some extent to the private gardeners and to amateurs. They should pay more attention to walls, erect more and utilize those they already have. I had a letter lately from Mr. J. A. Watson of Geneva, in which the writer describes a village church as being covered with Peaches and Nectarines, and goes on to state that the sexton gives a lecture on the subject now and then to the natives on Sunday mornings. I do not wish the example to be followed, nor the glorious old Ivy to be disturbed even for the luscious Peach; but we may do a good deal more than at present with our unoccupied walls. Probably many readers who live near Oxford can testify to the beauty and profit that results from the villagers covering their walls with Apricot trees. The same may be done in many parts of England where such a thing is not now to be seen; but in the case of cottagers and others the only thing likely to do good is example. If they see a specimen of success they need no other encouragement. And perhaps I may here suggest that a present of a few good kinds of trees, and perhaps a few minutes' advice from the gardener, would be more productive of benefit to cottagers than many other things given them in a charitable way.

As to our various other hardy fruits, including the Apple and Pear, there can be little doubt that it is to good

orchard culture we must look for the increase of our supplies.   The word orchard is familiar enough in our ears, but
a really good orchard is as rare round country seats as if it
were not a British institution.   There are farmers and
market gardeners and fruit growers who have the finest
orchards ; but at the country seat, with generally every
opportunity to select a good site, it is surprising how rarely
even a presentable thing of the kind is attempted.   Indeed,
in some parts of the country it is never thought of—the
ordinary type of kitchen garden being considered sufficient
for all attempts at fruit growing.   I say attempts advisedly,
for what do half the gardens in the country show ?   The
surface cannot be devoted to standard trees, as they hide
the light from the necessary crops, and the walls and dwarf
trees, if such there be, are those upon which the gardener depends.   Now good wall culture is not common, even in
places where a regular staff of gardeners is kept ; and in
hundreds of cases where there are not, the trees are " abandoned to themselves."   But supposing that the wall culture is
good, and that the most is made of the space, it is hardly
sufficient to yield a crop of fruit such as one would like for
eating, cooking, preserving, and presents.   If the walls
supply a good dessert for a reasonable length of time, it is
as much as is expected of them, and more than they generally do.   They who secure a good crop of winter Pears,
who can command really eatable specimens of this fruit
during the winter and spring months, are luckier than most
persons in possession of garden walls.   The walls can only
supply a portion of the choicest fruit—chiefly of those kinds
which require the additional heat of a wall for their perfect
development and flavour.

Let us next glance at the fruit trees in the garden itself.
Standards we see are not much grown ; they shade the
ground too much, and the crops are better when fully
exposed to sun and air.   In some places the culture of
bush and dwarf pyramidal trees is carried on successfully,
but in general it is so backward that nothing like a good
crop is gathered.   Besides, all dwarf closely pruned and
accurately trained trees require considerable expense and

time; and it is sheer folly to bestow these on kinds which will produce as good a result if grown as standard trees, requiring hardly any attention, and actually permitting of as good a crop of some things being gathered from under them as if the trees did not exist. Perhaps there may be a few espaliers in the garden; but they are usually so very few, and so very badly managed and ugly, that little fruit is got from them. I look forward to the time when the well trained espalier, on its cheap, neat, and permanent trellis of galvanized wire, will run along within a few feet of every garden walk; but little can be reaped from such as we have at present. It follows, then, that in private grounds there is as a rule no source from which an abundant stock of the better kinds of hardy fruit may be gathered.

Most of our fruits are wholesome and delicious food, or capable of being made so. They should be much more abundant than they are at present, and might form part of the daily meal of every Briton. But if the country gentleman, to whom the production of these fruits should be a matter of the greatest ease, does not lead the way, how are we to improve? The chief thing necessary is to plant an orchard, carefully choosing the site, and, above all things, selecting the very best kinds, all perfectly hardy, and such as ripen their fruit every year, be the season what it may. Such an orchard would be very convenient near the garden, and in fact might form part and parcel of it; but as the care required is nearly none, except the pleasant one of gathering the fruit, it would not matter much about its position. The first consideration should be the selection of the most suitable soil at the owner's disposal. Not an inch of space of the whole need be lost. All the trees should be allowed to grow as standards, and the crops to be gathered from them would soon put to shame the few dozens that are considered a wonderful crop on the wall or dwarf tree. All the wall, dwarf, and espalier trees might then be exclusively kinds that require some additional heat or attention, or that the shelter and support of the espalier and the cordon are an advantage to. As protection of some kind might be provided for most of these carefully trained

trees, it would of course be wise to include among them all
the sorts most liable to be injured by spring frosts. And
such kinds are so abundant that all the walls and espaliers
might well be devoted to them.

The apparent utility of such an orchard is so great that
to speak of its beauty can hardly be necessary; and yet we
question if those who ought to be most interested in the
matter have the least idea of this. It is difficult for those
who do not live in a good fruit growing or orcharding district to have any notion of what an ornamental as well as useful
thing a good orchard or fruit garden is. I have never anywhere seen in gardens of the usual type such a picture as I
did during the past year in the well managed orchards or fruit
gardens of a west London market gardener—one who devotes
about sixty acres to fruit culture. His groves of Cherries,
Pears, and Plums were superb—the Plum trees, densely laden
with their purple eggs, being as attractive from colour alone
as many ornamental plants are when in flower. The produce is enormous, compared to what we are accustomed to
from the garden managed in the ordinary way. Of course
such a scene is a garden in the best sense of the word. An
acre or two planted after this fashion would be productive
of more satisfaction than any other attempt at fruit growing,
though it is by far the most inexpensive of all.

The only points to be attended to as regards pruning,
are an occasional winter pruning to open them up to the
full influence of light, and a thinning of the fruit buds to
concentrate the energies of the trees, and thereby much improve the value of their produce. These operations performed once every second or third winter will do much good.
It is true that without them the pear may be profitably cultivated; but I know of one instance near London in which a
grower of the Pear as a standard, or orchard, tree on an extensive scale has doubled the market value of his fruit by well
thinning the buds and branchlets—operations which are
carried out in winter, when time can be most readily
spared for such work. If this were generally done by
orchardists it would lead to much improvement. The
orchard once planted it would not prove much addition to

the labour of the gardener, and the abundant crops might often save him from the grumblings that are sometimes known to accompany large garden expenditure and a scarcity of vegetables and fruits.

Having said so much in favour of good orchard culture it behoves me to give the names of the kinds of hardy fruit that do best as standard orchard trees :— *Pears :* Jersey Gratioli, Doyenné du Comice, Citron des Carmes, Jargonelle, Williams's Bon Chrétien, Aston Town, Beurré de Capïaumont, Louise Bonne of Jersey, Suffolk Thorn, Thompson's Pear, Beurré d'Amanlis, Swan's Egg, Croft Castle, Doyenné d'Eté, Comte de Lamy, Knight's Monarch, Althorpe Crassane, Marie Louise, and Beurré Superfin. *Apples :* Borovitsky, Early Harvest, Irish Peach, Joanneting, Summer Golden Pippin, Lord Suffield, Keswick Codlin, Adams's Pearmain, Blenheim Pippin, Cox's Orange Pippin, Early Nonpareil, Golden Pippin, Ribston Pippin, Sykehouse Russet, Bedfordshire Foundling, Hawthornden, Yorkshire Greening, Golden Noble, Court Pendu Plat, Golden Harvey, Sam Young, Sturmer Pippin, Beauty of Kent, Dumelow's Seedling, Royal Pearmain, Tower of Glammis, and Pitmaston Nonpareil. *Plums :* Pond's Seedling, Early Rivers, Orleans, Gisborne's, Victoria, Prince Englebert, and Damson. *Cherries :* May Duke, Early Prolific, Bigarreau, Late Duke, Knight's Early Black, Belle Agathe, Rival, and Mammoth. *Apricots* (for standard trees in the southern counties): Breda, Brussels, Turkey, and Moorpark. *Figs :* Black Ischia, Brown Ischia, Brown Turkey, and Courcourcelle Blanche. These would be better grown as shrubs, with low sweeping branches, and buried in the ground in winter to save them from the frosts, as the French do about Argenteuil. *Medlars :* The Nottingham is the best kind. *Nuts :* Lambert's Filbert (Kentish cob) is the best; Purple Filbert, Pearson's Prolific, and Cosford also good. Of the Quinces the Portugal is the best. The Berberry is rarely cultivated, though worth that trouble. Where the fruit is much in request, it would be a good plan to inclose the orchard with a dense hedge of this shrub. The stoneless variety is the best, but it is not easy to get the true kind.   The smaller fruits are so well known and abundantly grown that it is needless to speak of them.

Of the various waste spaces where good fruit might be grown the most conspicuous are the railway embankments. Here we have a space quite unused, and on which for hundreds of miles fruit trees may be planted, that will after a few years yield profit, and continue to do so for a long time with but little attention. I am not aware that any attempt has been made to cultivate fruit trees on these places in England; but learning that one had been instituted in France, I went to see the experiment which has been made for a distance of eight leagues or so along the line from Gretz to Colommiers—Chemin de Fer de l'Est.   The French see the great advantage of utilizing spots at present worthless in this way, and are beginning to work at it; but to all intents and purposes they are nearly as backward as ourselves.   It is true you now and then hear of somebody becoming a rentier by planting a barren mountain side with Cherries, but on the whole they have nearly as much to do as we have with regard to fruit culture in waste and profitless places.   However, they have commenced, and it is most likely the first trial will be a profitable one, though by no means so inexpensive as like ones might be made.

A cheap fence of galvanized wire runs on each side of the line, and on this Pear trees are trained so that their branches cross each other; and they are, though only in their fourth year, at the top of the fence. In some parts they are trained in like manner on the slender but very cheap and slight kind of wooden fence, so common in France. By training them in a way to cross and support each other, before the time the fence decays the trees are perfectly self-supporting, and form a very neat fence themselves. This is a plan well worth adopting in many gardens where neat dividing lines are desired. Judging from appearances, these trees will bear abundantly for many years to come.   But this, although something in the right direction, does not occupy more than a mere thread of the space on each side of the line, and I cannot but think that much more might be done on the remaining surface by planting small trees.   It would be a great point gained if we could have dwarf productive trees without having to go to expense for fixing or training them—if we could make them self-supporting, in fact.

It is quite possible to train espaliers of the choicest varieties of pears so that they shall be perfectly self-supporting, as shown by the figure, or in some like way. Established trees that I saw crosssed in this way were not allowed to get into a rough hedge-like condition, but, on the contrary, trained as neatly and perfectly as ever trees were on trellis or wall. No flaying of the branches re-sulted from their being inter-laced.

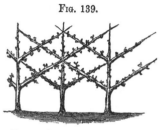

FIG. 139.

Young line of self-supporting Pear Trees.

A shoot was taken along the top so as to act as a finish and tend to hold all tighter, and the whole looked much firmer and neater than the ill-supported and ill-trained espaliers that one too often sees at home. Other examples of self-supporting espaliers are figured elsewhere in this book.

A mere line of trees, however, trained along a railway, will not effect the improvement we require. Why not plant pyramid or bush trees in such positions? Why not the Fig in the southern counties? By covering nearly all the sur-face of those sunny banks—in many cases of excellent soil—there would be enough work to do to make it necessary and profitable to have men in charge of comparatively short lengths of the line, and these men would be able to better protect the fruit. On the French railway in question the fence of fruit trees is carried along, no matter what the soil or situation. A more rational system would be to adopt the kind of tree to the soil, and simply take the more desirable spots at first.

CHAPTER XV.

THE first thing we have to settle is, What is a cordon? There has been some little discussion on this point—discussion that was utterly needless, and even mischievous, as tending to prevent the public knowing exactly what the term is used for. It simply means a tree confined to a single stem; that stem being furnished with spurs, or sometimes with little fruiting branches nailed in, as in the case of the peach when trained to one stem. Some contended that it meant any form of branch closely spurred in; but

FIG. 140.

The Apple trained as a Simple Horizontal Cordon, grafted on the French Paradise Stock, and in full bearing.

this is quite erroneous. The term is never applied to any form of tree but the small and simple stemmed ones. The French have no more need of the word to express a tree trained on the spur system than we have, and they have trained trees on that system for ages without ever calling them by this name. Before it was given to the forms of Apple and Pear and Peach-trees shown in this chapter, or rather before they came into use, it was chiefly applied to a mode of training plants horizontally—each plant resembling what we call the bilateral cordon. (See the engravings illustrating Vine culture at Thomery.) However, to settle the use of the term, I wrote to Professor Du Breuil, the leading professor of fruit culture in France. His reply was thus alluded to in the *Gardener's Chronicle* :—" What a vast proportion of controversy and dispute might be saved, would people only agree as to the meaning to be attached to words. Just now, as it appears to us, a great deal of unnecessary discussion is raised as to the word 'cordon.' A wrangle about words is about as satisfactory as an argument to prove a negative. It may serve, perhaps, to stop this futile wordy debate to give the opinion of M. Du Breuil himself on the matter. This renowned horticulturist, in a recently written letter, which has been submitted for our inspection, says that he applied the word 'cordon' to trees

Fig. 141.

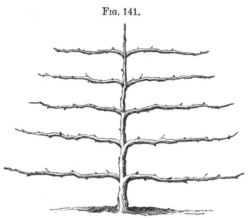

Tree with horizontal branches. This form, very commonly seen in our gardens, has been called a "cordon" by some writers, but has nothing whatever to do with that form.

consisting of a single branch, bearing fruit-spurs only, and never allowed to ramify. When there are two such branches, M. Du Breuil applies the expression 'double cordon.' In order to be quite accurate, we subjoin M. Du Breuil's letter verbatim et literatim :—

'Le mot 'cordon' dérive en français de cord ; j'ai employé cette expression pour désigner les formes d'arbres dont la charpente se compose seulement d'une seule branche qui ne porte que des rameaux à fruit.' "

Professor Du Breuil states distinctly that, struck with the long period it took to cover a wall by means of the

larger forms of trees, he adopted those quick-rising simple-
stemmed kinds to cover the walls rapidly and give an early
return.   Now it is clear that if we call a fan, or horizontally trained tree, a "cordon," we not only misapply the
term, but prevent the inventor's very clear idea from being
understood.   Notwithstanding this, some persons have
actually figured the old forms of fruit trees common in
our gardens for ages and called them cordons.   To show
how erroneous is the impression that the term applies to
any kind of tree with the branches closely pinched in, I
have merely to state that the cordon Peach trees in French
gardens are not pinched in in this way, but have the
wood regularly nailed in, just the same as the common Peach
trees on our garden walls.   However, the figures in this
chapter will give a correct idea of what the cordon system is.

A simple galvanized wire is attached to a strong oak post
or rod of iron, so firmly fixed that the strain of the wire may

Fig. 142.

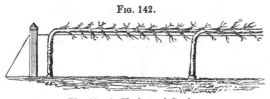

The Simple Horizontal Cordon.

not disturb it.   The wire is supported at a distance of one
foot from the ground, and tightened by one of the handy
little implements described elsewhere in this volume.   The
raidisseur will tighten several hundred feet of the wire,
which need not be thicker than strong twine, and of the
same sort as that recommended for walls and espaliers.
The galvanized wire known as No. 14 is the most suitable
for general use.   At intervals a support is placed under
the wire in the form of a piece of thick wire with an
eye in it, and on the wire the Apple on the French paradise
is trained, thus forming the simplest and best and commonest kind of cordon, and the one so extensively employed
for making edgings around the squares in kitchen and fruit-
gardens.

Cordons are trained against walls, espaliers, and in many ways, but the most popular form of all, and the best and most useful, is the little line of Apple trees acting as an edging to the quarters in the kitchen and fruit garden. By selecting good kinds and training them in this way abundance of the finest fruit may be grown without having any of the large trees or those of any other form in the garden to shade or occupy its surface. The bilateral cordon is useful for the same purposes as the simple one, and especially adapted to the bottoms of walls, bare spaces between the fruit trees, the fronts of pits, or any low naked wall with a warm exposure. As in many cases the lower parts of walls in gardens are quite naked, this form of cordon offers an opportunity for covering them with what will yield a certain and valuable return. It is by this method that the finest coloured, largest, and best French Apples sold in Covent-garden and in the Paris fruit shops at such high prices are grown. I

FIG. 143.

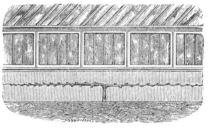

The Cordon on low sunny wall of plant-house. In this way Calville Blanc, Reinette du Canada, the Lady Apple, Melon - Apple, American-Mother, Newtown Pippin, and all the finer and tenderer French, American, and British apples may be grown to perfection.

have seen them this year in Covent-garden and in Regent-street marked two and three shillings each, and M. Lepère fils, of Montreuil, told me when with him last summer that they have there obtained four francs each for the best fruit of the Calville to send to St. Petersburg, where they are sold in winter for as much as eight francs each ! Why should we have to buy these from the French at such a high rate ? Considering the enormous number of walled gardens there are in this country, there can be no doubt whatever that by merely covering, by means of this plan, the lower parts of walls now entirely naked and useless, we could supply half a dozen markets like Covent-garden with the very

z

choice fruit referred to, and be entirely independent of the French.

Doubtless many think that these very fine fruit require a warmer climate than we have for them. But by treating them as the French do we may produce as good or a better result, and may, in addition, grow tender but fine apples, like the Calville Blanc, that do little good when grown as standards. The climate in most parts of England will be found to suit them quite as well as that of Paris, if not better, because the sun in France is in some parts a little too strong for the perfect development of the flesh and flavour of the apple. There is no part of the country in which the low cordon will not be found a most useful addition to the garden—that is, wherever first-rate and handsome dessert fruit is a want. So great is the demand

Fig. 144.

Young Cordon of the Lady Apple trained as an Edging.

in the markets for fruit of the highest quality that some-times the little trees more than pay for themselves the first year after being planted. In any northern exposed and cold places where choice apples do not ripen well it would be desirable to give the trees as warm and sunny a position as possible, while the form recommended for walls should be used extensively. In no case should the system be tried except as a garden one—an improved method of orcharding being what we want for kitchen fruit, and for the supply of the markets at a cheap rate.

When lines of cordons are perfectly well furnished the whole line is a thick mass of bold spurs. Some keep them very closely pinched in to the rod, but the best I have ever seen were allowed a rather free development of spurs, care being taken that they were regularly and densely produced along the stem. If anybody will reflect that as a rule the full vigour of the ordinary espalier tree flows to its upper line of branches, he will have no difficulty in seeing at a glance the advantages of the horizontal cordon, particularly if he bears in mind that the system as generally applied to the apple is simply a bringing of one good branch near the earth, where it receives more heat, where it causes no injurious shade, and where it may be protected with the greatest efficiency and the least amount of trouble.

The system is simply an extension of the best principles of pruning — a wise bending of the young tree to the conditions that best suit it in our northern climate. The fact that by its means we bring all the fruit and leaves to within ten inches or a foot of the ground, thereby exposing them to an increase of heat, which compensates to a

FIG. 145.

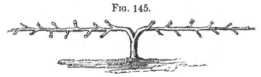

The Bilateral Cordon.

great extent for a bad climate, will surely prove a strong argument in its favour to every intelligent person.

The form is so definite and so simple that anybody may attend to it, and direct the energies of the little trees to a perfect end, with much less trouble than is requisite to form a presentable pyramid or bush. It does not, like other forms, shade anything; and beneath the very line of cordons you may grow a slight crop. They are less trouble to support than either pyramid or bush; always under the eye for thinning, stopping, and pruning; easy of protection, if that be desired; and very cheap in the first instance.

A few words are necessary as to the best method of planting and managing the Apple trained and planted around the quarters or borders. In a garden in which particular neatness is desirable it would be better to plant them within whatever edging be used for the walks; but in the rough kitchen or fruit garden they may be used as edgings. The

z 2

reason for supporting the cordon at one foot from the surface
is to prevent the fruit getting soiled by earthy splashings.   By having something planted underneath which
would prevent this, we might bring the cordon lower down;
but, though I have thought of several things likely to do
this, none of them are very satisfactory.   Doubtless, however,
we shall yet find something that may be cultivated with profit immediately under the cordon so as to prevent splashings,
and thus be able to bring it within six inches of the earth.
In gardens where it would not be suitable as an edging,
the best way would be to plant it ten inches within the
Box or whatever kind of edging was employed.   In planting, keep the union of stock and scion just above the
surface of the ground, to prevent the Apple grafted on the

Fig. 146.

Reinette du Canada trained as a Cordon.

Paradise from emitting its own roots, and consequently
becoming useless for such a mode of training.   The trees
should never be fixed down to wire or wall immediately after
being planted; but allowed to grow erect during the winter
months, and until the sap is moving in them, when they
may be tied down.   Some allow them to grow erect a year
in position before tying them down.   They should in
all cases be allowed to settle well into the ground
before being tied to anything.   For general plantings, the
best and cheapest kinds of plants to get are those known as
" maidens," *i.e.*, erect growing trees about a year from the
bud or graft.   These can be readily trained down to the
wire, or to the wall, in spring.   In training the young tree,
the point with its young growing shoot of the current year

should always be allowed to grow somewhat erect, so that the
sap will flow equably through the plant, drawn on by the
rising shoot at its end. To allow gross shoots to rise at any
other parts of the tree is to spoil all prospect of success.
If the tree does not break regularly into buds, it must
be forced to improve by making incisions before dormant
eyes.

A chief point is not to pinch too closely or too soon.
The first stopping of the year is the most important one,
and the first shoots should not be pinched in too soon ;
but when the wood at their base is a little firm, so that the

Fig. 147.

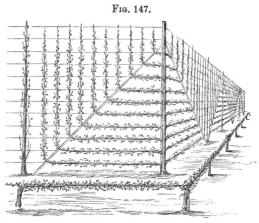

Edging of Simple Cordons three years old in French fruit garden.

lower eyes at the bases of the leaves may not break soon after
the operation. Stop the shoot at five or six leaves, as the
object is not to have a mere stick for the cordon, but a dense
bushy array of fruit spurs quite a foot or more in diameter,
when the leaves are on in summer. All the after pinching
of the year may be shorter, and as the object is to regularly
furnish the line, the observant trainer will vary his tactics
to secure that end—in one place he will have to repress
vigour, in another encourage it. About three general
stoppings during the summer will suffice, but at all times
when a strong soft "water shoot" shows itself well above
the mass of fruitful ones, it should be pinched in, though

not too closely. I have even in nurseries seen things called
" cordons" with every shoot allowed to rise up like a willow
wand—utterly neglected and on the wrong stock; and I
have in other cases seen them so pinched in as to be worthless sticks. Of course success could not be expected under
the circumstances; and I must caution the reader against
taking such things as examples of the cordon system, or
placing any reliance on the opinions of their producers.

As the Paradise keeps its roots quite near the surface of
the ground, spreading an inch or two of half decomposed
manure over the ground, or in gardening language mulching
it, could not fail
to be beneficial.
The galvanized
wire support (No.
14) is neatest and
cheapest, and in
fact, the only one
that should be used.

Fig. 148.

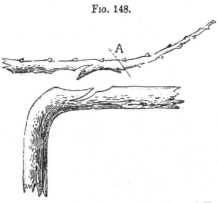

The cordons are
usually planted too
close together in
France. In Decem-
ber last I finished
an experimental
plantation of 500
at six feet apart,

Grafting by approach, to unite the points of Horizontal Cordons. The apex of each Cordon is cut at
A, when firmly united to its neighbour.

but anticipate taking every second one up after a year or two.
When the cordons overtake each other it is common to graft
them one to another—a very simple operation. If when
all are united they should grow too strong in rich ground,
the stem of every second plant may be cut off just beneath the
wire and the trees will be nourished by the others. When
the line is well trained and established the wire may be taken
away altogether; but it is so very inexpensive that it is
scarcely worth while removing it. If the plantation be
made on a slope all the trees should be planted so as to
grow up the incline.

Finally, in winter, the trees will be the better for being

looked over with a view to a little pruning here and there ; taking care to thin and regulate the spurs when the plantation is thoroughly established, to cut in objec-tionable stumps, and to firmly tie the shoots along the wire. These should never be tied tight-ly, so as to prevent their free expansion ; but they may be tied firmly with-out incurring any such danger.

As the system is chiefly valuable for the production of superb dessert fruit, only the finest kinds should be selected ; but, as some apples are of high value both for kitchen and dessert, some of the finer kitchen apples are included in the following list of such as will be found very suitable : Reinette du Canada, Reinette du Ca-nada Grise, Reinette Grise, Reinette de Caux, Reinette d'Espagne, Rei-nette très Tardive, Belle Dubois, Pomme d'Api, Mela Carla, Calville St. Sauveur, Coe's Golden Drops, Newtown Pippin, Calville Blanc, Northern Spy, Melon-Apple, Cox's Orange Pippin, Duke of Devonshire, Kerry Pippin, Lodgemore Nonpareil, White Nonpareil, American-Mother, Early Harvest, Lord Burleigh, Beauty of Kent, Bedfordshire Foundling, Lord Suffield, Cox's Pomona, Hawthornden, Tower of Glammis,

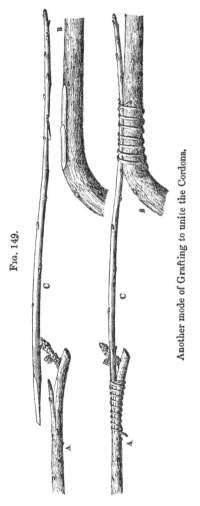

Fig. 149.

Another mode of Grafting to unite the Cordons.

Winter Hawthornden, Betty Geeson, and Small's Admirable. Some of the best of the above are valuable keeping apples. I have seen the Reinette Grise in fine condition in the markets at Rouen in June, and Reinette très Tardive is good in July. Those who wish to plant good early apples might try Borovitsky, and a few of the best early kinds; but it is best to devote most of our horizontal cordons to the growth of the finer, later, and most valuable fruits. Of the above selection the Calville Blanc, the Reinette du Canada, and Mela Carla must be grown on a warm wall; Newtown Pippin, The Mother, Melon, and several of the other later and finer apples will also be grateful for the same protection.

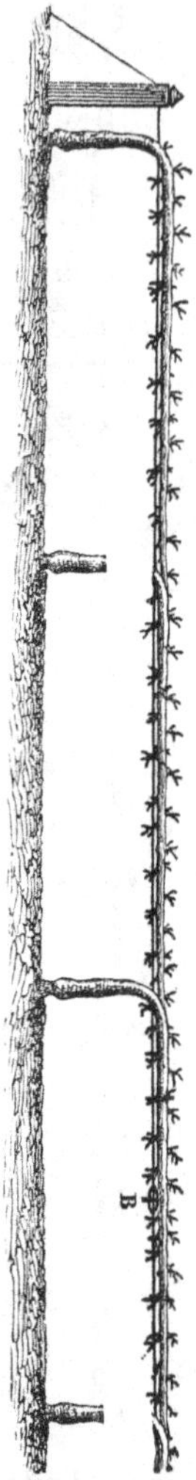

As our stock of apples on the Paradise are at present chiefly obtained from France, there seems little chance of our securing the finer English kinds for a while. But more than one of our nurserymen have assured me of their intention to plant and graft this stock largely, and I should advise all large fruit nurserymen to do so. They should be offered as cheaply as possible by the hundred, as they will be required in greater numbers for planting than any other form. Of the kinds of French apples that may be had grafted on the Paradise abundantly in nurseries all over France the following are among the best:—Calville Blanc, Reinette du Canada, R. d'Angleterre, R. Grise, R. de Caux, R. très

The Horizontal Cordon trained as an edging Originally the trees represented here were planted too thickly, and after all had been securely grafted together, every second stem was severed. B shows the position of the Raidisseur.

Fig. 150.

Tardive, R. de Bretagne, R. d'Espagne, Pomme d'Api, Belle Dubois, Belle Joséphine, Calville St. Sauveur.   Doubtless ere long we shall have an abundant stock of the best English kinds on the right stock.   As a great number of trees are required for this mode of planting ; as the apple on the Paradise occupies but a small space in nurseries compared to other trees, and as it is very likely there will before long be a large trade in this form of tree, it is to be hoped that our nurserymen will offer suitable kinds at ·a very low rate by the dozen, score, or hundred, as is the case in France.   I have no hope of the perfect and general success of the system till this is done.

I have recently received the following on the raising of the Apple on the Paradise stock in nurseries from my friend M. Jean Durand, of the well-known fruit tree nurseries at Bourg-la-Reine, near Paris, and have much pleasure in giving it, particularly as it is desirable that our nurserymen, and even in some cases private growers, should raise it for themselves :—

" If the Apple tree is to be grown in the form of the horizontal cordon, it must be grafted on the variety known to horticulturists as the Paradise.   This variety, which loves a fresh, damp, clayey soil, cannot be grown from seed, but must be propagated by means of layers or cuttings, which are obtained in the following manner :—Having chosen soil of the proper description, it must be well dug and manured.   Trenches, six inches deep and a yard apart, are then opened, and the stocks, which have been procured previously, planted in them.   They should be pruned down to twelve or fourteen inches in height, and placed in the trenches at a distance of four inches apart, and in such a way that about six inches of the top appears above the ground.   The trenches are then filled in and the ground levelled.

" In the following spring, as soon as there is no longer any danger from frost, the stocks are cut down level with the ground.   The object of this operation is to develope a number of shoots : these are earthed up about June or July by covering them with a small quantity of earth taken from the trench on each side of the line of plants, so as to cover their bases to the depth of four inches or so.

" In the following November these buds will have taken root, the plants from which they take their origin will consequently be alluded to in future as old stools, and will give every year a certain number of young plants.

" Every year during the month of November the young plants should be stripped from these old stools. It is necessary above all during the first year to use a strong sécateur for taking them off in order not to injure the stools; later on they may be simply broken off. Immediately after this operation the wounds left in the trees should be covered over with earth. They will perform the same service for a great length of time—from five to ten years, according to the care taken of them—and the young plants thus obtained will serve for grafting in the nursery.

" For this purpose the ground which is destined to receive them should be well dug and then divided in lines distant from each other two feet or two feet six inches. The operation of planting in beds requires great care. The young plants should be well trimmed both at top and bottom, so as to give the branches a uniform length of sixteen inches. They should then be planted sixteen or twenty inches apart and three or four inches deep, and the ground hoed frequently until the month of August, so as to destroy the weeds and break up the ground. At this period the young plants are ready for budding, each subject receiving a bud at about four inches above the surface of the ground. Immediately afterwards, particularly in dry years, it will be well to give them a good hoeing to prevent the ground from caking together, and to preserve it in properly moist condition.

" During the winter the young plants that have been budded should be stripped of all the shoots that have grown on them to within a height of three inches above the bud, and the plant itself should be pruned down to this height. The following spring a certain number of small shoots will make their appearance all over the pruned plant. When they have reached a length of an inch or an inch and a

half, they should be pinched down to favour the growth of
the graft which will have grown as large as the other buds.
The bud which has not been pinched will naturally soon
surpass in size those which have been.   From this time
it will attain sufficient strength to absorb the sap, and
it will be necessary to destroy all useless shoots. This little
pruning operation requires considerable caution, and is performed by the aid of the knife.   Care should be taken not
to confound the graft with the other shoots on the plant.
The stem of the stock above the bud being thus deprived of
its shoots, serves as a stake to which to tie the young growing
scion. Tied thus loosely, it is preserved from accident during
its growth.   This natural stake having served its purpose is
cut away at the end of the year, and the graft having attained its proper size is ready for sale as the scion or graft
of one year, and may be trained into any form the grower
may think desirable.

.'' The Apple thus grafted on the Paradise is, as is well
known, a great success throughout all parts of France and
the adjoining countries. _ In proportion to the space it
occupies, it furnishes a great quantity of the finest fruit.
It is not rare to count seventy or eighty apples upon a
little tree whose arms together are not more than seven feet
long. This form is due to M. J. L. Jamin, of Bourg-la-
Reine. This nurseryman used to sell dwarf fruit trees of all
kinds in pots in the Paris market, and amongst them the
now well known cordon. The form was much appreciated
and promptly spread abroad, and after having had some
success at a horticultural exhibition held at the Louvre, it
was definitely adopted in kitchen and fruit gardens under
the name of the cordon horizontale.

" To establish the growth of cordons in the nursery a
line of galvanized iron wire is stretched along the ground
at fourteen inches from the surface, and firmly fixed at
each end. The young trees are then cut down nearly to
the level of the wire, and when they start in spring two
opposite buds are chosen for the formation of the two arms,
and allowed to grow during the summer, the buds on the
stem below these being pinched within an inch or so of

their base. During the summer the two shoots ought to form a strong cordon fit for sale in the following winter. The simple cordon with one branch is formed in the same manner, except that one bud only is allowed to grow. There are many other methods of procuring these forms, but I like the one above indicated better than that of bending the shoot directly to the wire either in the first year of plantation or the second year, as recommended by some professors."

Since the discussion which took place in the *Times* and the gardening journals concerning the merits of this mode of apple-growing, and since the first part of this chapter was written, I have seen miles of cordons on the Paradise in many parts of France and in Switzerland, and sought everywhere to ascertain its merits and defects; and the result is that I am more than ever satisfied of its great value. Before stating my reasons, let us first devote a few minutes to the more important objections to the system. I have been active in proclaiming its merits. It should be equally well known that some of the most distinguished horticulturists in the country have condemned it. Some have considered that a late frosty season would be fatal to low cordons, and that our climate is too conducive to the growth of gross wood, whereas on the Continent it becomes ripe and stubby, and the trees may be preserved within bounds. If such were the fact, this objectionable tendency should be more developed in the warm parts of north-western France near the sea coast; but this is not the case. Grafted on the true French Paradise stock, the tree is always good, and keeps perfectly within bounds in parts of France and Belgium as cold as southern England. In a note from Professor Morren, of Liége, he says : " The culture of the Apple as a cordon on the Paradise stock has been extensively tried in this country, and is rapidly extending, particularly near Liége, Verviers, Huy, Namur, and in all the Flemish districts. Hedges of horizontal cordon Apples and of Pears are now formed along the sides of the railway between Brussels and Louvain. These plantations were made two years ago, and have proved very successful. The fruits attain a considerable size, and the experiment promises so well that preparations are being made to greatly extend it." Is there magic in the air, that there should be so much difference in the behaviour of trees separated by a few miles of sea? In many continental districts where frosts are quite as severe as here, the cordons escape yearly without injury; and besides, no form of tree is so easily protected in spring, it being so very low.

One distinguished horticulturist attacked the system by declaring that he had tried it alongside of grass walks; that the shoots grew as big as broom-handles, and the slugs ate any fruit that happened to occur in such unlikely covert—one can hardly call it fruit wood. But in this case the error is clear. He planted a Crab or a Doucin stock, which grew too much, and which it is perfect folly to plant in the hope of having a satisfactory result as a horizontal cordon. The shoots from trees grafted on the Paradise stock never grow as above described, and may be kept within bounds with very ordinary attention.

In addition to the objections above stated, some are good enough to observe that the cordons may, under certain circumstances, be desirable for amateurs, but that practically they are to be regarded as toys. If, as I believe, they will supplant our present mode of cultivating the Apple as a standard, half-standard, pyramid and bush tree, they will prove toys only in the sense in which a guinea is a toy compared with a penny piece. I have urged the advantages of improved orchard culture so much that it is needless to renew my commendation of it here; what I admire in the horizontal cordon is that it is the simplest mode of doing away with the gouty old Apple trees which now in multitudes of cases shade our gardens and haunt them with ugliness. Moreover, as people rarely let them have their own way as when grown in orchards, they form a lasting puzzle to the pruner, who, in cutting them in annually, merely makes them uglier, more vigorous, and less useful.

As to my reasons for being more than ever convinced of the merits of the system I advocate, after hearing all arguments against it, and travelling many hundreds of miles to have full opportunities of studying it, I would urge :—

1. The fruit is larger and finer than that borne on any large form of tree.—2. The tree comes into bearing much earlier—in fact, often bears freely the second summer after being grafted.—3. The growth is dwarfer and much more compact than that of the apple on any other stock, and the tree may, without root pruning, be kept in a more compact and fruitful form than Apples on the Crab may be with that troublesome attention.—4. The fruit being held at an average of one foot from the ground it is in consequence benefited by a greater degree of heat; and from the compactness of the form the leaves and wood enjoy a greater amount of sun than is the case with high trees: it need not be said that these are great advantages.—5. The tree being confined to a single stem, and stubby fruit-spurs held near the surface of the ground, there is in consequence no injury to the fruit from wind or the swaying about of branches; besides, the fruit, if it does fall, is not injured.— 6. The trees may be more readily protected than any other form whatever, should protection be considered necessary.— 7. They may be more easily attended to in pinching, pruning, and thinning the fruit, and the desired shape attained more readily than any other form of trained tree.—8. Being little taller than a neglected Box edging, they shade no garden crop.—9. They take up but little space, and the positions best suited to them are those that hitherto have been made little or no use of.—10. They will enable us entirely to do away with the ugly and gouty old Apple trees now so common in gardens.—11. The apple on the French Paradise grows to its highest perfection on stiff loamy and clayey and wet soils, those which are often most inimical to fruit culture in these islands.—12. By planting it against low walls we may grow for ourselves the fine winter apples now supplied to the capitals of Europe from northern France and from America.

The reader, knowing my views on this system—knowing perhaps the contemptuous opinions which many persons hold of the cordon in England, and understanding that it is highly appreciated by the best judges on the Continent, may like to learn how its merits are appreciated by one or two English horticulturists who have had some opportunity of examining its worth. The first opinion quoted is that of one of the most able and experienced horticulturists in England :—

"As Mr. Robinson's writings induced me to see for myself the horizontal cordons on the true Paradise stock, my opinion may be as useful on this question as that of some who have never seen them at all, or have not got beyond the erroneous idea that they are espaliers with one branch, being ignorant of, or ignoring the fact, that the stock is entirely different. I have seen a good many espaliers in my time, but never one that bore a crop like those little cordons that I saw at Ferrières, Versailles, and amongst the French fruit growers. The espalier on the Crab stock, no matter how big and ugly was the trellis you put it upon, was always with difficulty kept within bounds, always pushing its vigour to the top branch, whereas the little trees I saw in France growing on very stiff moist loams, were in the stubbiest and neatest condition that could be desired; and everywhere I was told that they were scarcely any trouble, a little pinching now and then, and some attention to see that the spurs were equally distributed along the line, being all that was required. Why, the trouble is worth incurring for the sake of having such a pretty garniture to our walks in spring and autumn, even if the great Apples were of wood, and not of the finest flavour. The pinching and training would be pleasant employment for ladies and young folks, in their few hours' garden rambles, affording both profitable and amusing exercise. So many tortured forms of trees have been presented to the public that I do not wonder at those rejecting them who cannot see the undeniable merits which have been claimed for these cordons; but when once they are seen well done, and in working order (we cannot expect they will be in England for a year or two), everybody interested in a garden will be charmed with them, and the plan will, I venture to say, be adopted in the largest as well as the smallest gardens in the land. Every operation connected with the culture of these trees will be agreeable in consequence of its simplicity; and it will be a pleasure to have the little trees under the eye, from the unfolding of the rosy buds in spring to the gathering of the fine fruit in autumn. It is to me very surprising that none of our great fruit growers, pomologists, and others, who are, I believe, in the habit of travelling in France every year, and some of them for the past thirty years, did not spy out and introduce this system long ago, and more surprising still, that it is but recently that we have learned from Mr. Robinson the real value and nature of the stock (others who have mentioned it have always recommended the Doucin or English Paradise), and no doubt but for his exposition of the matter, we might have gone on for many years without knowing anything of value about it, as we have already lived without such knowledge for many years, notwithstanding the proximity of the fruit gardens of northern France and southern England, and the abundant intercourse between the two countries. We have brick and tile edgings in all sorts of fancy forms, pebble, stone, slate, and wooden edgings, also Grass, Box, Thrift, and many other living edgings; but when once fairly understood, the little edging of choice Apple trees will prove the most popular, profitable, and useful of them all for the fruit or kitchen garden. Apart from edgings, the plan of planting the cordon on the ends, fronts, and low walls of plant-pits, and glass houses, low walls and fences, small vacancies or spaces between fruit trees on walls of any aspect—indeed, on any kind of blank space on walls—is another distinct improvement; and, when we have it in full operation, the specimens of the finer and tenderer fruits grown on this method will be such as we have not yet had the pleasure of producing in this country."—Mr. JAMES BARNES, of Bicton, in the *Gardener's Chronicle*, February 27, 1869.

The next is that of one who has had excellent opportunities of studying fruit culture in the country round Geneva.

"M. Vaucher, the President of our Horticultural Society, began fruit growing at Chatellaine, a mile from Geneva, three years ago. Knowing that he had made large plantations of the horizontal cordons, I paid him a visit early in July with a view of ascertaining their condition. The plantation is not more than three years old, the garden having been a grass field three years ago. In entering the garden the first things that catch the eye are the very neat lines of these little trees running around the borders, and at about one foot from the margin of the gravel walk. The space between the cordons and gravel is planted with the finer kinds of Strawberries. The borders margining each square are cut off from the body of the square by alleys, and these are also edged by cordons in the same way. In most cases two lines of cordons are employed, one above the other,—the fruit of the lower line sometimes coming within three and four inches of the ground. The effect of the whole is neat, and such as would make a tasteful gardener use them for edgings, even if the result they yield be ever so problematical. But as regards the Apple, with ordinary care there is nothing problematical about it, for the most dense crops already adorn these beautiful little trees. Here are my notes and measurements of a few of them:—Calville d'Hiver, eighteen inches from the ground, seven feet six inches long, thirty-seven fine promising fruit; the same kind, seven feet long, seven inches from the ground, twenty-four fine fruit; Pepin d'Angleterre, six feet long, the fruit fifty-seven in number, hanging at an average of fifteen inches from the ground; Reinette d'Espagne, three feet six inches long, twenty-four fruit; the Lady Apple, six feet long, 110 fruit. These were some of the best examples I saw; and I need not remind your readers that the fruit, instead of being too thin, is much too thick. I may safely say that if properly thinned as fine fruit as ever grew will be gathered from these young cordons—so neat to look at, and at the same time such a luxury and profit. I particularly observed that the fruits on the lower line of wire, at an average of about six inches from the ground, were quite as good and fertile as those on the upper wire, at an average height of about eighteen inches—although, perhaps, at some disadvantage from being exactly under the higher line. To the above I may add, that there are many gardens about Geneva in which these cordons on the Paradise are a perfect success, that they give little trouble to the gardeners, who are always fond of them, be the gardens or the 'help' large or small."—J. A. WATSON, Château Lammermoor, Geneva, Switzerland, in a letter to the *Gardener's Chronicle*.

It is not merely in the way it is at present practised in France or elsewhere that the cordon system is interesting and instructive to all taking an interest in the culture of hardy fruits. To me it seems to offer a means of training trees so that we may readily give them that protection in spring, the want of which is in nine cases out of ten the cause of all our failures in fruit crops. Hitherto the best course to pursue with the borders along our fruit walls has always been a disputed point: some contending that they ought not to be cropped at all; others that salads and small vegetables might be grown upon them.

Let us crop the borders with trees trained on the horizontal cordon principle as suggested in the accompanying figures, and in this way dispose of the much debated question as to what is best to do with the fruit borders. By so doing we should collect such a valuable lot of fruit trees immediately in front of each wall as would render it convenient and highly desirable to protect efficiently both walls and borders, and by the same means. The low cordons will no more shade the wall than a crop of small salading, will prevent all necessity for disturbing the border, and will utilize every inch of its space. Indeed, I can conceive of no greater improvement in our fruit culture than devoting to fruit trees those excellent sunny borders that usually lie at the foot of our fruit walls. By this plan we should, it is true, sacrifice some of the more suitable spots for our early vegetables and salads, but we should gain very much more, and the change would be in every way conducive to the beauty and utility of our gardens. When the wall trees are being attended to the cordons cannot be forgotten, and the whole will be under the eye at a glance.

Fig. 151.   Fig. 152.

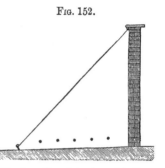

Narrow border in front of fruit wall, with two lines of horizontal cordons, protected in spring by wide temporary coping and rough canvas.

Peach wall and border, with five lines of cordons, the whole protected in spring.

Fig. 153.

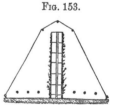

Double espalier of pears, with three lines of cordon apples on each side, the whole to be protected in spring as suggested in the illustration.

The Pear may be grown thus, and the Apple to the highest degree of perfection; so much so that I have no doubt whatever that the splendid Apples which may be grown in this way would, if put to the market test, more than pay for the expense of protecting cordons and wall trees at the same time, by means of the plan shown above. Other

A A

fruits will probably be found to submit to this mode of culture as well as these, and all sorts should be tried by those with opportunities for making experiments in fruit culture, kinds of a fertile and compact habit being selected for trial. Should we in time find varieties of our other hardy fruits conform as readily to the cordon system as the Apple on the Paradise, this way of covering borders as well as walls with fruit trees will prove a gain in the culture of our choice hardy fruits, the importance of which it would be difficult to over-estimate.

Efficiently protect borders and walls from the time of flower-

FIG. 154.

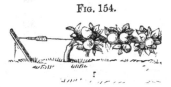

Simple wooden support for Cordon, the wire attached to a stone in the ground.

FIG. 155.

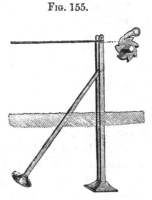

Iron support, with ratchet wheel at the top.

FIG. 156.

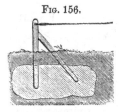

Iron support let into stone.

ing till the fruit is beyond all danger, afterwards expose all to the refreshing summer rains, and then there will be an end to all but mere routine work till the protecting season comes again. Every hundred feet in length of such well protected wall and border would be equivalent to a well-managed orchard house; and how attractive the borders would be considered from an ornamental point of view! The fact of the borders being thus covered with fruit trees will make it almost imperative to protect the wall and border at the same time; and without efficient protection at flowering time, we can hope for but very little success with the finer hardy fruits in this country.

Although the cordon is so simply supported, it is desirable to know the best means of doing so in a permanent and ready way. The simplest way of all is to drive a tough post in the ground to the required height, and attach the wire to it. The post may be driven in obliquely, as in fig. 154, or erect, as is the custom. An iron support with a stay let into a block of rough stone, as suggested in fig. 156, would be as satisfactory as any other, because so permanent. The strongest, best, and dearest kind I have ever seen was in M. F. Jamin's garden at Bourg-la-Reine ; fig. 155 is a representation of it. The stay is bolted between two vertical irons, and the galvanized wire strained between them at the top.

THE PARADISE, DOUCIN, AND CRAB STOCKS.—The stock is to the cultivated fruit tree as important as the foundation is to the bridge ; if we have not the right stock, all is wrong. The Paradise stock (the French Paradise stock, it must be remembered—for the term " English Paradise" is a misnomer) is the only one that should be used to form the cordon, except on the very poorest and driest of soils. Of quite an opposite opinion, Mr. Thomas Rivers, our great authority on fruit growing, says :—" It is exceedingly dwarf in its habit, and too tender for this climate, unless in very warm and dry soils !" But in fact it is as hardy as the hardiest tree of the forest, not perishing even if thrown with its roots exposed on the surface of the ground, and allowed to remain there through a rigorous winter ; and the soils above all for which it is peculiarly unfitted are those that are hot and very dry, while it flourishes with the most satisfactory vigour, fertility, and health on rich, moist loams, and even bad clays—the very soils which often present the greatest amount of difficulty to the British fruit grower. As will be readily seen, this is simply a matter for experiment, and I appeal to the horticulturists of Britain to settle the question by direct trial, a thing they can so readily do. The " English Paradise " recommended by Mr. Rivers for this purpose is, according to his own statement, the Doucin

—one that as regards vigour is intermediate between the Crab and the Paradise, well fitted for neat standards, pyramids, and large bushes, but growing too vigorously to furnish anything but disappointment if planted as a low cordon, except on very light calcareous or " burning " soils. To plant the Apple on the common or Crab stock, and expect to form a dwarf fertile tree, is simply folly.   By mutilation and removals you may secure a crop, and keep the Doucin or "English Paradise" within bounds ; but what we want is a stock that will furnish a dwarf and fertile growth, without any root-pruning or attention whatever, beyond that of pinching in the shoots two or three times in summer, according to their luxuriance. This we have exactly in the Paradise stock, grown by millions in the nurseries around Paris, and in many other parts of France.

We have next to determine what is this Paradise stock. It need scarcely be said that a plant like this, which exerts so marked an influence on the trees grafted on it, and is so truly valuable for our gardens, deserves to be at least as well known as any one kind of fruit, however good.   Yet this is so far from being the case that but very little is known about it.   To most of the French botanists its origin is involved in obscurity.   I failed to find perfect fruit or flowers in any garden in the neighbourhood of Paris or London, but have had some young trees of the Paradise and Doucin planted with a view of allowing them to fruit.

As regards the origin of the trees, apparently the clearest account is that of Professor Koch of Berlin, who has paid a great deal of attention to the origin of all our fruit trees. He says :—" The name Malus paradisiaca appears to have been first used by Ruellius in the year 1537.   It is a native of South-Eastern Russia, Caucasus, Tartary, and the Altai Mountains.   I have often seen this shrub in the Caucasus, and near the Don and the Volga, where it forms shrubs and dwarf trees, frequently accompanied with suckers."

Without attempting to throw any light on the origin of the Paradise, M. Carrière of the Jardin des Plantes has studied its characteristics, compared them with those of the Doucin, and described both in the Flore des Serres :—

PARADISE.

"Roots much ramified and tidy, short, remaining near the surface, and never tap-rooted. Shrub, bush-like, much branched, the branchlets rather long, and with a lateral tendency, the adults covered with a smooth bark of a reddish colour; lightly pubescent in the case of the young shoots. Leaves lanceolate, elliptical, of a light green above and velvety beneath, finely denticulated, acuminate at the ends, but principally at the base. Petiole broadish and channelled. Calyx, with divisions acuminated and recurved, often contorted, as long as the peduncle. Petals straightly elongated at the base, faintly keeled, borne on a thin base, prolonged into a sort of keel. Ovary on a slender base, pubescent. Fruit higher than broad, lightly ribbed, skin white, flesh sweetish, almost insipid; ripening in July." It flowers more abundantly, and eight days earlier, than the Doucin.

DOUCIN.

" Roots rather long and strong, tap-rooted. Tree not much ramified, straight in its growth, with branchlets short, large, in adult specimens covered with a deep dull brown bark; very tomentose, and whitened in the case of the young shoots. Leaves broadly oval or nearly oboval, lightly blistered, shining on the upper and pubescent on the lower surface, rather broadly denticulated, scarcely acuminate at the apex, abruptly contracted and round at the base. Petiole broad, scarcely channelled. Calyx with divisions usually horizontal, occasionally recurved, rather large. Petals sub-oval, nearly blistered, keeled, borne on a base short and rather broad. Ovary on a stout support, covered with a tomentose down, white and thick. Fruit depressed, broader than high, not ribbed, the skin of an intense green, marked here and there with brownish spots: flesh of a high and agreeable flavour; ripening in August."

The Paradise stock has been known in France for between 200 and 300 years. The Doucin would appear to be not quite so ancient, but has been known for at least 160 years. It is used to form neat low trees, pyramids, wall, espalier, and even standard trees less vigorous and more suitable for gardens than those grafted on the Crab, or commonest stock, and occasionally for cordons on bad and very poor and dry soils. It is most probably a vigorous and deep-rooting variety of the same species as the Paradise, healthy everywhere, and succeeding well on some very dry and poor soils, where in consequence of its habit of surface rooting the Paradise would suffer and prove useless. Apples grafted upon it come into bearing earlier than upon the Crab, and it is admirable for all forms of garden trees in size intermediate between the very dwarf cordons and bushes, and the tall and vigorous orchard trees.

The Crab stock it is needless to describe. It is the stock on which our Apples have been grafted for ages, and which is the only one employed in the majority of British gardens. It is the natural stock for the Apple, and that on which it grows with greatest vigour; but it takes a much longer time to come into bearing, and the attempts to keep it of a size suited to gardens by pruning, pinching, and root-pruning which may be seen everywhere, are all efforts thrown away.

Thus it will be seen there are three distinct stocks, each suiting distinct purposes, and that those who experiment upon or criticise the cordon system of Apple growing without acting upon or bearing these facts in mind as the greatest and most important of all in connexion with the subject, may be likened to an individual attempting to study the moon by gazing through the wrong end of his telescope.

Of these three stocks, the one which has been most abused and least known, but which will yet prove the most valuable of all as a garden stock for the Apple, is the true French Paradise. When it is fairly tried it will prove to be of all stocks yet known the hardiest, most dwarfing in its effects, and most powerful in inducing early fertility. This stock, which has hitherto been characterized in England as a thing quite worthless and contemptible—only fit for growing in pots, and such toy gardening—will, if planted in the coldest and wettest of soils, instead of sending long and hungry roots down into the sour bowels of the cold clayey earth, like the Crab, and in a lesser degree the Doucin, keep its wig-like mass of small roots near the surface, and without root-pruning bear fruit long before the others, even if they receive every attention. The above is the way to best test its powers of withstanding cold, and the other merits I claim for it : on all ordinarily rich and cool soils it will be found to succeed perfectly without root-pruning of any kind. Amateurs and gardeners throughout the length and breadth of these islands have only to try it to prove that instead of sickening and dying in our cool climate, and on our moist soils, its general adoption will lead to one of the greatest improvements our garden fruit culture has ever undergone. It is necessary to observe that in trying this stock healthy plants should be secured to begin with. It has been ascertained that some of our nurserymen who have tried this stock import the Paradise from France in a very small state, and then graft it soon after it arrives. The consequence is that the little trees have no power to push forth a healthy graft. If imported in this way they should be allowed one year's growth before being grafted.

THE PEAR AS A CORDON.—Having said so much about the Apple as a cordon, we will next turn to the Pear trained as an oblique cordon on walls. It does not, as applied to this fruit, offer a distinct and economical way of producing a better class of fruit, as in the case of the Apple. Its advantages are simply quick growth, early fertility, and a considerable number of varieties from a limited space. Figure 162 will fully show the appearance of a wall covered with Pears on the cordon system. The plants at each end, which display a fuller development, show the means by which the ends of the wall are covered. As will be seen, the trees are placed very close together, which makes the plantation costly. They, however, soon run up to the top of the wall, and yield a quicker return than the larger forms. Then if one fails it is easily replaced. But are these advantages sufficient to justify us in adopting this system to any extent for our wall Pears?

We may secure handsomer trees, less distortion, longer life, and more fruit, by adopting such simple and easily conducted forms as those figured in the account of the Imperial garden at Versailles, and other medium-sized and simple forms. Those forms are handsomer than the wall or espalier cordon for the Pear, yield a great number of kinds from a comparatively small space,

FIG. 157.

Pear Tree trained as a Vertical Cordon. This form is best suited for very high walls, &c.

and moreover, allow of a somewhat free and natural development. We all know how comparatively few are the varieties of first-class Pears which succeed to perfection in any one place, and that the necessity of planting a new kind at every eighteen inches along the wall does not exist. For the fruiting of seedlings and testing of new kinds, it is however a good plan, and if the object be to cover a wall in a short time and get a quick return, it is certainly the best way. In this case it enables us to attain our ends in the shortest space of time, and with the least possible waste of space.

Some of the leading teachers of fruit culture in France adopt the oblique cordon as the surest way of getting a quick return, and plant extensively the finest varieties trained

FIG. 158.   FIG. 159.

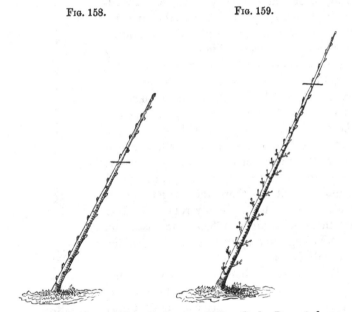

The Pear as a Simple Oblique Cordon.   Oblique Cordon Pear. 2nd year.
1st year. Showing the first pruning
after planting.

in this way; but others ridicule the planting of trees as closely as one would coleworts, and laugh at the system as only profitable for the nurseryman. My opinion is that for the finest kinds of winter pears and high walls it is well worth adopting, provided the trees can be got at a low price. To plant high priced trees so thickly would be ruinous. It would be better to wait for years and allow the larger forms to cover the walls in their old-fashioned way. But where the cultivator can graft and raise his own trees, or procure first-rate kinds cheaply as " maiden," or very young plants, then he will probably find the system an improvement. None but the very best kinds should be planted, and to begin with, it would be desirable to plant a goodly number of one kind known to succeed well in the neighbourhood

Fig. 160.

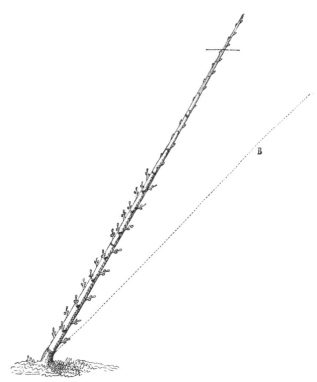

Oblique Cordon Pear.   3rd year.   B is the position which the tree will
eventually occupy.

rather than a variety of sorts.   The kinds known to do best in this contracted form are *Beurré Superfin, Flemish Beauty, Beurré Giffard, *Louise Bonne of Jersey, *Marie Louise, Beurré Clairgeau, Duchesse d'Angoulême, *Easter Beurré, and Beurré d'Anjou.   Of these a beginner would do well to take those marked with an asterisk.   As regards the

training of the Pear in this way, it is too simple to require
description here. The tree is simply treated as we treat a
single branch of a fan-shaped tree, and requires none of
the careful pruning necessary to form the more elaborate
shapes. Healthy young plants, a year from the graft, are
chosen, planted at from 16 to 18 inches apart, and
pruned as explained in the accompanying figures.

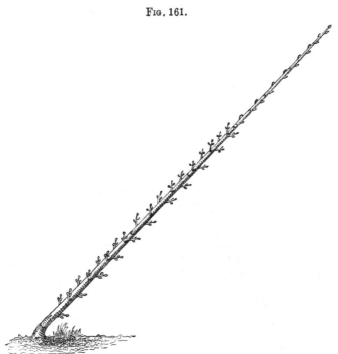

Oblique Cordon Pear.   4th year.

Sometimes the Pear is trained as a vertical single or double
cordon.   Of the two forms the single is preferable, and it is
chiefly suited for very high walls, the gable ends of out-
houses, and the like.   It need scarcely be added that the
trees should be on the Quince stock.

The Pear may also be trained as a horizontal cordon on
low walls, the fronts of glass houses, and as an edging like
the Apple.   But generally the Pear pushes too vigorously

to be trained in this way, while the pendulous habit of the fruit renders it more liable to be soiled. I once saw Uvedale's

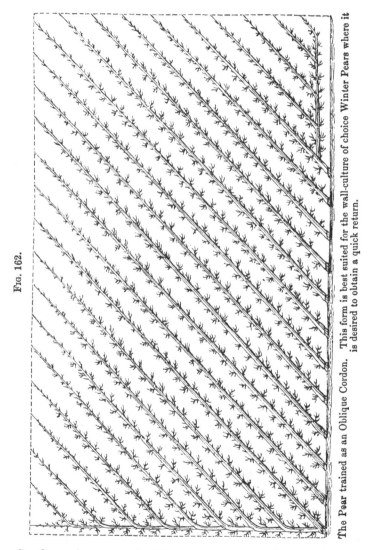

FIG. 162.

The Pear trained as an Oblique Cordon. This form is best suited for the wall-culture of choice Winter Pears where it is desired to obtain a quick return.

St. Germain grown in this way, the great fruit sitting on the ground, and quite encrusted with earthy splashings.

I have frequently seen plantations of the Pear as a horizontal cordon, but never one that I could call thoroughly satisfactory. The disposition to form a neat compact line of spurs so abundantly manifested by the Apple when well

Fig. 163.

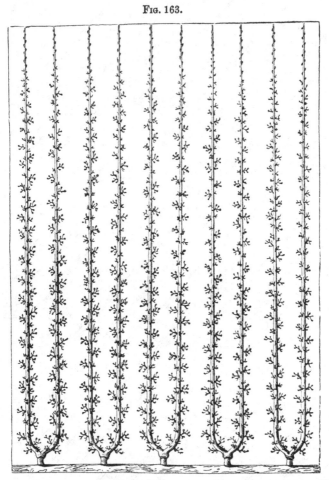

Pear Tree trained in U form. This has no special merits, and is only useful for very high walls.

trained on the Paradise is very rarely shown by the Pear. Nevertheless a few varieties, such as Louise Bonne and Beurré Giffard, might be tried; allowing them to attain a greater length of stem than the cordon Apples, and placing them a little higher from the ground. As regards the Pear as a horizontal cordon, Mr. Watson of Geneva wrote as follows to the *Gardener's Chronicle* :—" I question if there exists elsewhere a more extensive collection of Pears trained on the horizontal cordon system than may now be seen in M. Vaucher's garden near Geneva. There are hundreds of them, consisting of every good sort that M. Vaucher could buy.

Fig. 164.

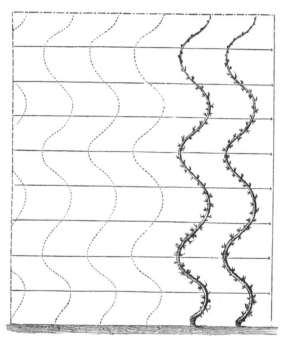

The Spiral Cordon against walls. This form is not to be recommended.

After carefully examining them, I came to the conclusion that Mr. Robinson's conclusions are right—namely, that the Pear as a rule does not conform to this mode of culture like the Apple, and can report no such success. Still some sorts are all that could be desired. Beurré Noirchain, four feet six inches long, had twenty-three fruit ; Beurré Giffard, six feet six inches long, twenty-two fruit. The fruit of the last-named kind are hanging about four inches from the ground."

Of the various modifications of the cordon system the
spiral cordon deserves note. The merit
claimed for it is that a greater length of
stem is secured, and consequently that the
tree is not so likely to suffer from being
confined to a single stem. Trained as
shown by Fig. 165, it is pretty, but against
walls it has not even that merit. The
isolated spiral cordon may be trained on a
galvanized iron support like that shown in
the illustration, or round a circle of stakes
inserted in the ground. The first way is
certainly the neatest and the best. It is
quite easy to train trees round this spiral
support, which seems best fitted for adoption where a thorough system of protection is carried out, in consequence of the
number of trees that may be packed into
a small space. It also seems worthy of
attention for orchard house and pot culture.

Fig. 165.

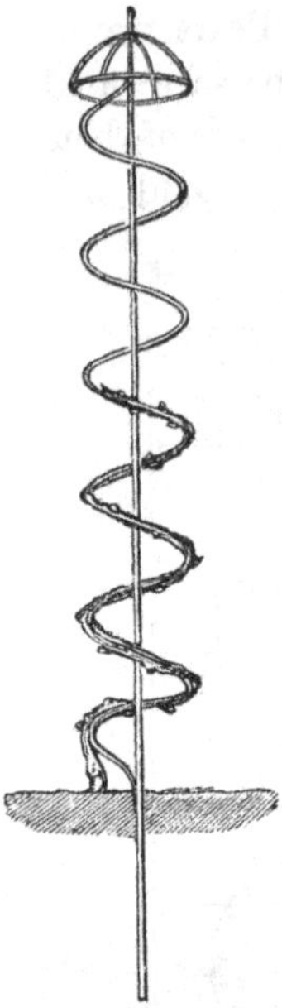

The Spiral Cordon.

THE PEACH AS A CORDON.—With the
Peach as an oblique cordon, a good result
is attained, the wall being covered very
rapidly ; and the neat laying in of a great number of shoots
on each side of the simple stem does away with the
crowded and unnatural appearance which a plantation of
cordon pears assumes when old and the stems are thickened.
But instead of the wood being closely pinched in, as people
might suppose in England from reading of the method of one
M. Grin, it is nailed in at each side of the branch, ay,
more so than if that branch were part and parcel of one of
the older and larger forms of tree. I once saw an excellent
result afforded by this system against the high back wall of
a vinery in the establishment of M. Rose-Charmeux, at
Thomery. By its means he perfectly covered his wall in a
short time, and gathered a great variety of fruit from a
small space. Out of doors I have seen it afford beautiful
results, and that not unfrequently. It is well calculated for
high walls, and it may be adopted for low ones by training
the trees at a more acute angle with the earth.

Considering the time usually required to furnish walls in the ordinary way, there can be little doubt that this mode of training the Peach is a real improvement, where a considerable number of varieties are required from a small space. Apart from that, however, the facility and simplicity with which walls may be covered by this method, and the readiness with which a diseased or otherwise objectionable tree may be replaced, will doubtless prove a sufficient recommendation for cultivators who are not restricted as to space.

Fig. 166.　　　　　　　Fig. 167.

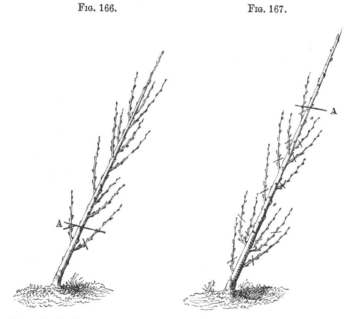

Young Peach Tree trained as an Oblique Cordon. 1st year. A shows the first pruning.

Peach Tree trained as an Oblique Cordon. 2nd year's pruning. The leading shoot is cut at A, and the side shoots at the cross-marks.

It should, however, be borne in mind that on very good soils where the Peach grows very vigorously, it will not suit so well as on poor ones where it grows slowly, and that medium-sized forms may be adopted for the Peach as well as for the Pear. The following is a description of the mode of forming it after M. Lepère :—

" There are two modes of growing this form. One, which was recommended by a professor of arboriculture, and frequently put in practice by many amateurs, but which I consider faulty, consists in planting the trees just as they come from the nursery, and training them at once in the oblique form. The inconvenience arising from this method consists in being obliged to place the tree close to the wall, which crowds the roots too much, preventing them from affording sufficient nourishment to the tree. Besides this, on account of the inclination of the tree, part of the roots are directed towards the surface of the earth or placed in an unnatural position, thereby preventing their full development. By-and-by, the trees that have been planted thus are cut to half the length that they were when they came from the nursery, having a number of weak, useless branches on the lower part, a condition which, as every one knows, is always unfavourable. The second method differs from the first, inasmuch as the plant from the nursery is cut down instead of being planted in an oblique direction. To obtain the oblique form without planting the tree in a crooked position, the stem is cut at eight inches from the graft, and placed in the hole in such a position that the base of the stem is four inches from the wall, with its extremity just touching it. The roots are well spread over the hole and drawn as much as possible towards the border in which the tree is planted. Care is taken to leave a well-placed bud on the side where the oblique branch is to be formed, and its development must be encouraged by ruthlessly pinching off all useless shoots. Under these conditions, the tree grows as long during the first year as the one planted obliquely and allowed to be of its full length from the first. This method is also to be preferred, because the shoot thus obtained the first year can be left intact and allowed to attain a development equal to that of the tree planted according to the first method. Besides this, the shoot is calculated to grow faster in consequence of its bark being less hardened, and each year the terminal point may be allowed to grow without cutting back. Sometimes the terminal bud does not develope owing to its having been killed with the

PLATE XLIII.

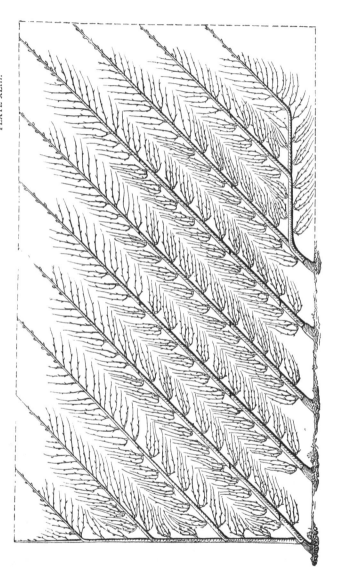

THE PEACH TRAINED AS AN OBLIQUE CORDON.

cold. In such a case a stronger eye is chosen lower down to make the desired prolongation.

" As in the case of other forms of training, the branches of the Peach cordons are allowed to grow in a more erect position at first than they are finally intended to occupy. I should advise this cordon form to be adopted in the case of gardens whose walls are on the incline, as often occurs in certain localities, and for soil of inferior quality where the Peach tree grows slowly, because under such circumstances it never attains its full development. The plan does not answer where the ground is flat and the conditions are such as favour the rapid growth of the tree."

Fig. 168.

Some fruit growers think that there is no occasion for resorting to this simple cordon in the case of the Peach, any more than in the case of the Pear. My friend M. F. Jamin, of Bourg-la-Reine, plants in his fruit garden a form of tree with three vertical branches, and if he wants a great variety of fruit from a small space, works a different variety on each branch. This figure shows, on a small scale, the appearance of one of his young specimens, trained on this principle. The U and double U forms, described in the Chapter on Montreuil, are also extensively adopted in preference to the oblique cordon by many growers.

Peach Tree with three stems, a different variety being grafted on each.

THE SHORT PINCHING SYSTEM APPLIED TO THE PEACH.—The system generally known among us as that of M. Grin is confounded by some writers with the cordon system, from which it is entirely distinct. It has not in the least influenced the old way of growing the Peach in France, and a commission of first-rate fruit growers sent to examine it, reported that the system pursued at Montreuil is still much the best. It may be shortly described as an attempt to do away with nailing by a system of close pinching, and that alone is sufficient to condemn it for our gardens, and also for those of the French, for the wood to be well ripened must be nailed in, and the pinching required to keep the shoots from running away from the wall is something prodigious. As the French fruit growers say—the cultivator who pursues this method had better provide himself with chairs, and place one before each tree to accommodate the person who has to see that the pinching is done at the proper time! The report of the commission sent to examine this method is as unfavourable to it as anything can be. I translated it with a view of giving it here, but space prevents my doing so, and therefore I sum up its statements in a few words. "This system, which is an attempt to do away with nailing in of the shoots, presents on the whole no advantages over the one in common use, but, on the contrary, certain drawbacks." Having read so much about the doings of M. Grin, I was astonished at the very ordinary aspect of his trees, and the by no means remarkable result attained. The individual who pays his penny to see the "blue horse captured in the Black Sea by Captain Jones of the ship *Adventurer*—the most extraordinary monster ever seen," &c., in the New Cut, and finds the blue horse to be a puny young seal, could not have been more disappointed than was I at the aspect of the trees in this garden. For when one reads of a method as being about to supplant everything else, it is quite natural to expect that it must at all events possess some merits over the older one; but in this instance such is not the case. Of course I speak of this mode of pinching as a system.

FIG. 169.

Peach Shoot of the current year bearing a number of secondary shoots—*bourgeons anticipés*.

It has one merit, however, and may be used incidentally with any system of summer-pruning. It should be remarked that M. Grin commenced by simply adopting a method of very short pinching in of the shoot. He now depends chiefly on pinching the stipulary leaves, as shown at A. This is the best feature of the system, and chiefly in dealing with the little lateral *bourgeons anticipés* that sometimes push forth on the current year's wood. By pinching the leaves of these little buds just when they push, as shown in the figure, the development of the shoot is not interfered with ; but a sufficient check is given to cause the eyes near the base to fill and become fruit buds, as shown at A, Fig. 171. This not done, the young shoot pushes away, and is often quite naked of buds at its base.

To think of adopting the system of Grin in its purity would be folly. As to training Peach cordons on this

Fig. 170.

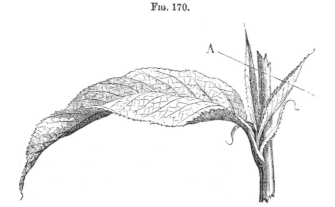

Portion of Shoot of Peach Tree, showing the pinching of the stipulary leaves.

principle, it is simply nonsense ; as well might we think of repressing the flow of the tide as hope to succeed with trees confined to a single stem, and pinched in quite close. It is by no means a success even where large forms are adopted ; but with the cordon there is no outlet. The cordons trained after this method in the public garden at Chartres must have exhausted the patience of the cultivator, for their shoots had started right away from the wall, and grown as much as eighteen inches long !

"This," said I, "will never do for England." "Nor

for France !" added an eminent Parisian fruit grower.   The
only chance of success with the Peach as a cordon is by
laying in the side shoots regularly.   Since the above was
written a report on Grin's method has been presented to
the Imperial Horticultural Society of France, and in this
also a very unfavourable account of the system is given.
There can be no doubt that as a system of pruning it is
ridiculously bad.   From reading a book on Peach pruning,
by the Government reporter on fruit trees at the Paris Ex-

FIG. 171.

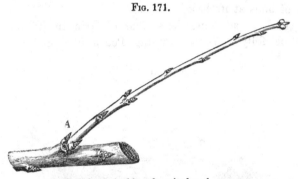

Result of pinching the stipulary leaves.

hibition, which was first published in one of our horticultural
journals, it is possible that some people interested in the
subject may suppose that this system of M. Grin, so highly
praised therein, is extensively practised and adopted instead
of the old way of laying in the shoots.   The interests of
fruit culture compel me to declare that it is not practised
by any intelligent French fruit growers.   It should by no
means be confounded with the true cordon system of growing
the Peach as it has been in the writings above alluded to.

CHAPTER XVI.

TRAINING.

TRAINING is very much better understood in France than in the British Isles. In France the commonest labourers frequently possess a knowledge of pruning and conducting a tree, which we might look in vain for anywhere in this country; and by way of illustrating their skill in this way, we cannot do better than examine their means of forming two of the most popular forms of fruit trees—the Palmette Verrier and the Pyramidal Pear trees—chiefly after Du Breuil. The Pear will serve to illustrate training and pruning as well as any other tree, or better, and the principles laid down will apply equally to other fruit trees.

THE PALMETTE VERRIER.—Wherever large wall trees are grown, the simple and beautiful form known to the French as the Palmette Verrier is sure to obtain a place among them. It is indeed the finest of all large forms, and is preferred by many of the best French cultivators to any other. They use it for other trees besides the Pear; and by far the finest Peach tree I have ever seen was trained after this method near Lyons. The English reader may think it impossible to attain such perfect shape as is shown in the accompanying plate, and such perfect equalization of sap as it suggests; but I have seen several trees even more beautifully finished than the one represented. This figure also shows the advantages of the kind of support used in France for espalier trees as compared with our own ugly method of using rough wooden and iron posts and strong bolt-like expensive wire. It will be seen that the tree differs radically from the usual form of Pear tree that we are in the habit of placing against walls, and it is easy to

point out its advantages in securing an equal flow of sap to
all the branches.

In the common horizontal form strength and fertility
are apt to desert the lower branches, in consequence of
their not possessing a growing point to draw the sap through,
and particularly when constant care is not taken to repress,
by summer pinching, the upper portions of the tree.   The
form here figured, in common with all very large wall and
espalier trees, takes a long time to complete.   Given a wall
10 ft. or 12 ft. high, and 20 ft. or 24 ft. long, to be covered
with a tree of this shape, it would require fifteen or sixteen
years to form it.   By adopting a more contracted form
based upon the same plan, we may cover the wall or trellis
more quickly.

The Palmette Verrier is named after the fruit gardener
at the Ecole Régionale de la Saulsaie, with whom it was first
observed.   To form the tree, we have in the first instance
to plant an ordinary young plant of a desired kind, and of
course that should be of the primest kind, both as to
quality and constitution, as so much care is about to be
exercised to make it a handsome and long-lived ornament to
the garden and valuable aid to the fruit room.   In forming
this as all other trees, the usual and most economical custom
is to choose plants about a year old from the time of grafting,
or what are usually called " maiden plants," and which when
planted are cut down to within about a foot of the surface
of the soil.   Three well placed buds are allowed to remain
and form three shoots.   The two side ones go to form the
lowest and longest branches of this handsome form of tree,
and at the second pruning the young trees would have
somewhat the appearance of fig. 172.   It is quite easy to
buy trees a little more advanced to make the same form
more quickly ; but they will be more expensive the further
they are advanced beyond what is called the " maiden"
stage.   The young trees should be allowed to remain a year
or so in their positions before being cut, so that they may
have rooted well.   At the first pruning the young tree is
cut down to within a foot or so of the ground, and just
above three suitable eyes, one at each side to form the two

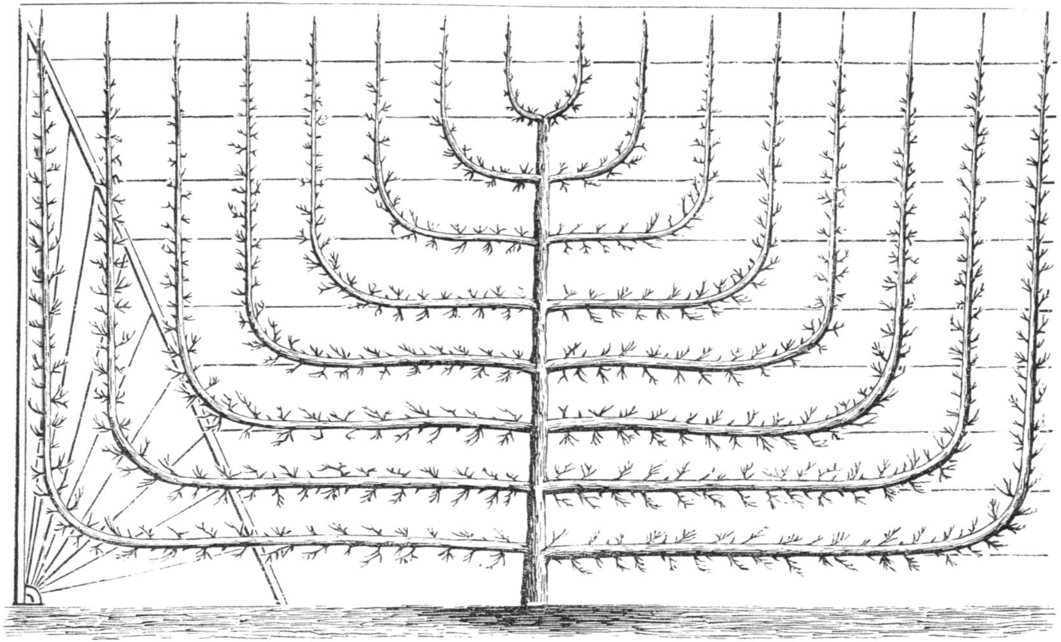

PLATE XLIV.

PEAR TREE TRAINED AS A "PALMETTE VERRIER,"

On trellis ten feet high, supports of T iron, horizontal lines slender galvanised wire (No. 12), wires united in strong ring at base to secure rigidity in end supports.

lowermost branches, the other a little above them and in front to continue the erect axis. Of course all the eyes, except those that are to send forth the three first shoots, must be suppressed in spring. Now, although the tree in the plate looks so very exact and regular in its lines, and the branches appear as if they had been " bent in the way they should go" at a very early stage, it is not so ; they are at first allowed to grow almost erect, and are afterwards gradually lowered to the horizontal position. During the first year of the young tree possessing three shoots, care must be taken (as at all times) to secure a perfect equilibrium between them. If one grows stronger than the others, it must be loosened from its position on the wall and lowered.

FIG. 172.   FIG. 173.   FIG. 174.

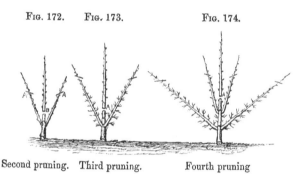

Second pruning.   Third pruning.   Fourth pruning

The Palmette Verrier.

This will divert the sap so as to strengthen the rest. Nothing is more easily conducted than the sap when we pay a little attention to it ; if not, it soon rushes towards the higher points, and spoils the symmetry of the tree.

We then, at the second pruning, have to cut them at B, and also cut off about a third of the length of the side shoots, as at A A, Fig. 172. If one side branch happens to be stronger than the other, cut the stronger one somewhat shorter. In cutting and pruning wall trees the cut should be made above a front bud, so that the wound made by the knife may be turned towards the wall, and away from the eye, from which, of course, it soon will be effectually hidden by this front bud pushing into a shoot, and

thickening at its base.   During the second year no more
branches must be permitted to grow, simply because the
trainer desires to throw all the strength he can into the
lower branches, which are to be the longest.   Sometimes,
however, the strength of the lower branches will permit the
second stage of branches to be made during the second year
of training.   At the third pruning the trees will present
somewhat the appearance of Fig. 173, the central stem
being cut at six inches or so above the previous incision,
which is indicated by a slight ring, and a third part of the
new growth of the side branches cut off, as shown on the

FIG. 175.

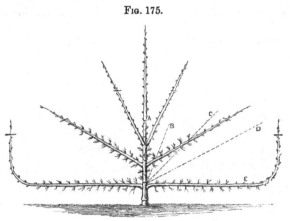

The Palmette Verrier.   Fifth pruning.

side branches of Fig. 173.   Here, again, we cut above and
inside of three promising eyes to obtain a new set of
branches, and each succeeding year add another series
until the tree is formed.   Fig. 174 represents the aspect
of the young tree at the fourth pruning.   At the end
of the following growing season the specimen will have
grown sufficiently to allow the lower branches to be turned
up towards the top of the wall, and begin to look shapely.
Fig. 175 is an exact representation of what it ought to be
at that stage—A, and the cross marks, indicating where the
cuttings are to be made.   Above all things is it necessary
to keep the growth and flow of sap equal, not only for the

sake of symmetry, but also to insure perfect health and
fertility ; for if one part be allowed to grow grossly at the
expense of another, an awkward state of things will soon
take place.   Sometimes, when the vegetation is very
vigorous, time is gained in the making of this form by
pinching the central growth at eight inches or so above
the highest pair of opposite branches.   It then breaks
again, and care is taken to secure two side shoots and one
erect one.   Thus, with care, and in good soil, two stages of
branches may be secured in the same year, but this must
not be attempted till the proper formation of the two lower

FIG. 176.

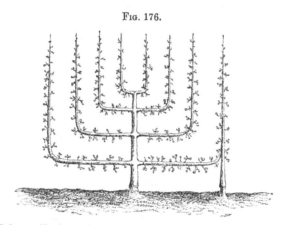

Palmette Verrier, with weakly outer branch completed by grafting.

branches is secured.   The dotted lines in Fig. 175 will
show the positions that have been successively occupied by
the branch E, when in course of formation, and that it is
by no means necessary to train a young branch from the
beginning in the exact position it is required to take.   In
fact, this form is only to be well and easily perfected by
allowing the young shoots to first grow and gather strength
in an erect or oblique position.   The branch E kept company when young with the central branch, and was at B ;
then it was lowered to C, next year to D, and finally to its
horizontal position.   Some care is required to make the
bend of the shoots equal and easily rounded.   If the tree

be trained on a wire trellis, it is best to place two bent rods
in the exact position necessary, and before we require the
shoot to be bent. They must be placed at exactly equal
distances from the main stem, and be equal in curvature.
Then it is an easy matter to gently attach the growing
shoot to them; it will soon harden to the desired bend.
Against a wall it will be easy to direct it with shreds and
nails; if the wall be wired the bits of bent twig may be
applied, as on the trellis. Like care should be bestowed
upon the other bends, as they require to be made; but of
course the outer and lower one is of the greatest importance.
As this form is not at all presentable if the outer branches
be incomplete, grafting by approach is sometimes employed
to repair this defect, as shown in Fig. 176.

The reader will observe that, in the formation of this
Palmette Verrier, the custom is not to attempt training the
young shoot in the position it is finally destined to occupy;
but, on the contrary, to permit it first to grow sometimes in
an erect, or at least in an oblique direction, so that the sap
may flow upwards without check. Nothing is easier than
taking down the shoots from time to time, as they become
strong and well developed. Now this is a principle almost
unknown, and certainly not practised in this country; being
applicable to many forms of training, I can strongly recommend it, having frequently witnessed the good effects produced by carefully carrying it out.

PYRAMIDAL TRAINING OF THE PEAR TREE.—This culture
is, considered from the stand-point of beauty alone, as desirable as any with which amateurs interest themselves. I
have seen in the gardens of even very humble French amateurs pyramidal pear trees, which, if they never afforded
a fruit, would be beautiful objects; and I have met with
few avenues that afforded me more pleasure than a short one
of pyramidal pear trees leading up through a little town
garden within the walls of Paris. We will begin, then,
with the fully formed pyramid, and in addition to its
symmetry will be observed the straight clean growth
of each branch, springing at regular intervals from the
main stem, which is so erect and well furnished. From

the summit to the base such a tree ought to be garnished with nothing but branches well set with fruit spurs. The greatest breadth of the pyramid should equal about one-third of its height. Pyramidal trees may be purchased in all stages; but trees ready-formed are costly, and as many would prefer training their own, and as those who plant on a large scale will find it economical to begin with trees a year from the graft, we will commence at the beginning with a "maiden tree," letting it grow one year in the ground before pruning it.

Fig. 178 represents the first pruning of this young tree, and its appearance one year after being permanently planted, or two years from the graft. B shows the union of stock and scion; and the terminal bud A just below where the shoot is cut should be placed on the side opposite to that on which the scion was inserted, as shown in the figure, so that the stem of the tree may rest perpendicularly on its base.   It is by attending to such little points

Fig. 177.

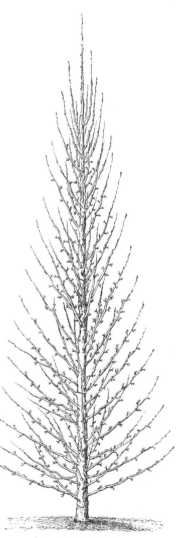

Pyramidal Pear Tree.

as this that the French get that perfectly equal distribution
of sap which is so essential to the satisfactory management
and prolonged fertility of trained fruit trees. The summer
following the first pruning, the young trees push with great
vigour, and their shoots should be thinned when a few

FIG. 178.

Pyramidal Pear Tree.
First pruning.

FIG. 179.

Top of Young Pear Tree. B, the
leading shoot. A A, shoots requiring to be pinched.

inches long, removing every
shoot from the base of the
stem to a height of about one
foot, and thinning out those
above this point to six, seven,
or eight shoots; reserving of
course the best placed shoots,
and taking care to have them
arranged as far as possible at
regular intervals.   Should they
in the course of the year assume an irregular development, pinching with the finger and thumb must be resorted
to.   This is shown in Fig. 179.   The shoots, A A, have
pushed too much; and one of them rivals the leading shoot
B; they therefore must be pinched, merely taking an inch
or so off.

In the spring of the following year the young trees should present the appearance of Fig. 180; the cross marks showing how the pruning is to be performed. This second pruning has for its object the production of a new set of lateral branches, and the further development of those already obtained. It is evident that to secure a beautiful tree, the branches must spring forth regularly from the main stem, which they are not likely to do if the tree is left to itself. Fig. 181 shows the way in which the careful cultivator furnishes his stem, as regularly as could be desired. The eyes which he desires to break strongly have an incision made above them, as shown in the figure. This is particularly desirable as regards the lower part of each successive growth of the erect stem; the vigour of the rising current of sap often pushing towards the higher buds, and causing the lower part to be poorly furnished. These incisions, A, A, A, must be carefully performed on the young branch: deep enough to penetrate the sap wood, and yet not so deep as to hurt the slender rising point. The top of this shoot, instead of being cut off, has been barked for some portion of its length above the bud that has been selected to continue the growth of the coming summer.

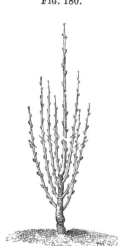

FIG. 180.

Pyramidal Pear Tree.
Second pruning.

FIG. 181.

Leading shoot of Pear Tree, showing incisions A, A, A, made above the buds required to break strongly.

To this the young shoot is trained, and a perfectly vertical growth for what we may term the pillar of the tree is thereby secured. The bark is neatly cut round above the upper eye; the branch is cut off at about four or five inches above

FIG. 182.

FIG. 183.

A, the best position at which to prune for the terminal bud.

FIG. 184.

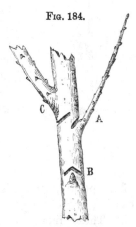

A, part of old leading shoot barked and left to tie the young shoot to. It is cut at B when the shoot is arrived at maturity.

A, B, C, incisions made above and below branches and buds to check their irregularity.

that point, and then the bark is taken clean off. When the young leading shoot is long enough, it is fastened to the bare portion of stem, as shown at Fig. 182. The portion A is cut off at B at the next winter pruning. This process may be prolonged as long as necessary or convenient.

In pruning the tree considerable judgment is required, so as to get the base of the specimen well furnished, and secure fertility in the fruiting branches.   Fig. 185 shows how this is performed, and several of the following figures well explain the principle.   It is to cut them of the greatest length at the base of the tree, and gradually shorten them as we reach the top.   The nearer they spring to the soil, the longer they must be left, or, to be more precise, only a third must be cut from the points of the lowest branches; half the length may be taken from those situated between summit and base; and lastly, three quarters may be cut from the most elevated.   In cutting-in the lateral branches, the directly oblique direction which it is desirable they should take must be borne in mind in the pruning, and the terminal bud of each left as far as possible, as at A in Fig. 183. In case of a very irregular development among the laterals, incisions are made above a weak branchlet to encourage it, as at A, Fig. 184, and below a strong one, as at C, to

Fig. 185.

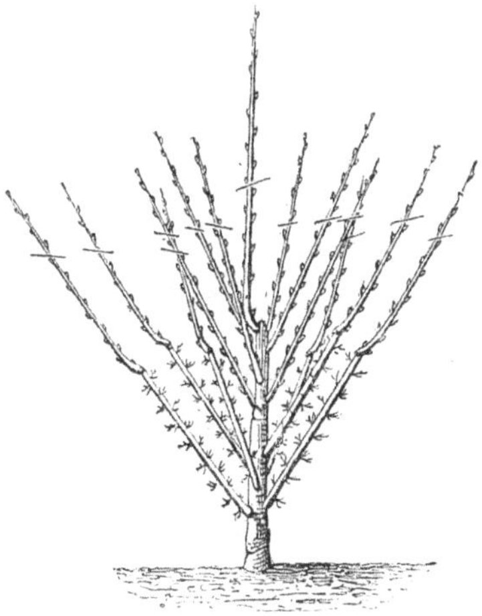

Pyramidal Pear Tree.   Third pruning.

retard it until the equilibrium of the branches is established.   At B this incision is made before a dormant bud that has failed to become developed into a lateral. This figure also shows the relative proportion to establish in pruning irregularly developed branches springing from a main stem that we wish to be equally balanced in all its parts.   The weak shoot is not cut, or but very little; the strong one is cut to below the level of the one it is desired to encourage.   These incisions should be performed with a little saw, so that the cuts may not soon heal over.   The incisions should penetrate sufficiently into the layer of young wood to well intercept the sap vessels. If with all these precautions there are objectionably bare spaces on the stem, they furnish them by grafting by approach, or in other words, turn back a vigorous branch to the main stem, and graft it on to the bare space; and if this cannot be done, insert a short ordinary graft in the stem. This, however, with good management will rarely be necessary.

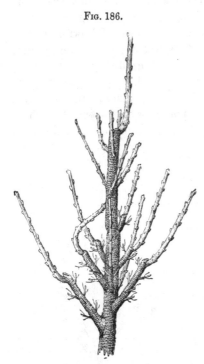

Grafting by approach, to cover bare spaces on Pyramidal Trees.

Having trained the branches straight, the next point is to see that they follow the desired oblique line; and it will be seen by the cuts that the disposition given them is better than the one they assume under a less careful system. The light enters freely to the stem, and illuminates all; the more important part of the tree is under the command of the eye and hand, and the top is prevented from running away. This, however, is more owing to the fine formation of the lower branches than to the position they assume, though certainly such free and straight outlets for the rising sap are very effective in preventing a gross development above, and consequently in keeping the tree

FIG. 187.

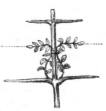

Grafting by approach as applied to Wall and Espalier Trees.

in the desired condition. During the summer following the
second pruning, the operations for maintaining the lead
with the vertical branch, and equality among the lateral
shoots of the new growth, must be carried out as before
described. The third pruning is shown at Fig. 185, and it
will be seen that here again the young lateral branches of

FIG. 188.

Pyramidal Pear Tree.   Fourth pruning.

the preceding summer are cut in much shorter than the
lower ones to favour the development of these.

At the fourth pruning the lower branches are not cut
nearly so long as in the previous pruning, because they have
now attained to almost the desired length and sufficient
vigour. The new branches of the second series are left
somewhat longer, and the pruner looks more to the top
structure, so to speak. The wisdom of well forming
the base at first will be seen at a glance. During the

summer following the fourth pruning before described, attention should be given to the young branches at the top of the pyramid, while the side ones will also require it. As the lower branches will have attained to nearly their full length, a too vigorous growth of the terminal shoot of each must be prevented by pinching.

FIG. 189.

Pyramidal Pear Tree.   Fifth pruning.

Fig. 189 shows the aspect of the tree at the fifth pruning, and how the pruning is performed. As is well seen by glancing from B to A of Fig. 189, the new growth of the lower branches is cut very short, while the higher the remaining superior branches are, the longer they are cut. A careful glance at Fig. 189 perfectly explains all this. The succeeding prunings differ nothing in principle from the preceding, future development taking place principally in the middle and higher parts of the tree. Care should be taken to guide in the desired direction by means of twine, and sometimes slender stakes, any branches that may have deviated from it. Thus the pruning is carried on till the tree becomes a large and perfect pyramid, the laterals being well pinched in, and in

FIG. 190.

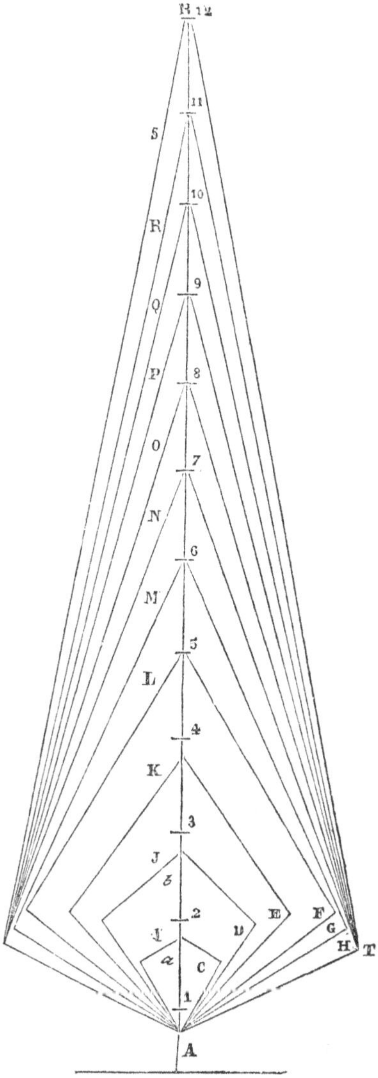

Figure theoretically indicating the Mode of forming a Pyramidal Pear Tree. (See p. 389.)

FIG. 191.

Young Pyramidal Pear Tree.

every case a free terminal shoot being allowed to proceed from each, so that the tree may be kept equally balanced

FIG. 192.

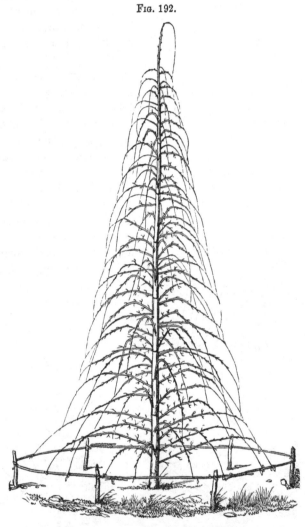

Pyramidal Pear Tree with bent branches.

and the sap freely conducted through each branch. They may of course be cut back well every year: always, however, at a bud likely to furnish a good shoot for the following season.

FIG. 193.

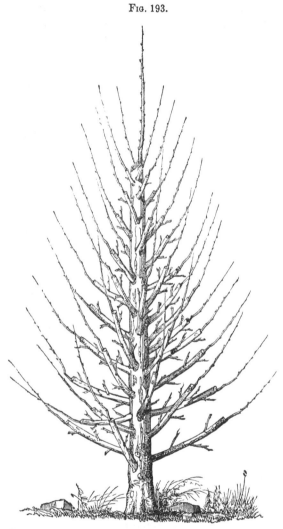

Pyramidal Pear Tree of an inferior sort, cut back and regrafted
with a good variety.

Fig. 190 shows theoretically, first, the central stem A to
B, and its successive cuttings back, 1 to 12; secondly, the

position successively occupied by the lower branches during the first six years, during which they were successively lowered and elongated from the point C to T; and thirdly, the lines from I to S show the lines of each year's pruning.

It is very questionable if the mathematically designed pyramid here alluded to be so desirable for gardens generally as a flatter and less pointed form. For example, the pyramid as represented at the time of its fourth or fifth pruning is in outline preferable to the tall and finished pyramidal tree depicted in Fig. 177, and a style somewhat like that shown in Fig. 191 will prove easier to form to those who have no time to spare for the niceties of training.

FIG. 194.

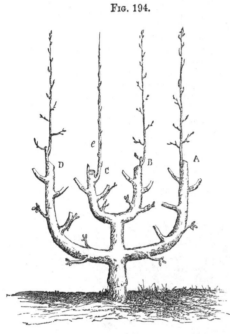

Wall Pear Tree regrafted. On each of the branches A, B, C, D, a graft has been placed. The graft at C failed, and consequently a shoot, *e*, is allowed to ascend; so that it may be budded the autumn following the grafting.

Occasionally the pyramidal Pear has its branches bent downwards, as in Fig. 192, some thinking that this induces a more fruitful habit. I never saw any clear evidence of this, and believe the form to be no better than the simple pyramid.

The excellent practice of cutting in pyramidal and other trees that happen to be worthless varieties, and regrafting them with superior kinds, is much recommended by the French growers.

This system is quite as applicable to wall trees as to pyramids or standards. In numbers of our gardens great good might be effected by regrafting with good varieties, and doing away with the worthless ones, so very common.

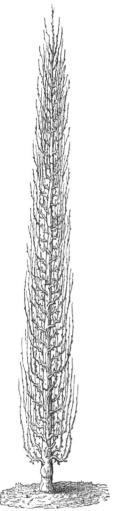

Fig. 195.

Fig. 195 represents a mode of training to be seen here and there in France. The woodcut shows a fully formed tree before the winter pruning takes place, and, as will be seen at a glance, it is an erect stem densely furnished with short fruiting branches. This form is considered better than the pyramidal one, where saving of space is a consideration, and where we do not wish the trees to much shade the crops between them. They are also well suited for small gardens where space cannot be afforded for a large number of varieties if trained in the usual way. I have thought it worthy of a figure, but except on the Quince in suitable soils it is not likely to present many advantages ; for if on the Pear and confined thus closely to a fastigiate bundle of shoots, it would in all probability run too high to permit of proper annual pruning or of the crop being gathered with convenience. Judging by the strength and thickness displayed by our old horizontal wall trees grafted on the Pear stock, what should we arrive at if we adopted a contracted form like this with trees worked on the Pear ? Why, in a few years, and

Pear Tree trained in the Columnar form.

especially with the cordons, we should have objects more like rustic gate-posts than trees.

It is not uncommon in English gardens to train the
branches of the pyramidal Pear in a pendulous fashion; and

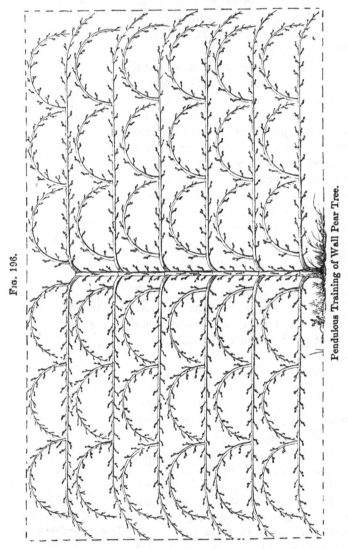

FIG. 196.

Pendulous Training of Wall Pear Tree.

it is a system admired by some, though somewhat more
troublesome to form than the simple pyramid.   Fig. 196
represents a mode of applying a modification of the same

principle to the ordinary horizontally trained Pear tree. I do
not say that it is as good as it is graceful in appearance, be-
lieving as I do in simple easily-conducted forms, but as these
smaller arching branches may be established on kinds that
bear better on the young wood, or on trees with the branches
thinly placed, it may prove useful.   The mode of formation
is so simple and so easily established that no further description is needed.   However, I cannot say too often that the
simple and quickly-formed trees, described elsewhere, are as
excellent for walls as for trellises, combining as they do the
advantages claimed for the cordons with a not too contracted, repressed development.

When the exact system of training described in this
chapter is well carried out, well furnished branches and
fruitful spurs are the rule.   Should it not be so, the
growers frequently resort to grafting fruit-buds on the bare
spaces, as shown by the following figures :—

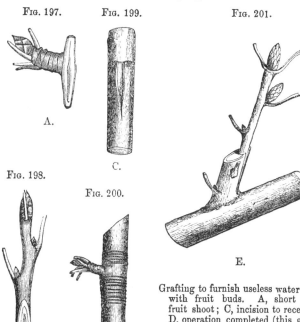

Fig. 197.   Fig. 199.   Fig. 201.

A.

C.

Fig. 198.

Fig. 200.

E.

B.   D.

Grafting to furnish useless water shoots
with fruit buds. A, short lateral
fruit shoot ; C, incision to receive A ;
D, operation completed (this graft is
performed in August, the buds fruiting the following year) ; B, terminal
fruit branch; E, crown grafting of fruiting shoots on gross unfruitful ones.

The greatest attention is paid to the proper and neat pinching and pruning of the shoots, as shown by the following figures :—

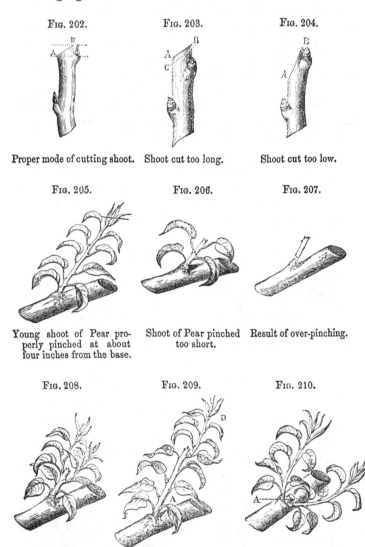

FIG. 202.

Proper mode of cutting shoot.

FIG. 203.

Shoot cut too long.

FIG. 204.

Shoot cut too low.

FIG. 205.

Young shoot of Pear properly pinched at about four inches from the base.

FIG. 206.

Shoot of Pear pinched too short.

FIG. 207.

Result of over-pinching.

FIG. 208.

Another result of over-pinching.

FIG. 209.

Pinching of the *bourgeon anticipé*, or second growth of the Pear.

FIG. 210.

The stipulary shoots forced into growth by the removal of the principal shoot, A.

CHAPTER XVII.

FIG CULTURE IN THE NEIGHBOURHOOD OF PARIS.

FIG culture as practised in the neighbourhood of Paris is very peculiar and interesting, as well as successful, and is, I believe, perfectly well adapted for the southern counties of England. As I have seen the Fig bearing well as a healthy standard tree at Arundel and elsewhere in Sussex without any attention, there cannot indeed be a doubt that the Parisian mode is perfectly applicable in sunny spots in the south. It might even be carried out on the railway embankments. It may not be amiss to state that the culture is founded upon the habits of the Fig in the climates of Paris and London. In hot countries the Fig is an evergreen tree, growing and bearing almost perpetually. In cold countries the Fig loses its leaves in winter, and becomes in fact a deciduous tree. Then the rudimentary figs borne at the end of each branch, instead of falling off prematurely as most other fruits would do, seem to rest stationary; in the spring they recommence their growth, and ripen off into the large succulent and well-flavoured figs supplied to the Paris market in summer. The French call those figs that require part of two years for their development figues-fleurs; those formed in spring and which ripen during warm autumns are known as secondes figues, or figues d'automne. These ripen but rarely in the climate of Paris, and it is to the care of the figues-fleurs, or figs formed in the preceding year, that all attention is given. To protect them and the young branches, the trees are trained in long sweeping shoots pretty near the soil, and in such a form that they may be readily interred in the ground when the winter and its dangers come. The frosts are often of great severity in the neighbourhood of Paris; so great indeed that the Fig would have little or no

chance if left exposed. So in autumn the sagacious cultivators throw the branches into four bundles, make a little
trench for each, and cover as shown by Figs. 214 and 217,
with small sloping banks of soil, protecting the crown of the
root by means of a little cone of earth, which merges
gradually into the four little ridges that protect the
branches. When the plantation
is made on deeply inclined
ground a somewhat different
system is followed, as is also
shown by the figures.

FIG. 211.

Branch of Fig Tree bearing the
Figs formed during the preceding year, D; those formed
during the current year, A; and
rudimentary Figs, C.

For the details of the culture
we will refer to Du Breuil's
"Culture des Arbres et Arbris-
seaux." In the climate of Paris
the Fig tree is grown as a low
shrub, with free sweeping branches arranged in single lines
or planted all together on a piece
of ground devoted to the purpose, and which for a better
name may be called a "figgery."
The branches of these tufty
trees are not allowed to grow
longer than from six to nine feet,
so that the tree may be conveniently buried in the ground
during the winter. Those varieties which produce rudimentary
figs in autumn in abundance
are the only ones grown, as
the figs of the current year
very rarely arrive at maturity. Argenteuil and La Frette
are the two most famous localities for the cultivation
of the Fig tree in the neighbourhood of Paris. Before
the southern railways were constructed, these two villages
used to supply the whole of Paris with all the green
figs that were seen in the markets. The introduction of
the Fig tree into Argenteuil appears to have taken place

about two centuries ago. It is cultivated in orchards in deeply dug and richly manured land, the soil of which is of a siliceous, calcareous, and clayey nature, well sheltered from the north and north-west winds, and open to the south and east. The cultivation of the Fig extends over a space of 130 acres, the production being somewhere about 400,000 figs per annum. The variety grown in this locality is the Blanquette or white Courcourelle, and the method of growing it is as follows :—

Layers raised in baskets or in the ordinary way are planted in the month of March in holes about four feet

Fig. 212.

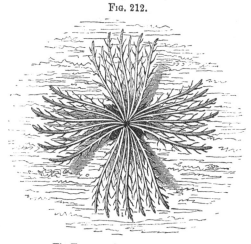

Fig Tree growing on level ground.

six inches in diameter, and one foot eight inches deep, filled with well manured mould. The planting is performed in such a way that the roots of the layer are buried from ten inches to one foot deep, and that the stem which springs out of the earth in an oblique direction should be covered with from three to four inches of earth. To form the stool more quickly two layers may be planted in the same hole instead of only one. In this case the two layers are placed in lines parallel to those of the plantation at eight inches' distance from each other, and in such a way that the stems are opposed to each other in the direction of this line.

The surface of the hole should be at least a foot below that of the surrounding soil. The rest of the soil is arranged slantwise round the stem of the layer, so that the rain-water may be easily retained round the roots of the young trees. The trees are planted five or six yards apart, the lines being separated by a space of about four yards, so as to form a kind of quincuncial arrangement. These young plants are left to themselves during the whole of the summer, care being taken to keep them from drought by means of frequent watering and careful covering. During the first half of November, when the first cold days set in and the trees are completely bare of leaves, a dry day is chosen when the ground is not too damp, and the young branch is carefully bent downwards until it reaches the bottom of the trench. It is then covered up with a layer of earth a foot deep to preserve it from the cold. Towards the end of February, as soon as the weather has become mild, the branches are uncovered and the trench is arranged the same as it was before earthing up. The development of the young plant is again allowed to proceed during the summer, after which it is once more earthed up in November.

FIG. 213.

Section showing Fig Tree growing on level ground.

The third spring after planting, a fine day is chosen to-wards the middle of March—the young stem is cut at from six to eight inches from the ground so as to favour the pro-duction of a large crop of shoots, which will afterwards form the principal branches of the tree. These shoots are allowed to grow through the summer, and are earthed up in autumn. This process is performed according to the following directions :—A dry day is chosen when the soil is in a friable condition, so that it will fill the spaces between the branches without leaving any empty places. The soil used should be free from leaves, grass, and straw, which if they were allowed to come in contact with the buried branches would cause them to become rotten. It is also necessary to pull off the half-grown autumn figs, which would rot in the earth, and cause the same mischief as any other decomposing vegetable matter. These precautions having been taken, the branches of the tree are divided into four equal bundles, each being tied together with string. As many trenches as there are bundles of branches are then dug in the ground. Each trench commences at the foot of the tree, and is made of sufficient depth to contain the bundles of branches. They are dug in different directions, according as the ground is inclined or horizontal. In the former case they are dug all in the same upward direction as in Fig. 217; when, however, the ground is horizontal, they radiate equally

Fig. 214.

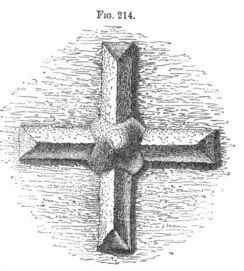

Showing the Mode of burying the Fig Trees cultivated on level ground, to preserve them from being destroyed by frost in winter.

from the centre. The earthing up of the branches being accomplished according to these directions, each bundle is covered with mould to the depth of eight inches, a small cone being piled up exactly over the root.

Towards the end of February of the fourth or fifth year of planting, a damp warm day is chosen for uncovering the buried Fig trees. The sooner this operation is accomplished the more forward will be the growth of the tree, and the ripening of the fruit; but the early fruit is often destroyed by the late frosts.   For this reason some growers prefer to defer this operation until the end of March, although the trees frequently suffer from being thus suddenly exposed to the heat of the sun, and the fruit does not ripen so well.   Others uncover one-half their trees at the end of February, and the other half at the end of March. By this means a better average crop is insured both in quality and quantity.   The branches are separated from each other by equal distances so as to avoid confusion, as

FIG. 215.

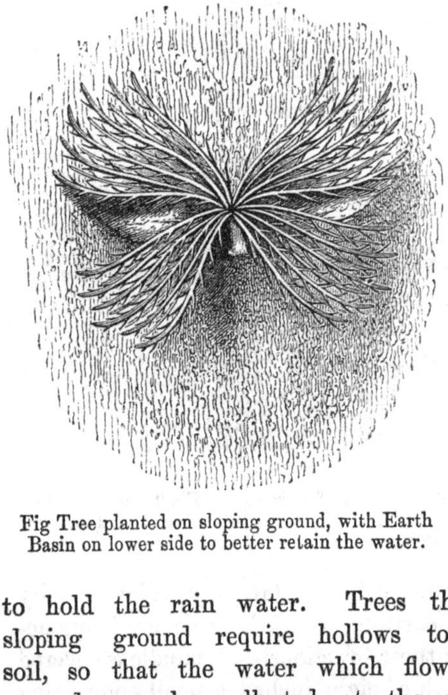

Fig Tree planted on sloping ground, with Earth Basin on lower side to better retain the water.

well as to prevent the leaves from rubbing against the fruit, which would have the effect of blackening them, and render them c o m p a r a t i v e l y worthless.   Those branches that are too near the ground are also held up by means of forked pieces of wood. The soil is carefully levelled where the ground is horizontal, a little hollow being made round the root of the tree to hold the rain water.   Trees that are planted on sloping ground require hollows to be made in the soil, so that the water which flows from the higher ground may be collected at the root of each tree. In this way a proper degree of moisture is insured during the whole of the summer, besides which the soil is prevented from being cut up by the rains. This plan would seem to be peculiarly well suited to plantations on the steep slopes of railway embankments.   Henceforward the young shoots growing from the stock are carefully cut off; otherwise they would weaken the larger branches. These precautions are taken during the fifth year.

In the spring of the sixth year the oldest branches are of the form shown in Fig. 218. The operation of nipping off the buds at the end of each branch is performed as soon as the uncovered trees begin

FIG. 216.

Section showing Fig Tree planted on inclined ground, with Earth Basin to retain the water.

to show signs of springing into leaf, that is to say that on some fine day the bud on the end of each lateral shoot is pinched off so as to favour the formation of buds on the wood lower down, as well as for the encouragement of any young figs that already show signs of making their appearance. About one-half of the buds on the side branches are also pinched off, choosing those that are nearest

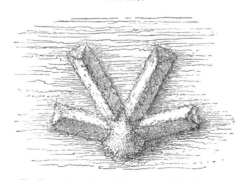

FIG. 217.

Fig Tree planted on sloping earth buried for the winter months.

to the young figs. Two, however, are always left on near the base of each branch, and one towards the tip, in order to draw up the sap. The end shoot of each branch is treated in the

D D

same manner but with this difference, that the bud immediately below the one at the end is allowed
to remain on, as well as one or two more
for the purpose of producing side branches,
which ought to be left about a foot from
each other on each stem. As soon as the
young shoots attain the length of about
two inches, the shoots on all the lateral
branches and on the end branch are nipped
off—a fine day being chosen for the purpose.
Of the former only a single shoot—the
one nearest the base of the branch—is
allowed to remain so as to replace the one
which bears the fruit of the year. The
shoot at the end of the terminal branch is
allowed to remain, and some of the lateral
ones intended to bear fruit in the following year. These last are spaced out so
that they may receive an equal amount of
sunshine without being interwoven or rubbing their leaves against the fruit. As
soon as the proper number of branches
that each stool ought to bear is reached,
all new shoots growing on the parent stem are nipped off.

FIG. 218.

Stem of Fig the sixth
year after planting.
The points of the
shoots A, A, A, A,
are pinched off in
spring to favour the
development of the
Figs, and also of
wood-buds at the
base of the shoots.

Although the figs which make their
appearance during the current year
ripen with difficulty, a certain number
may be grown in favourable years.
To hasten their ripening you must
proceed in the following manner.
Those branches which appear to be
most prolific are allowed to retain
two shoots at their base instead of
only one. The one nearest the base
is intended to produce the young figs
for the following year, the other the
autumn figs. In order to force
these latter into rapid growth the
end of the shoot must be nipped

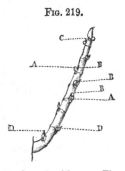

FIG. 219.

Fig branch with young Figs.
The lateral side buds,
A, B, B, B, are pinched in
spring,—two at the base,
D, D, and one at the apex,
C, being left.

off as soon as it has attained the length of four or five inches. As the process of forcing autumn figs to ripen makes the trees weaker and less able to produce the buds, or "fig flowers," for the next year, only those which are strong and vigorous should be chosen for the purpose.

Should the late frosts destroy the crop of figs, an event which may take place before the middle of May, summer pruning must be resorted to, that is to say, each lateral branch is pruned back to near the stem. This operation causes the sap to flow to the old wood and produce a large crop of shoots. This circumstance is taken advantage of to fill up empty spaces, of course taking care to leave only those shoots growing that are really useful. The shoots are thinned out according to the method already described.

After the figs are gathered, each branch bearing fruit presents the appearance shown in fig. 220, or that in fig. 221, if the shoots have been retained so that we may have autumn figs. Towards the end of August a dry day is chosen for cleaning the figgery. The portions of the shoots which

Fig. 220.

Branch of Fig Tree after the gathering of the crop. Should the year be unusually warm, some of the lower Figs may ripen; if not, they are removed. The shoot that has borne figs is cut at B.

have borne fruit are cut off as shown at fig. 220, and useless shoots are taken away just above the lowest eye. If this eye should develope the succeeding year it is disbudded in its turn. Withered branches are also removed quite close to the stem, care being taken to cover the open places with grafting wax. Some growers leave this operation until the spring of the following year, but prunings made at that time give rise to a much greater loss of sap, and the wounds made in the tree heal up with great difficulty.

In the spring of the seventh year the lateral branches of each stem are treated like those of the preceding year. The other operations are similar to those already described. The principal branches are allowed to grow longer every year, taking care to allow the fruit-bearing shoots, which are replaced from year to year, like those of the Peach tree, to remain at regular intervals. When the branches have grown to the length of from six to nine feet, their growth should be checked, otherwise the sap will desert the fruit-bearing branches at the sides, and so cause them to wither away. When sufficiently long the principal are treated in the same way as directed for the side branches.

FIG. 221.

Branch of Fig Tree after the gathering of the crop. C bears the young Figs for the coming year's crop; D is pinched back to help the ripening of some of the Figs of the current year; and the fruit has been gathered from the naked shoot, which is cut at B.

The earthing up to which the branches of the Fig tree are subjected every year causes them to grow in a horizontal direction a foot or eighteen inches from the ground. This is an element of success, for on the one hand the fruit nearest to the ground receives the greater part of the heat and ripens readily, and on the other the sap is more evenly distributed amongst the different side branches. The Argenteuil Fig trees begin to bear when they are six years old, and are in full perfection at ten years. They live a long time, but it is necessary to renew the long and old stems, which wear out every twelve or fifteen years. For this purpose the requisite number of shoots are allowed to grow on the parent stem to replace those which are cut away in the

August following. The soil round the trees should be dug up every year in the spring after having unearthed the branches and before covering in the trenches at the foot of the tree. They should also be well watered several times during the summer and manured every three years. The practice of putting a drop of fine oil into the eye of the fruit just as it colours and shows signs of opening, to hasten its maturity, is employed about Paris, especially during cold summers.

Preserving Grapes through the Winter without letting them hang on the Vines.

The preservation of grapes through the winter with the least amount of trouble is one of the most important of all matters to the British grape grower. Every cultivator, young or old, knows to his cost what a task it is to keep grapes hanging all the winter after they are ripe, especially in places where there are a good many houses devoted to vines. The latest books on the vine give directions for regulating the vineries so as to preserve the grapes on the vine after they are ripe, and every calendar of operations tells us how to manage them in that respect, though I fear the directions are not always intelligible. Here, for instance, is an extract from a recent issue of a leading garden paper:—" Those who wish to keep grapes hanging as fresh and plump as possible to the longest possible period, must take care not to afford them too much heat, as an excess of this, no matter how dry the structure may be, or how favourably treated otherwise, is sure to cause them to shrivel. more or less prematurely. Give only just such warmth to the pipes or flues as will insure sufficient buoyancy to any humidity (!) which may arise in the house as to enable it to make its escape. Independently of the ill effects caused by actual heat, a too warm atmosphere, even in the driest house, will cause a correspondingly excessive evaporation and consequent condensation." Then of course we must have fire heat and give air when foggy days occur, " as," says Mr. Thompson of Chiswick—" the mean temperature of this month (November) is on the average little above 40°, and the air is generally saturated with moisture. When this is the case, moisture will be deposited on all substances exposed to the air, if they are not warmer than it is. Grapes that are ripe should therefore be kept warmer than the air, otherwise they will be liable to damp. The application of fire heat would effect this : but if it were applied suddenly, and without air being given at the same time, the heated air would deposit moisture on the berries ; for although these would ultimately acquire the same temperature as that of the air surrounding them, yet for a time they would be colder, and so long as this is the case they would act as condensers of the moisture in the warmer air in contact with them. The more rapidly the air is heated, the greater for a time will be the difference between the temperature of the fruit and that of the air, and of course the slower the heating the less at any time will be the difference. Give therefore in damp weather, a little fire heat in the morning and admit air. If the nights are cold, the temperature of the house should not be allowed to fall lower than 45°."

Here then are nice operations and a lot of trouble to bestow on perhaps half a dozen houses during the winter months ! If the vineries are shaky and badly heated, the task is most difficult and annoying ; in the best constructed ones it is a great and needless labour. The trouble of regulating the atmosphere, the expense for fire heat, and the necessity of keeping the house almost entirely devoted to the grapes, must render any improvement very acceptable. Several times during the spring of 1867 I noticed grapes hanging from branches the ends of which were inserted in vases of water—grapes which the exhibitors described as having been for a long time so preserved in a fresh state. From such few specimens I did not derive sufficient confidence in the method to speak with certainty of its merits, but having since then visited a good many gardens in which the method is practised, I found that it is accepted as a great boon by some of the best gardeners in France, and their system of keeping grapes has been altered accordingly.

The best example in a private place was in the gardens of Ferrières, the magnificent country seat of Baron Rothschild. Here they have constructed, in addition to very fine and well filled fruit rooms, a grape room, which is filled with stands thickly hung from top to bottom with all kinds of grapes. M. Bergman, the manager, was cutting down all his grapes in harvest fashion, and would in a few weeks, as soon as the latest houses were ripe, have his many and well managed vineries to do as he pleased with: ripen the wood, prune and clean the vines, or utilize the cleared space of the houses for any purpose that might be con-

Fig. 222.   Fig. 223.   Fig. 224.

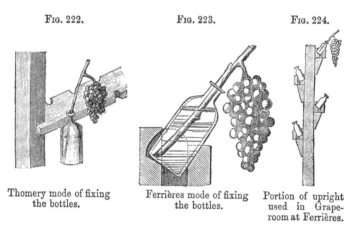

Thomery mode of fixing the bottles.   Ferrières mode of fixing the bottles.   Portion of upright used in Grape-room at Ferrières.

venient, not fearing as we do to spill a drop of water or make full use of the house.

The grapes are cut with a considerable portion of the shoot attached, much as if one were pruning the vine ; the shoot is inserted in a narrow-necked and small bottle containing water, and these little bottles are fixed firmly along, so that the bunches hang just clear of each other. In the first instance two pretty strong uprights are erected, each supported on three legs. Then from one to the other of these, on both sides alternately, are nailed sets of strong laths, two for each line of bottles. These laths are kept an inch and a half or so apart by a piece of wood at each end ; in the inner one are made incisions, into which the bottom

of each little bottle fits, and then the outer lath has a con-
cave incision in which the side of the bottle rests, so that,
caught in the inner and leaning firmly on the outer lath, it
holds the stem and stout bunch quite firmly.   I thus particularize it from having seen other ways of doing the same
thing less neatly and simply than this.   Walking space was
left between the walls of grapes ; for six or seven rows were
arranged one above another on both sides of each support.

Charcoal is mixed with water, allowed to stand for some
time, and then the water is strained off to fill the bottles.
But there can be no doubt that to put a pinch of animal
charcoal in each bottle would prove a better plan of guarding
the water from any impurity from the slight deposit of or-
ganic matter that might be expected ; at least, it does not
seem very clear how charcoal removed from the water before
the vine-stem is put in can have much effect in keeping it
pure.   However, this is not an important matter, and it is
certain that a pinch of animal charcoal, which is very cheap,
will keep the water quite sweet.   One cultivator who keeps
grapes on a large scale by this method, never uses any charcoal at all, but simply fills his little bottles almost full with
water, and then inserts the branches, which nearly close the
necks of them.   He appeared quite as well satisfied with
the plan as those who had taken more pains to keep the
water sweet.   In case evaporation should cause the water
in the bottles to fall below the bases of the shoots it is
simply necessary to add a little more.

Of course it will be understood in a moment that with
one-tenth the amount of expense and trouble that is now
necessary in large grape-growing places, we may in a grape-
room like this maintain conditions infinitely better calculated
for the preservation of the fruit than the atmosphere of any
vinery can possibly be.   We may keep the fruit dark, preserve the necessary amount of dryness in the atmosphere,
and keep up a temperature constantly equal—all of which
are essential to the well-being of fruits, and none of which
can possibly be attained in the house in which the grapes
are grown.   It would of course be wise, in arranging a room
of the sort, to have hollow walls and other contrivances to

attain the conditions under which fruit is known to keep best.

M. Rose-Charmeux, the great grape grower of Thomery, was, I believe, the first to try this plan. Now, as we grow by far the best and largest quantity of hothouse grapes of any country, this method will prove of far more use to us than to the French. I was told by experienced French growers who have adopted the system, that they keep the fruit as long this way as upon the vine, with fewer mouldy

Fig. 225.

Interior of Grape-room in which the System described is carried out.

berries, and almost without trouble; and it is not likely that a man would cut down half a dozen houses of fine grapes at the beginning of October unless he had already proved it to be a good system. The advantage of having all the stock of grapes safely housed and away from the attacks of vermin and other interlopers, is another of the many presented by this plan, which I now leave in the reader's hands for trial, confident that it will prove a great boon to the grape grower, and tend to make that fruit—

every day growing in popularity—a great deal more enjoyable and obtainable in the winter and early spring months. For if it be a process requiring much care in large well-conducted gardens, how much more difficult must it be for the large class of amateurs and small gardeners to preserve their fruit in good condition? In places where the stock of grapes is not sufficient to require a special room for their keeping, part of the fruit room might be adopted, or even a dry cellar or store-room.

The above was written previous to visiting M. Rose-Charmeux, with whom the system originated. I have since seen his grapes stored for the winter; the method was in full working order, and even more simple and effective than could have been supposed. He began by having a stove and a couple of chimneys to try to regulate the atmosphere of his large grape-room; but finding that the grapes keep very much better without this, he now simply devotes to his winter stock a large room in his house, fitting it up in all parts to accommodate handily the little bottles before spoken of, padding the inside of the windows so as to exclude light, and obviate, as far as possible, changes of temperature. The grapes are cut in October, and preserved in good condition until April, when his earliest are ripe. He has frequently shown them in May, and even later, and has kept them till August; but of course the quality cannot be expected to be good after such very long keeping, which is merely done for the sake of show. A small room in M. Rose-Charmeux's house illustrates to a nicety the fact that a similar one in most houses may be made to answer the purposes of keeping grapes. It has no windows, and scarcely any means of ventilation. The house is heated by hot air; but while there are openings in the floor of the passages and other rooms to admit this, there are none in this little room in which the grapes keep perfectly. Thus it is clear that the ordinary dwelling house will present suitable conditions for the long preservation of grapes. The system was attractive enough when it was considered necessary to construct a room specially to carry it out; it is much more so now when it has been proved that not only is it not necessary to take any special means to warm or ventilate the structure, but that the grapes keep much better without that trouble. The first result of the method was a gain to the village of Thomery, which is almost wholly occupied with grape culture, of from 100,000 to 150,000 francs per annum. The system enables the cultivators to keep their grapes much later than of old, and thus to add considerably to their revenue.

Since the above was written this system has been tried and favourably reported on by Mr. Hill of Keele Hall, the famous grape grower, and by other practical men. Mr. Whittaker of Crewe Hall sent some to a meeting of the Horticultural Society, but he had taken unnecessary trouble by corking and sealing the bottles. The insertion of the shoot into a bottle of water is all that is required, and as the bottles used are little more than wide enough at the neck to admit the shoot, the evaporating surface exposed is very small. It was urged against the method at this meeting that the grapes " lose their sugar." This is not the case unless the fruit is kept a very long time. The French in carrying out their experiments have kept some of their grapes as long as they could, and have frequently shown them in a nice plump condition long after they ripen their early grapes—just for the " honour of the thing." In these instances a loss of sugar was no doubt perceptible; but what kind of flavour would berries possess if left hanging on the vine till the summer months when the Frenchmen exhibited their grapes ? The necessity for keeping the grapes till they lose their sugar does not exist. In most of our large gardens grapes are forced early, and would be ripe before the fruit of the previous year had lost its virtues in the least degree. And in our comparatively small gardens, containing perhaps a vinery or two, how many bunches of grapes are left after the consumption of the winter months ? To be able to clear the vineries of grapes for two months before the ordinary time would be a decided gain to thousands of gardeners in this country.

" About the 15th of April," says Mr. Thompson, " the sap began to rise in the vines, and some of the berries that were a little shrivelled suddenly got plump, while others that had shown no signs of shrivelling burst their skins, and the sap of the vine that had forced itself into them began to drip from them !"   Surely even in such a case as this it would be a gain to the grape grower to cut his grapes a few weeks before any danger of such a thing existed, and thereby keep them a little longer from bursting their skins and giving forth what cannot be very rich in sugar !   The expense and care required to keep grapes during the dull and cold months of winter in the ordinary way is very considerable, and the inconvenience and loss of space are great. The latest writers among grape growers recommend the surface of the "interior borders to be kept perfectly dry and to remain so all the winter, care being taken that as little sweeping or raking takes place as possible, for by this means dust is raised which settles on the bunches."   Practically speaking, houses treated in this way are nearly useless for anything except keeping the grapes, consuming fuel, and wasting labour.   Remove the necessity of keeping grapes on the vines long after they are thoroughly ripe, and the houses may be filled at a season when every inch of the room in vineries is wanted for storing plants.

The Culture of the Vine at Thomery.

At first I had intended to say nothing whatever about the grapes grown against walls in the open air, but further consideration has shown me that the culture of the grape in this way may be attempted with profit over a large part of the southern and midland counties of England, and therefore an account is given of the successful and highly interesting culture of the Chasselas Grape near Paris, where it must be grown against walls as well as with us.   Respecting this grape, it may be well to notice that when well ripened against walls the French think it the best grape ever grown, and superior to our hothouse grapes, fine as they look.   Here I am simply stating an opinion without endorsing it, merely adding that this estimate is not solely confined to those who have no opportunity of judging both sides of the question, but was held by the late Baron Rothschild, who grew all our finest grapes.   Grape culture is often successful against houses with us when it receives mere chance attention from cottagers and others.   By selecting the soil and position, and really paying some attention to protecting and cultivating the Vine, we may grow capital grapes against our walls, even in many places where ground vineries are now resorted to.   Should any person doubt the possibility of cultivating the Chasselas and others of our best hardy grapes in the open air, I have merely to

FIG. 226.

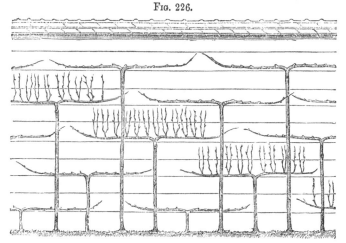

Wall of Chasselas at Thomery, showing the Vines trained as Horizontal Cordons, and both in a pruned and unpruned condition.

refer him to the horticultural papers for the autumn of 1868. They contain abundant evidence that even with the rough treatment grapes now receive in the open air, it is quite possible to grow them of good quality on walls.   Grapes are already grown well in the open air in a few places— by Mr. Darkin, at Bury St. Edmunds, for example ; and by Mr. Fenn, in the Rectory Garden at Woodstock ; so that there can be no doubt about the possibility of ripening good grapes over a considerable portion of England and Ireland.

It is necessary to observe that the plan is only recommended for warm soils and positions, for gardens not having much glass and yet some wall space, for covering cottages, out-offices, &c., and not in any way as a substitute for Vine culture indoors. It may, nevertheless, be added that I have never yet tasted the Chasselas de Fontainebleau or Royal Muscadine nearly so well flavoured as when grown in the open air, and that all who admire this grape would do well to attempt its culture on warm walls. The culture of the

FIG. 227.

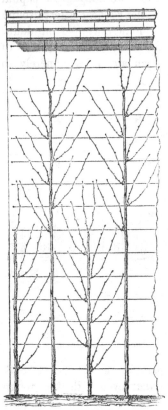

FIG. 228.

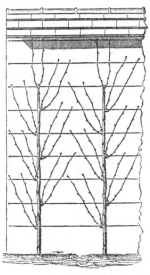

Rose-Charmeux's System of Vertical Training. The Vines are planted at sixteen inches apart.

Vines trained Vertically with alternated spurs, wires nine inches apart on wall; Vines about twenty-eight inches apart.

Chasselas de Fontainebleau at Thomery and other places in the vicinity of Paris is the best example of open air culture anywhere to be found; and this variety, more generally known in England as the Royal Muscadine, is also far the best for culture in the open air in this country.

An account of the grape growing at this place from the pen of M. Rose-Charmeux is likely to convey the most practical information on the subject, and the following is translated from his " Culture de Chasselas :"—

Fig. 229.

Low Double Espalier, and Mode of Protecting the Vines.

•" At Thomery the soil is of a sandy and clayey nature, and mixed with pebbles in those parts which are near the river. The soil is at all times easy to work. Near the Seine it lacks depth—so much so, indeed, that before cultivation it has to be dug and trenched so as to remove some of the stony subsoil. Everywhere else the layer of vegetable mould measures from four feet six inches to six feet in thickness. This layer lies on a reddish clay of about the same thickness, and beneath the clay is a broken - up stratum of building stone filled with fissures. This building stone is easily extracted. The grapes ripen a fortnight earlier in the flinty districts than in those parts in which the soil is deeper and richer.

Fig. 230.

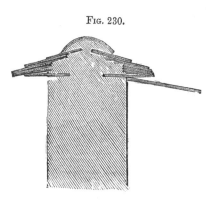

Section of top of wall at Thomery, showing the projection of the temporary coping.

" The gardens at Thomery, taken altogether, present much the appearance of those of Montreuil-sur-Bois. There is nothing but walls in all directions, distant from

each other about forty feet, and ten feet high. This
height has only obtained during the last fifteen years,
before which period they were rarely higher than six
or seven feet. The change has been advantageous for
two reasons; first, the grape growers have been able to
increase the space required for their purpose by taking possession of a larger portion of air instead of having to buy
fresh ground; and secondly, the high walls
are found to improve the appearance and
quality of the grapes. The walls are built of
hard stone quarried in the neighbourhood,

FIG. 232.

FIG. 231.

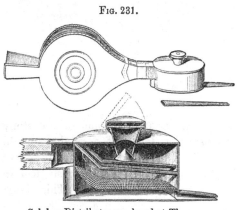

Sulphur Distributor employed at Thomery.

the stones being laid with mud only. The
face of the wall is then covered with a
mortar made of lime and sand, and is
finally covered with the same material
thinned to a cream.

Pruning to obtain
the two arms of
the Cordon.

" Every wall is topped with a roof of pantiles, surmounted
by a row of gutter tiles. These roofs project about ten inches,
and below them are fixed at every yard iron rods, inclined
slightly downwards. These supports project about twenty
inches beyond the edge of the tiles, affording altogether a
support of at least two feet six inches wide. Upon this is
fixed, when occasion requires it, a coping of bituminized felt,
or, where economy is necessary, a piece of thin plank. The

bituminized felt is stretched on frames of wood, about ten feet in length by eighteen inches in width, the felt being fastened to them by means of small nails. These frames are chiefly used when the grapes are perfectly ripe, which is generally about September 15, or when there is danger of the fruit being spoilt by heavy rains. Formerly, before these methods of shelter were employed, large quantities of grapes were continually lost through becoming rotten with the wet; since their adoption, however, there is no fear of such a result. The size of the temporary copings to be used is always de-

Fig. 234.

Fig. 233.

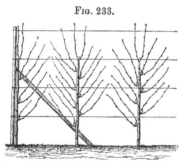

Low Espalier of Vines trained vertically, four feet high.

Layer of Vine raised and planted in basket.

pendent on the aspect and height of the walls. With walls facing the south and ten feet high, frames containing felt at least thirty inches in width ought to be used. With a western aspect, they ought to be even wider, in order to avoid all danger from the heavy rains. With the old low walls, frames twenty-four inches wide for the south, twenty-eight inches for the west, and sixteen inches for the east, were found to be quite sufficient."

"After selecting a proper position and soil, the most important point is the sulphuring to prevent the Oidium.

E E

Sulphur is the effective cure for this pest, and it should be applied directly after the first pinching of the shoots, at a temperature below 96° Fahr. in the open air.   If the heat is too great, the young skin of the grape is liable to become decomposed.   In full sunshine at noon the fruit would be burnt up in an hour's time.   Sulphuring may be carried on while the dew is falling.   There is no fear in this case of soiling the grapes.   The operation should not be deferred until the Oidium has made its appearance.   The second

FIG. 235.

Moveable Scaffold used for thinning the Grapes.

sulphuring should be performed when the grapes are about as large as a pea, or even earlier if the Oidium has appeared at all. It would be preferable to sulphur while the vines are in flower. The operation is performed with sublimated sulphur, blown upon the vine with a pair of bellows (Fig. 231) specially contrived for the purpose.   It may be effectively done without the operator standing an instant in one spot, but passing quickly along the line.   In these latitudes heavy rains destroy in part the effect of the sulphur, and it is nearly always necessary to repeat the operation three or four times.

If the grapes themselves are attacked, it is on them that the
flowers of sulphur should be applied.   It has been remarked
that under sunshine the Oidium may be totally destroyed in
one hour, a result that may be attributed to the speedier
disengagement of sulphurous acid gas by the heat of the sun,
but it is dangerous to apply it if the sun is too strong."

The pruning of the vine is so well understood in England
that it is needless to give it here in the full detail with which
it is honoured in M. Rose-Charmeux's book, the "Culture
du Chasselas." Indeed, after having translated his direc-

FIG. 236.

FIG. 237.

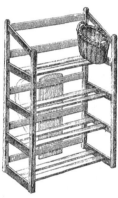

Shade to protect the Grape-thinners from
strong sun.

Frame for carrying small baskets filled with Grapes from
the walls to the store-rooms :
four feet high at back,
thirty-one inches wide, and
ten inches and a half deep.

tions their painful and unnecessary minuteness and great
length have obliged me to omit them.   The system as shown
in Figs. 227 and 228 is simply the well known spur pruning
practised in nearly every English vinery.   There are indeed
several modifications of training ; but this as everybody knows
is of no real importance.   In this case, as with the vine
indoors, the selection of a proper medium for the roots is of
far greater importance than anything else, while the
simplest form and the best system of pruning are without
doubt the same as those seen in our vineries—an erect

stem with the side shoots annually pruned in. At Thomery
the vine is frequently trained as a horizontal cordon line
over line ; but to execute this form well requires time and
skill, which only cultivators who devote themselves specially
to it can afford, and it may be safely said that letting the
vines run straight up the walls and with their spurs at each
side is better than any less simple mode. The really im-

Fig. 238.   Fig. 239.   Fig. 240.

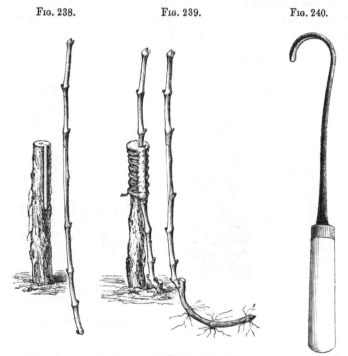

Mode of Grafting the   Mode of Grafting the Vine, by   Gouge   used   in
    Vine at Thomery.   approach, practised at Tho-   Grafting the Vine:
                            mery.   ten inches long.

portant points to bear in mind are—first, the warmer the
exposure is, the better for the grape ; second, that the walls
are white, or nearly so, as the vines get more heat on such
walls than they do on dark ones, and are maintained in
better health ; third, that wide and efficient copings are
used to permit the fruit to thoroughly ripen in autumn,
and prevent its being spoiled by heavy rains ; and that the

higher walls are found to possess an advantage over the lower ones. The plants are frequently raised in rough baskets for convenience of removal and sale. Several of the appliances here in use are sensible ones, which might be found useful in other ways than that of vine culture. I allude to the moveable scaffold to facilitate the labours of the women who attend to the walls in summer (Fig. 235), the shade to shield them from the sun (Fig. 236), and the frame for conveying a number of small baskets laden with grapes from the walls to the grape room (Fig. 237).

Grafting is frequently performed, and chiefly to replace a bad by a good variety, or to hasten the fructification of a new one. The plant is cut down to within nine or ten inches of the soil, and with the gouge (Fig. 240) an incision is made on the smoothest side, a corresponding cut being made in a scion or in the stem of a young plant, both of which methods are shown in Figs. 238 and 239. The grafting is performed as soon as the sap begins to move in spring, and the grafts are tied and covered with grafting wax, as shown in Fig. 239.

FIG. 241.

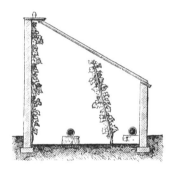

Small Pit used for Forcing the Vine.

A particularly noticeable feature in the cultivation, is that the young vines are as a rule planted at a considerable distance from the wall—say a little more than three feet, and the stem laid into the ground to near the base of the wall. Sometimes the stem is allowed to rise some distance from the wall, and in the following year when it has grown a little it is again lowered and taken to the wall. This method is obviously pursued to secure a number of vigorous roots spread over a large surface. Where the ground is stony and poor it is probably a good plan.

As regards the forcing of grapes at Thomery, I need hardly say there is little to note of any importance to the British grape grower, who is certainly in advance of all others as regards the indoor culture of this old and ever popular fruit. Nevertheless, M. Rose-Charmeux's garden exhibits such an advance on the ordinary style of forcing grapes around Paris that it deserves a few words. " The walls of the pits are of brick; the highest, towards the north, measures about five feet in height; the front wall being only about two feet high. The width of the hothouse at its base between the walls is about four feet six inches, and the length indefinite. The higher wall is covered on the top with a deal board a foot wide and

Fig. 242.

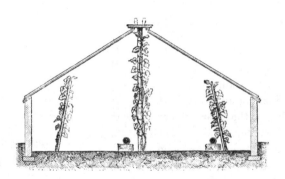

Small span-roofed house for forcing the Vine : ten feet five inches wide, and five feet five inches high.

projecting towards the south; the lower wall is covered in the same way with a board five inches wide. The walls ought to be rough cast, and kept perfectly white like those of the gardens. Bars of iron serve as supports to the frames, and to keep the walls in their places when the frames are taken away, and rods provided with holes are placed in the middle of each frame so that they may be opened to different heights according to circumstances. A copper hotwater pipe, four inches in diameter, serves to warm the structure, and an entrance-door is constructed at each end. Grape forcing begins from the 15th to the 25th of December, in order to have ripe fruit by the end of April. During the first fortnight the heat is not allowed to rise above from 58° F. to 65° F. The fortnight after it is allowed to rise to 78° F. or 80° F., from which time until the grapes are ripe the heat is maintained at from 80° F. to 88° F. The time of flowering requires a great deal of attention, for on it depends entirely the success of the result. In order that fecundation should take place under the most favourable circumstances, and that the grapes should be well formed, it is absolutely necessary that the temperature should be maintained between 78° F. and 88° F.; also that the vine should have plenty of light and dry air."

" The low span-roofed house is constructed in the following manner :—On the east and west are built two small brick walls twenty-eight inches high, and in the centre of the enclosed space are placed strong posts about five feet high, and distant from each other about three feet. A plank fourteen inches wide, nailed on the top of these posts, ties them together solidly and forms a sort of coping. This plank is covered with sheet zinc, and bars of iron are carried from it to the walls serving as supports to the lights. At each end a door is constructed for the attendants to go in and out, and on each side is a thermometer for regulating the temperature. The interior of the hothouse is about ten feet wide at the base, so that the rows of vines are distant from the side walls about eighteen or twenty inches, and one side gets the effect of the sun in the morning, the other in the afternoon. Two rows of pulleys are attached to the wooden coping for working the straw mats, which ought to be taken off every morning and replaced in the evening."

Thus M. Rose-Charmeux speaks of his forced culture of the vine. In addition to the houses here figured and alluded to he employs a well constructed portable lean-to house—portable because the French yet believe in the virtue of the plan of alternately forcing and resting their trees, a system which we have long ago proved to be worthless.

CHAPTER XVIII.

THE IMPERIAL FRUIT AND FORCING GARDENS AT VERSAILLES.
—THE NEW FRUIT GARDEN OF THE CITY OF PARIS IN THE
BOIS DE VINCENNES.

THE imperial fruit and forcing gardens at Versailles form a
large establishment, not so costly nor nearly so fine as
Frogmore, but containing a few things novel and instructive
to the English visitor.　Generally the crops do not display
the high cultivation nor the surface the rapid rotation to be
seen in the market gardens round Paris, but in the culture
of hardy fruits there is something to admire.　It is a forcing,
culinary, and fruit garden solely, therefore there are few
pot plants to be seen, the houses being nearly all devoted to
the pine-apple.　Some years ago the culture of this fruit
was considered by some of our gardening authorities to be
better understood in France than in England; but though
very fine pines are grown in the neighbourhood of Paris,
our pine growers are on the whole the best.　Such growers
of the pine-apple as Mr. James Barnes of Bicton, Mr.
David Thompson, Mr. Rose at Frogmore, and many other
English gardeners, afford us the best example of how to
produce it in the highest degree of perfection.　The forcing
department is usually well-ordered and neat so far as the
more permanent houses go.　In them the back walls may
be seen very prettily covered with the two well-known
Vincas, alba and rosæa.　To cover the walls of all kinds of
glass-houses devoted to ornamental purposes is an object
with most people who possess such things.　It is very rarely
well accomplished, mostly from using a bad selection of
vigorous growing plants, which often get covered with insect
filth, and become a capital breeding place for it, or perhaps
never yield flowers.　If anybody possessing a stove, pine-

house, or intermediate house, or any other warm structure
with a back walk and a border against it, will plant in it
and train against the wall the two pretty subjects named
above, plant for plant, the result will prove strikingly
pretty. The plants are always glossy and full of flower,
may be kept at two feet or allowed to grow six feet high,
and are always free from insects or vermin of any
kind. They keep neatly to the wall with but, little trouble,
and bloom all over the surface, top as well as bottom.
They are in this state very useful for cutting, and the effect,
when you enter the house, is of the most pleasing kind.
Their culture in this way is far more satisfactory than in pots,
and in almost every warm stove or forcing house in France
you see them trained against the back walls. The system
of forcing grapes and early vegetables in very small rough
frames is extensively practised here.

The fruit growing department is undergoing a gradual
and complete alteration, especially as regards the choicer
Pears trained as espaliers. So satisfactory is the system
adopted, that I am certain if English cultivators generally could get an idea of its excellence it would lead to a
revolution in our fruit culture, and a great improvement in
the appearance of our gardens. I know of no way whereby
we may so highly improve the garden culture of the Pear
than by paying more attention to it as an espalier tree. This
is also the opinion of many of the best fruit growers in
Britain, who agree that there is no finer fruit than that
gathered from well-managed espalier trees. It is well
known that some pears lose quality by being grown against
walls. It is equally certain that a fuller degree of sun and
exposure than the shoots and fruit get on a pyramidal tree is
very desirable in many parts of this country, especially for
particular kinds. Many sorts grow beautifully as pyramids ;
others, to be had in perfection, must be grown upon walls ;
but by means of the improved espalier system the majority
of the finer kinds may be grown to the highest excellence.
If the French can teach us nothing else they can certainly
give us a lesson as to the improvement in appearance,
cheapness, and utility of the espalier mode of growing

fruit, especially as regards the finer varieties of Pear trees.

It should be borne in mind that the good opinion of espalier trees given by British cultivators has been won by them under great disadvantages, for nothing can be uglier or more inefficient than the usual mode of supporting and training espaliers in our gardens. It is generally so costly and disagreeable to the eye, that it has been done away with for these reasons alone in many gardens. I know some important ones near London, and indeed in many parts of Britain, where the espalier support is the most unworkmanlike and discreditable affair to be seen in the place. Great rough uprights of wood, which soon rot and wabble out of position, thick and costly bolt-like wire, cumbrous and expensive construction, and, in a word, so many disadvantages as would suffice to prevent the prudent cultivator from attempting anything of the kind. The form of tree used, too, is such that the lower branches become impoverished, and often nearly useless.

To support his espalier fruit trees the Emperor's gardener, M. Hardy, has largely adopted a system which is at once cheap, neat, and almost everlasting. Instead of employing ugly and perishable wooden supports he erects uprights of T-iron, and connects these with slender galvanized wire. These are tightened with the little raidisseurs before alluded to, and then there is an end of all trouble. He manages to erect this trellising nine feet high for less than a shilling a yard run; but it could not be done so cheaply in smaller quantities. Then, instead of adopting the common form of espalier tree, with horizontal branches, he more frequently uses trees of which each branch ascends towards the top of the trellis, and thus secures an equable flow of sap through the tree. The accompanying figure (243) will give a better idea of both trellis and tree than any description. There is no more important matter connected with our fruit culture than this very point, and therefore I should be much obliged to all my readers, both amateur and professional, if they will give the subject attention, as I am sure that by doing so they will be led to largely adopt it, and much improve their fruit

culture.   The finest stores of pears I have ever seen were in
gardens with a good length of tree trained in this manner;
and I know few places in France where the espalier
system is so extensively and so well carried out as here.
The form here represented is much better than the cordon
or single-branched Pear tree, because a more free and
natural development is allowed to the tree, and at the same
time the trellis is covered quickly, and a considerable variety

FIG. 243.

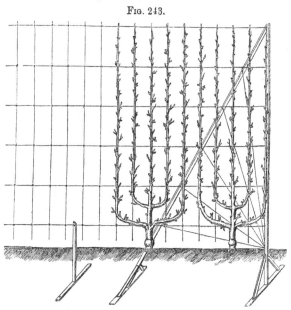

Trellis for Pear Trees : ten feet high. Uprights and stays of T-iron, horizontal
   lines slender galvanized wire ; vertical lines, pine-wood half an inch square
   and painted green : to these the ascending branches are trained.

of fruit may be obtained from a small space.   It is very
extensively adopted by M. Hardy, upon walls as well as on
the neat and elegant trellis, of which he has constructed so
much.   Of course the Palmette Verrier, the fan, or any
other form, may be trained on these trellises, but decidedly
the best are such as combine the advantages of quick
covering and early productiveness claimed for the cordon,
and the fuller development and more pleasing appearance

of the larger forms.   It should be borne in mind that
planting erect cordons close together, as they must be
planted, involves a great expense which is avoided by
using trees of a fuller development.   It takes a good many
years to form the large style of tree usually adopted,
and therefore I advise the general planting of these intermediate forms.

Nothing can be neater alongside garden walks than lines
such as these trained on the trellis alluded to.   There is
no shaking about of rough irons or wooden beams, no
falling down or loosening of the wires ; the fruit is firmly
attached and safe from gales, the wood is fully exposed,
and the trellis when well covered forms an elegant dividing
line in a garden.   The best way to place them is at from
three to six feet from the edge of the walk, and if in the
space between the espalier and the walk a line of the cordons elsewhere recommended be established, the effect and
result will prove very good indeed   In some cases where
large quantities of fruit are required, it may be desirable to
run them across the squares at a distance of fifteen or
eighteen feet apart.   The principle is quite simple, the proof
of which is that the trellises at Versailles were erected
by the garden workmen.   M. Hardy, the head gardener at
Versailles, is the son of the celebrated writer on fruit trees
of that name, and has had much experience in fruit growing.
" These trellises," says he, " are the cheapest as well as the
most ornamental that we have yet succeeded in making,
and the trees which I plant against them are of the form
that I prefer to all others, for promptly furnishing walls
and trellises, and for yielding a great number of varieties in
a comparatively restricted space."   The mode of employing
the uprights of pine wood painted green and reaching from
the top of the trellis to within six inches of the ground, is
not a common one, though very desirable where the erect
way of training the shoots is practised.   The reader will
readily perceive that this system combines the advantages of
the cordon and the large tree.   Of course many other
forms, or any form, may be used with this system of trellising, with slight modifications to suit different kinds of

trees or different forms.   The double trellis shown is simply
a modification of the preceding, and is not only desirable
where space is limited, but also for its economy, for one set of
uprights supports the two sets of wires simply by using cross

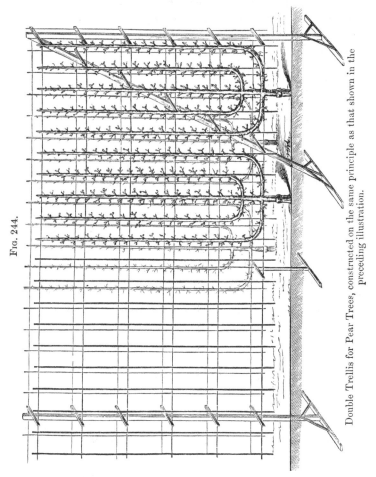

FIG. 244.

Double Trellis for Pear Trees, constructed on the same principle as that shown in the preceding illustration.

bits of iron about eighteen inches long, and at the desired
distance apart.   However, the engraving (fig. 244) shows
this at a glance.

The Pear as a low cordon is found to succeed very badly,
and to plant it as an oblique cordon at fifteen or eighteen

inches apart is considered much too close and very unwise. A white wall fourteen feet high covered with Easter Beurré Pears was very fine indeed. The trees were mostly on the Quince stock, a few on the Pear, but all bore equally well. They were all trained in the five-branched form usually adopted here, and had almost covered the tall white wall. The fruit-growing foreman insisted very strongly on the necessity of having white walls for fruit trees, and stated that dark ones injured both fruit and leaves, while white ones benefited both. White walls, apparently well lime-washed every year, are to be found in every good establishment, whether for peach, grape, or other wall fruit culture. The Easter Beurré may be seen here double-worked on the Curé. On one wall the trees are established and in good bearing; on another they had been budded last year only. The Curé is first grafted on the Quince and allowed to form five vertical branches before it is budded. The Easter Beurré is found to do best when double-grafted, though the trees directly on the Quince and Pear seemed to do well. The naked parts of the stems of fruit trees in this garden were in many cases protected from injury from a strong sun by being neatly covered with straight straw, tied with willow twigs. Neatly done, it seemed better than the commoner plan of placing slates or boards before them. Brackets to support straw mats in spring are placed on every wall at a little more than a yard apart.

There are a great many old and worn-out trees in the garden which have a bad effect on its appearance here and there, but the gradual adoption of the new trellises will much improve matters. The Pear makes as strong a growth here as I have ever seen it make in Britain, though some of our growers are continually saying that quite a different and very much more fruitful kind of wood is formed in the fine climate of France. There are a few specimens of forming letters with trees to be seen here, as in many other French gardens.

When I last visited this garden M. Hardy had commenced carrying out a system of protecting his espalier trees. The plan is simply to strain lines of galvanized wire above the top of the espalier, so as to form a low span when covered with rough canvas. The sides are not covered, but the protection at the top is sufficient to prevent radiation, and to throw off heavy rains when the trees are in bloom. If there is a wall running at right angles with the lines of espaliers, wires are stretched from it so as to form a light support over each espalier; if not, a post is driven in so as to support and stretch the wire in the firmest way. The lower of these two lines _____ may be sup-

FIG. 245.

Section of protection used for Es-palier Trees at Versailles.

posed to represent the top of the espalier, the upper a line firmly supported at a few inches above it. Wires are also stretched at each side of this, at about twenty inches from it, so as to form the outline of a very low span-roof of strained wire. It is a matter of little difficulty to stretch cheap canvas of some kind over these wires, letting it be an inch or two narrower than the breadth between the outer wires, so that it may be strained tight, say a yard for the canvas, and two inches more for the wires. The outer margins must of course be firmly threaded to the outer wires with twine or any convenient tying or rough sewing material. Here they simply use

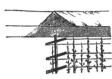

FIG. 246.

Side view of protection to double line of Espaliers.

the stems of the glaucous or Hard Rush (Juncus glaucus), which grows wild all over Britain, and find it answer admirably. A neat ridge is then arranged over each line of espa-liers, which throws off the rain and pre-vents radiation, thereby saving the bloom from frost and insuring a crop. The protection is put up before the buds are liable to be injured, and removed when the fruit is set, and all danger has passed away. Thus a very cheap and effective protection is secured. The old trellising used for fruit growing in these gardens is inferior compared to the new. The kinds of pears mostly grown here are Easter Beurré, by which several walls are covered; Duchesse d'Angoulême, of which there is a square of trellising in all nearly 600 yards long, and

about nine feet high; Beurré Diel, and Louise Bonne
d'Avranches.

The Peach is well-grown and trained in some parts of the
garden, a form with five main branches being adopted with
success. It is analogous to the form used for the pear in
the same garden, and is very readily made.

In addition to the trellises above described, the most remarkable feature of this garden is the presence of a vast
number of horizontal cordon Apple trees, both in single
lines and in superimposed ones of two or three stages, all
on galvanized wire. The trees are on the Paradise stock,
and nearly always confined to a single stem. These trees

Fig 247.

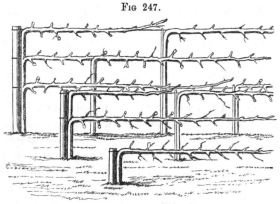

Border of Superimposed Cordons at Versailles.

bore an enormous crop during the year 1868, but the fine
apples were nearly all destroyed by the worm. At the end
of September, the display of fruit was quite remarkable, although much had fallen before that period, and the
year had been too hot for the perfect development of the
Apple. One border devoted to cordons is 300 mètres
(984 feet) long, and altogether there is 4000 mètres of
cordon apples in the garden. As the greater portion of this
length is composed of two and three lines of wires placed at
distances of a foot one above the other, there is really quite
8000 mètres, or more than five miles of horizontal (or
French) cordon Apple trees on the true Paradise stock, and

the plantations are being extended as often as circumstances will permit. It should be observed that though the cordons are often grown in lines one above the other, one plant does not furnish more than one line except at the ends. There, however, it is necessary to take several branches from one plant to fur- nish the two or three lines of wire starting from the same post. Here, as in many other gardens superintended

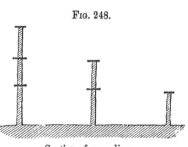

FIG. 248.

Section of preceding.

by experienced fruit growers, this mode of Apple growing is preferred to any other, but the enormous number planted best speaks of the estimation in which it is held. The cordons, though generally well-managed, are not quite so good as I have seen them elsewhere, and apparently from being too closely confined to the main stem. I have always noticed them best and most satisfactory when allowed to form a free and regular bush of spurs along the stem. The soil is as cold, stiff, and disagreeable for fruit culture as could well be devoted to that purpose.

The new Fruit Garden of the City of Paris in the Bois de Vincennes.

Not long since it was determined to make a new school of fruit culture for Paris, and in the spring of 1868 the first trees were planted. Naturally there is but very little to be seen as yet; but, nevertheless, a description of it can scarcely fail to be of use. As to plan and arrange- ment it is almost identical with that given overleaf, and recommended by M. Du Breuil for the north of France. It is situated near the Avenue Daumesnil entrance to the Bois de Vincennes. The first thing remarkable about the new garden is its walls; they are of felt, supported on a rough wooden framework. The felt is first nailed on frames of

F F

Fig. 249.

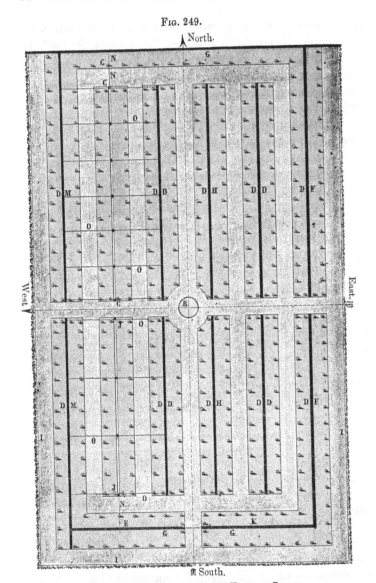

PLAN OF A FRUIT GARDEN FOR THE NORTH OF FRANCE.

B, Cistern. C, Double espalier for apricots. D, West side of walls planted with oblique cordon pear trees, 16 inches apart. E, Gooseberries. F, East side of wall, planted with peach trees as simple cordons, 16 inches apart. G, Cherries as oblique cordons, 16 inches apart. H, Plum trees, trained in like manner. I, Hedge. J, Summer pears, as vertical cordons on espaliers. K, Raspberries, cultivated at the foot of the wall. M, Wall of vines. The borders are surrounded by a cordon of apples, planted 7 feet apart, and 12 inches within the margin of the border. The black lines show the walls, and N and O wires stretched from the walls to support the espalier.

wood about six feet long by four feet wide, which are dropped into a groove made in the uprights, the stronger framework being based upon a few inches of masonry; the felt is whitened over, and the whole surmounted by a little ridge-like coping. This peculiar form of wall was erected in consequence of the objection of the authorities to have any walls of solid materials in the neighbourhood, which is so near the fort: but this merely helped to prove that in cold northern countries we may hope to grow good fruit by means of something less expensive than well-made brick walls. These walls are about nine feet high, except at the north end, where they are more than twelve feet high.

The garden, which is not a yard larger than is necessary for the purpose to which it is devoted, is in two divisions— one to illustrate the practical and profitable culture of fruit for market, the other all the important modes of fruit culture, the various curious and useful forms of wall and standard trees, and, in a word, most things necessary to know concerning the subject. The division devoted to illustrate the mode of culture best calculated to afford a quick and certain return is planted almost entirely with the finest of all winter Pears, Easter Beurré, and that well-known Apple the Calville Blanc, one of the best of all Apples for either dessert or culinary uses. The Pears are all cordons, either planted against walls or espaliers, and the Apples are all the low horizontal cordon, the form I have so often recommended. The most valuable and excellent fruits are the only ones cultivated. Most of the cordons against the walls are oblique (thus, / / / /), except at the high end wall, where they are vertical. The Professor's reason for adopting this form, is that the walls are more readily covered by it, and a much quicker return obtained; and of course he thinks these advantages compensate for the expense of planting so closely, or any other objection that may be urged against the system. Between three and four thousand trees of Easter Beurré, and the same number of Calville Blanc, are planted here in this small garden. The trees have done very poorly indeed, having been planted too late, and it is to be feared many of them will die, so that much in the way of healthy and fertile specimens will not be seen for some years.

One thing cannot fail to strike the British visitor who takes an interest in fruit growing, and to give him a valuable lesson at the same time; precautions to protect the trees effectually from wet and frost are taken, which are never seen or thought of in British gardens. All round the walls iron brackets project from immediately beneath the permanent wooden coping, to receive wide copings made of felt nailed on a cheap wooden framework, in lengths about six feet long and over two wide. These are slipped in under the short permanent coping, and rest on the bracket, the hooked point of which holds them in position. A small eye is at the under side of each, so as to thoroughly fix the coping by attaching each length with a piece of wire to another eye near the upper portion of the wall. Thus a most effective and excellent protection is afforded the delicate blossoms and fruit in spring. This is against the walls, where the British cultivator occasionally takes a little trouble to protect his trees from the cold rains and frosts of the budding and flowering season. Equal care is taken to protect the espalier trees—a thing which has never yet been attempted by British fruit growers, who, however, are not slow to contrast the difficulties they have to contend against with those of the French, for whom of course the climate is said to do everything. The protection for the espaliers is afforded by iron rods projecting from the top of the pine posts that are used to support the double espaliers, and running through them are six lines of galvanized wire, forming a sort of span over the trees. A little above these wires runs a stronger one, connecting the posts beneath it, and resting on the lower wires are two lines of

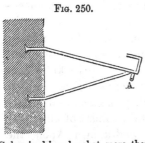

Fig. 250.

Galvanized iron bracket, more than two feet wide, for supporting a temporary coping of bituminized felt. A wire passes through at A for supporting curtains, where these are necessary.

neat thin frames of straw, each at least a yard wide. These are firmly fixed down to the wires, so that in spring the trees are placed under what may be called a neatly-thatched

FIG. 251.

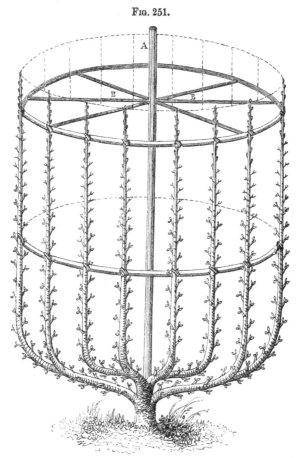

Fruit tree in the Vase form,—one of many forms more curious than useful. A, Stake; B, B, Crossed sticks to sustain hoop in position. To form a handsome tree of this kind, eighteen or twenty branches are required.

shed. No doubt some other material would look better than the straw, but it is cheap, and when nailed firmly between laths does not look untidy; and, moreover, it is the object of the place to show the cheapest as well as the best modes of protection, and also the best way of applying those most commonly in use; and the use of neat straw mats for protecting walls is very common in France. Posts of pine wood five or six inches in diameter are employed to support the espaliers, because they are cheap; and, to

Fig. 252.

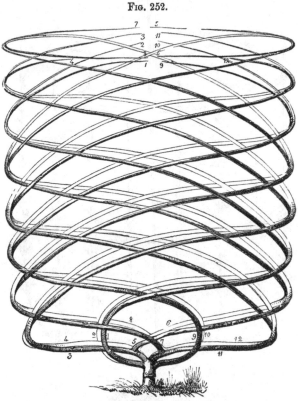

Pear trained in Vase form, with the branches crossed. The branches are grafted by approach where they cross each other, and the tree rendered self-supporting. It is somewhat better than the preceding form, and as easy to make.

secure their durability, they are thoroughly saturated with blue vitriol before being erected. This is a cumbrous and bad plan, the kind of fruit trellises employed at Versailles being neater, more durable, and in every way so superior that I am astonished that anybody who has seen the Versailles trellises could think of erecting such things as these.

In the garden devoted to teaching purposes only, all the lines which the branches of the wall trees are to pursue when fully formed are indicated by small rattan canes—accurately placed, so that as the tree grows the trainer has no hesitation as to the exact position each branch should take, but merely has to attach it to the rods so definitely laid down. The larger trees against the walls are mostly those I have figured as the Palmette Verrier. This is however occasionally trained " double," that is, it has two vertical stems instead of one. Useless as well as desirable forms are shown; for instance, trees formed like a goblet, with the branches crossed or ascending vertically, or sometimes like a goblet reversed. These are all useless for practical purposes, though they may serve to amuse an amateur; who, however, would do better to amuse himself with trees more beautiful, productive, and easy to train. The way of making a hedge of Pears—a hedge that when once made, and with its branches crossed and intertwined, will support itself—is also shown; and without doubt neat and productive screens may thus be made in any garden, and the trees kept quite as neatly as if supported by expensive trellising. Altogether the place will prove an instructive one after a few years to the British visitor, and particularly in convincing him of the necessity of protecting our finer wall fruits; but as a fruit garden it is quite unworthy of the city. There are amateurs' gardens about Paris better arranged and more instructive than this, specially designed to illustrate fruit culture in the capital of La Belle France, " the orchard of Europe!" Such was my impression when I visited it in September, 1868.

CHAPTER XIX.

THE PEACH GARDENS OF MONTREUIL.

The finest supplies of Peaches for the Paris market do not come, as perhaps many would suppose, from the sunny south or the balmy west, but from within a few miles of Paris, where they have to be grown on walls furnished with good copings, and receive in every way careful protection and culture. Approaching Montreuil the country is seen covered with good crops of vegetables and fruit to the tops of the pretty, low hills in the neighbourhood. All the crops, however, are divided into small plots, showing how each person has his own little portion, and has it moreover for ever if he so chooses—land being bought and sold here as simply as an overcoat is in England. But getting nearer still to the village, a great number of white walls, about eight or nine feet high, are seen, enclosing rather small squares of land, and almost entirely devoted to the Peach. As the walls are netted over many acres in some parts, the effect is curious when you look over them from a distance. In the squares are small fruit trees and all sorts of garden crops. To the visitor who takes a general look at the plantations here, it is quite apparent that it is not to the climate that the best growers owe their success. Among the two hundred and fifty cultivators having Peach gardens here, there are many with very shabby-looking trees on the walls, while those in some of the best gardens are perfect models of health, fertility, and skilful training. It will be seen by a glance at the cuts that the French mode of pruning the Peach tree is quite different from ours, inasmuch as they always aim at securing straight, well-formed, well-furnished, and equidistant branches, and always spur in the shoots rather closely in spring. The cuts showing their mode of pruning,

disbudding, covering bare spaces on the stem, &c., scattered
throughout this chapter, fully explain the regular and close-
pruning French system—any garden wall with a Peach tree
will illustrate our own.

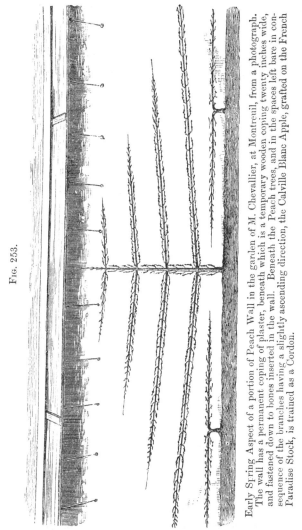

Fig. 253.

Early Spring Aspect of a portion of Peach Wall in the garden of M. Chevallier, at Montreuil, from a photograph.
The wall has a permanent coping of plaster, beneath which is a temporary wooden coping twenty inches wide,
and fastened down to bones inserted in the wall. Beneath the Peach trees, and in the spaces left bare in consequence of the branches having a slightly ascending direction, the Calville Blanc Apple, grafted on the French
Paradise Stock, is trained as a Cordon.

Fig. 257 shows a shoot that on its first pruning was cut back to
four or five inches, bore two good fruit, and furnished four shoots.

How these are to be dealt with is explained in the illustration,
and all other important operations in those that follow it.

The garden of M. Chevallier is less extensive than that
of the better known M. Lepère, but certainly displays
examples of cultivation not anywhere to be surpassed; and
no person interested in fruit-growing should visit the town
without seeing it. The first impression is very good, for the
outer side of the walls is covered with admirable specimens
of Peach trees, the narrow strip forming the border in which

FIG. 254.   FIG. 255.

Leaf of Peach Tree attacked by the
  Cloque, a disease caused by sudden
  cold at the commencement of
  vegetation.

Peach Shoot attacked by
Cloque.

they are planted being cut off from the road-side by a
fragile fence covered with vines. To merely walk along this
wall, without entering the garden at all, would repay the
visitor, so perfect are the trees in health, bearing, and
training. Overhead is a permanent coping of plaster, and
immediately beneath it, and at intervals of three or four feet,
the spokes of old wheels project eighteen inches; on these
are placed the temporary copings of boards or mats in spring
in this very paradisiacal climate. In the garden the same

admirable culture everywhere prevails. The walls are as white as snow,—they are whitewashed every year, with a view to the extermination of insects,—and the trees are of the brightest and healthiest green—quite a pleasure to look upon. The knuckle end of the leg-bone of a sheep projects from the wall, at intervals of a couple of inches only, and at about a foot and a half from the top of the wall. These are placed so as to firmly fix the temporary coping in spring. The boards or neat frames of straw are placed beneath the permanent coping and on the supports. The space between coping and brackets being very narrow, there is considerable support afforded the temporary covering, especially at the back part; by attaching an eye to its under surface, and firmly tying it with a twig of osier, wire, or strong twine, to the bone projecting so neatly and firmly below, it is perfectly secured from all danger of removal by winds. The cold rains which occur during the several months while the trees are in bud and flower, and all the time the shoots and newly formed fruit are tender, run off the plaster coping on to the temporary one, and from it safe beyond the trees, while radiation and consequently frost are effectually prevented from doing harm.

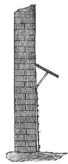

Fig. 256.

Small Wooden Coping used to protect young Peach Trees in spring.

To suppose that this thoughtful protection is merely necessary for the flowers and to secure fruit is a fallacy; a little temporary coping is improvised here even over quite young trees without a fruit on them, simply to guard their leaves in spring from the maladies consequent upon the extreme cold and many vicissitudes of the French climate at that season. This extemporized coping is simply formed by placing little wooden brackets against the wall at about four feet from the ground, and placing thereon a thin rough board. Such a thing is never thought of in England, where there is of course quite as much necessity for it. The effect of the sun on the stem and larger branches of the tree is also guarded against, pieces of bark or boards being placed before the short bole or base of the tree, the main branches on the upper parts being carefully shaded by training over them the young branches of the current year's wood.

The black marks seen on the white walls are lines which the main branches of the trees are to follow. In some cases they are quite simple vertical or horizontal lines, according as the form to be attained may require; in other places they form crowns, eagles, initial letters, flourishes, &c.; for though the cultivator generally prefers simple and definite forms, he is also proud of his skill in overcoming difficulties of training, and shows it by these curiously and very successfully trained trees against his walls. M. Chevallier is, however, a younger cultivator than M. Lepère, and has not his curiosities in this way perfect as yet, but there is every sign that his fancy trees will be even more elaborate and remarkable than those of M. Lepère. It is only just to state that these elaborately-trained trees bear freely and well;

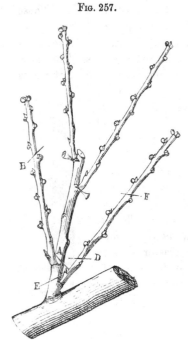

Fig. 257.

Second Pruning of Fruiting Peach Branch. F is cut at D above two wood-buds to furnish shoots for the following year; B remains to carry the fruit, and the shoot is cut at A. Cut E would only be applied if shoot B did not bear flower-buds.

but except for curiosity's sake or for show, they should not be attempted.

Branches of trees like that in Fig. 253, fifteen feet long, were three inches higher at the apex than at the base, a difference which scarcely removed them from the horizontal position, and yet sufficed to give an easy ascent to the sap, and prevent all tendency of the branch to shoot vigorously from any point near the base, as is sometimes the case with the branches when placed exactly in the horizontal position. Apart from this, the growing point of each main branch is allowed to push freely a little upwards, so as to encourage the sap to flow regularly through the branch, and not halt at any one point to the detriment of all. Grafting by approach is practised to cover naked branches. Four to five hundred fruit are gathered from the best trees, or an average of about ten fruit per metre of fruiting branch. Cheap and rather thin planks, about twenty inches wide, are preferred for the temporary coping; walls twelve feet high would be benefited by a few inches more. Cordons of Calville Blanc and other fine Apples are planted plentifully on the spaces between the trees; no matter how well the walls are covered, there is always space for cordon trees between them, in consequence of the branches having a very gradual upward inclination. M. Chevallier's garden is one of the most interesting and instructive I have ever seen, and the trees in it are models of beauty and of perfect training.

FIG. 258.

Mode of preserving the Lower Part of the Stems from the heat of the Sun.

We will next visit the garden of M. Lepère. It is large, and consists simply of a series of oblong spaces which are surrounded by Peach walls, both walls and ground being well covered and cropped—neat, clean, and in all respects satisfactory. The Peach is the favourite subject, but neat pyramidal and cordon Pear and Apple trees are also to be seen, and the place is altogether many degrees above the ordinary type of French fruit or kitchen gardens. There are two entrances to M. Lepère's establishment, and it may not be amiss to say that the finer examples of cultivation are those nearest the one approached by a narrow lane-like road, which is margined on each side by Peach walls. Outside the entrance of the walls there is a small corner of ground, where against a wall may be seen several capital examples of Peach trees, the finest being trained after what is called the Carré form. This is much admired by the best cultivators, but they prefer and generally adopt the Palmette forms, and say they are the best. The Candelabrum form is also to be seen in fine condition in

FIG. 259.

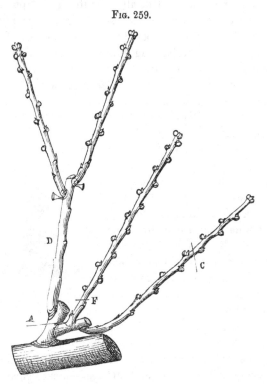

Fruiting Branch of Peach submitted to the third year's pruning. D, which has borne the fruit of the past year, is cut at A; the wood-buds below F will furnish fruiting shoots for the following year; and C bear the fruit of the coming summer.

this out-of-the-way nook; it is simply trained by raising vertical branches from horizontal ones running along near the bottom of the wall. But as to the form itself, it is not a matter of so much importance—the two chief points are covering the walls and the treatment of the fruiting branch. Yet it is interesting to notice the forms adopted by the most successful growers, who, however, are sure to have several trees most fantastically trained.   They will tell you that form is not a matter of so much consequence, but, nevertheless, certain forms are preferred, and certain principles strictly adhered to.

A very old man, dressed in a blouse, is moving along the walls nailing in the shoots here and there, and with him a dozen young men, his pupils.   This is M. Lepère, who has a class twice a week.   Incidentally I may say that the

FIG. 260.   FIG. 261.

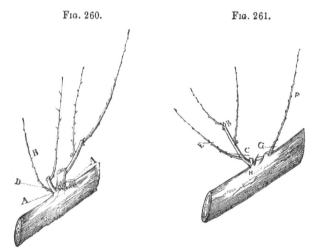

Pruning to replace old fruit-spur: wood-buds are developed at the base, and all the shoots are cut sharp off, as at A.

Result of the preceding operation.   B is cut at C; E bears fruit, and G wood-buds; and thus the spur is renewed.

principle of giving a full explanation of their system of doing anything well, animates all French gardeners more or less.   Did anybody ever hear of an unusually successful English market gardener or fruit grower calling a class round him at a low fee, or no fee at all?   The French, though proud of their success in this way, are careful to give it the fullest possible ventilation; and those who attend here cannot fail to learn the culture of the Peach as well as need be, if so disposed, for the master glides along the wall, and stops and nails in the shoots, and cuts out the foremost

Fig. 262.

The Napoleon Peach Tree.

branches here and there that are not wanted for next year's work ; and, in short, does and explains everything before his pupils.   He has been cultivating Peaches here for a couple of generations, and certainly has reason to be proud of the result.   He inquired as to the state of gardening in England, and I told him we could beat him in most things, but not with the Peach, and that he was indisputably the Emperor of Peach-growers.

Entering the garden, your eye for a moment rests upon the perfectly-covered walls, but presently the famous Napo-

FIG. 263.

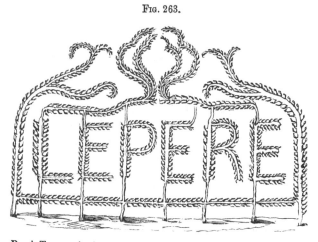

Peach Trees trained to form their Owner's Name against Garden Wall.

leon Peach presents itself. It is in good health, but looks a little weak about the central letters. It is, I need scarcely add, beautifully trained, and a striking evidence of what may be done by a skilful pruning.   Looking in another direction another specimen quite different from the Napoleon presents itself, and it takes the form of the letters of the owner's name—LEPERE.   It is against a high and very white wall, and at a long distance the letters stand out quite clearly, while, upon approaching the tree, the abundance of fruit and regularity of good wood are equally satisfactory.   The letters complete, a shoot is taken from the top of each, and these are united in a somewhat arching line above, and spread out again into a crown over the name, while on each side a single tree springs up, and, forming a border for the letters, spreads out above into a triple flourish on each side of the crown. It is a finer object than the Napoleon, and bears a splendid crop. The sketch gives but a very poor idea of the beauty of the tree, which I by no means figure here by way of recommending it or similar curious forms, but simply to show the mastery attained over the trees. Such a fanciful form is interesting in a great Peach garden, where the grower wishes to show his skill, but is useless for private gardens or for general purposes. It should be added that the formation of the LEPERE was much easier than that of the NAPOLEON tree, inasmuch as a plant is devoted to every letter in the former.

The well-made walls all run east and west, and are placed within about ten yards of each other. This proximity of the walls makes the scene quite different from what we have in England. It is done so that many walls may be accommodated on a comparatively small space, and they are also effective in concentrating the heat and for sheltering. The ground is thus divided into very long narrow strips, the white walls covered with the fresh green of healthy Peach trees, and the ground planted with fruit trees, Strawberries, and Asparagus. The soil is of a calcareous nature, and the long strips enclosed by the walls are generally about fourteen yards across. The syringe is rarely or never used, sulphur being the remedy for spider. The ground was in all cases mulched near the trees, a wide alley being left ; and for preparation of the border they simply trench and manure the ground a couple of feet deep, and about six feet wide. The trees are pruned on the spur system, and as for their shapes, they are many, in addition to the alphabetical ones alluded to above. The Taille en Candelabre is one of the handsomest and most useful. To form it two branches are taken to the right and left along near the bottom of the wall. From the uppermost, single shoots are taken at regular intervals to the top of the wall—the lower branch simply running along to the end and rising to the top of the wall, or in other words, forming a great oblong frame for the interior. Then there is the Taille à la Montreuil, a sort of fan-tail, but with the divisions somewhat far from the base in most cases, and several modifications of the common horizontal mode of training, which we employ so much for the Pear, but never for the Peach. These seem favourite varieties, and by their means the walls are perfectly covered—if indeed one can draw any distinction

Fɪɢ. 264.

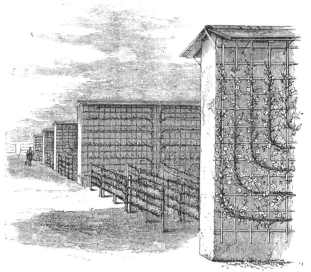

Spring aspect of Fruit Garden formed in North Germany by M. Lepère fils, on the same plan as the best gardens at Montreuil.

between the walls here, which are all as fresh-looking as a meadow in May.

A form presenting the advantage of the cordon, without its too confined and unnatural development, is very common. It is properly termed the U, bearing a considerable resemblance to that letter much elongated. Frequently this is doubled, and a tree with four ascending branches obtained. These forms are excellent for poor ground, or that in which the Peach grows with but little vigour. The number of fruit borne by the finer examples of trees here rusn from

four to five hundred, and this without injuring the tree in
the least degree. As to the pinching of the summer shoots
after they are laid in, it is done according to their strength;
but the greater number are pinched at from six to ten
inches, and the lateral branches that spring from these are
of course pinched also, while weak shortish branches are
allowed to grow to their full extension. The pruning is
distinct from ours in this: it is done on the spur, and not
on the cutting-out principle. We generally leave the shoots

Fig. 265.   Fig. 266.

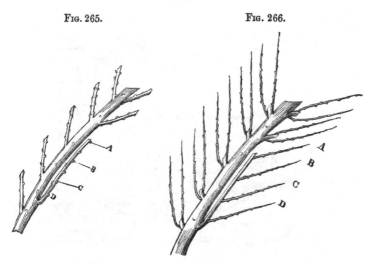

Mode of Pruning to cover bare
spaces on the branches of
Peach trees, first year. The
shoots arising from the buds
A, B, C, and D are allowed
to grow freely, and are nailed
in during the summer.

Result of preceding operation, second year.
A, B, C, D are the shoots developed from
the buds to which the same letters refer in
the preceding figure.   This figure shows
the appearance of the branches before the
pruning.

of the past year long, and cut away a good deal of the old
wood; here the branches are conducted in straight lines
and regularly spurred in every year, fruit and wood buds
being left at the base according to the judgment of the
cultivator. The wood of the current year is laid in against
the wall with nails and shreds just in our own way, only
thicker, as of course must be the case when a close array of
spurs along each shoot has to be obtained.   There can be

little doubt that this system is better than our own, and perfectly suited for our wants, provided we take care to protect the young shoots and flowers in spring, as common sense directs.

In passing along by the walls, grafting by approach may be seen in operation here and there, with the object of covering naked spots, strengthening shoots, and even adding a young shoot to the base of an old spur that has become too long. An interesting example of its utility was shown

FIG. 267.

FIG. 268.

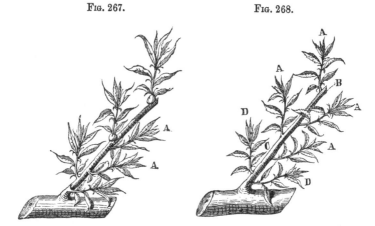

Summer management of the Peach. The shoots above the fruit are retained and stopped, A, A removed, and the two lower shoots furnish the fruiting wood for the following year.

Shoot of Peach without Fruit: the branches A, which would have been retained had the shoot borne a crop, become useless, and the shoot B is cut at C to favour the development of D, D, which will be the fruiting branches of the following year.

by the outer branch of a tree. It is considered very desirable that the lower and outer branch of a Palmette Verrier should be the strongest and highest of all, so as to secure a flow of sap to the lower parts of the trees, instead of allowing it to flow rapidly towards the higher parts, and thus spoil all. In one case, one of the outer branches was feeble and delicate, and did not seem to push much more than to the bend, from whence it ought to have grown strong to the top of the wall. A healthy and vigorous

shoot of a neighbouring tree was worked on it by approach,
and in the course of a single season the desired strength

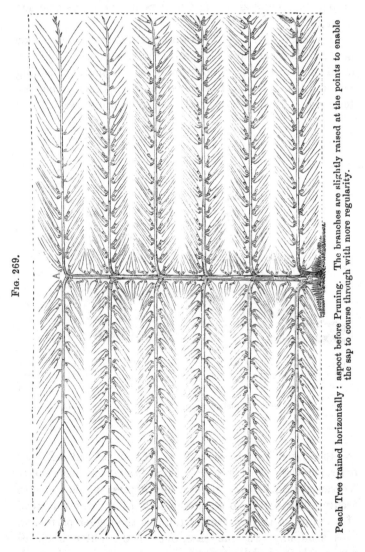

Fig. 269.

Peach Tree trained horizontally: aspect before Pruning. The branches are slightly raised at the points to enable the sap to course through with more regularity.

was obtained, and the shoot went vigorously to the top of
the wall. Not only are the pruner's best precautions taken
to secure abundance of vigour and sap in the lower parts

of the specimen, but slow-growing and not very vigorous
kinds are grafted a little above the middle of the tree,
so as to prevent in the completest manner the tendency
which the sap has to rush towards the higher points.   To
show the difference between cultivators, it is sufficient to
mention that M. Lepère considers this precaution indispensable; while another distinguished cultivator in the
same neighbourhood does not practise it at all, but pinches
the upper shoots and deprives them of leaves when too
vigorous, and thus preserves the most perfect health in his

Fig. 270.

Disbudding of the Peach, second year.
C and A are removed; B, B, furnish the wood for the following
year.

Fig. 271.

Disbudding of the Peach, second year.
If no fruit be borne on E, it is cut at
F, leaving G to furnish the fruiting
wood for the following year.

trees.   This repulsion of the sap to the lower parts of the
trees is also slightly effected by the use of the wide
temporary coping, which guards against frost and keeps the
growth down by partly excluding light from the upper
part of the wall.   When it is removed, and when all
danger of frost is past, the sap has flowed so freely
into the lower branches that but little trouble is required
to keep the tree in a perfectly equable state, all parts of
the wall doing a full amount of work.   I noticed some
walls alongside a road at Montreuil made of blocks of
plaster two feet long, one foot high, and five inches

thick, forming a strong and presentable wall. The blocks are sold at sixty francs per hundred. The walls are about nine feet high, and have a coping of plaster six inches wide. Plaster is very cheap in the neighbourhood, being dug up in quarries quite near to the gardens, and thus it is easy to form a neat and thin projection from the ridge of plaster which forms the top of the wall, by placing boards underneath till the coping sets. This protection is more necessary at the west and south than at the east, the cold rains being more feared than frost, and more difficult to guard against; for while a narrow coping will

Fig. 272.

Grafting by approach to furnish bare spaces on the main branches of the Peach Tree. The second spring after grafting, when the Graft has firmly united, the shoot D is cut at C, and B forms a well-placed shoot.

save the trees from frost, it is not so effective against driving cold rains. A finer crop could not be desired than was visible everywhere here on the day of my visit, 5th July, 1868. It is particularly noticeable that, no matter what form of tree is adopted, all the fruiting branches are higher at the apex than the base, instead of pursuing the horizontal line, as is the case with us. Perhaps to the passing visitor some of the trees in their full summer dress might appear to have their branches horizontally placed; but even in cases where there is most room for the

supposition, the outer ends of the shoots are several inches
higher than where they spring from the ascending axis.

Many cordons are to be seen in abundant bearing in
the garden, both against the walls and in the open.   The
Calvilles against the walls were very good, and were
not always confined to a single line, but were superimposed.
It is a better plan to confine them to a single stem, allowing
that to elongate as much as space will permit, that is, if
the space to be covered is a mere narrow strip of wall, as is
the case under these Peach trees, and the object be to secure
a crop of the finest fruit.   Some of the Calville and other
Apples to be seen here on cordons have nut-brown scars

FIG. 273.

Multiple Grafting by approach, to furnish bare spaces on the stems of
Peach Trees. A, A, A, ligatures of Grafts.

near the apex, showing where the destructive worm has
been cut out ; by taking it in time the fruit is saved,
and this attention, which would be ridiculous in the case
of ordinary fruit, is repaid in the case of the Calville,
for the very finest specimens of which four francs each
are sometimes received by the owner of this garden.
It need hardly be added that this price is for fruit quite
exceptional both as to appearance and size.   There
are specimens of the Peach trained as cordons bearing plenty of fruit, but they present few advantages
in this case that should make them be preferred to
forms that are more fully developed.   It is not with them,

as with the Apple on the Paradise stock, a union that induces a very dwarf development, but, on the contrary, in consequence of being confined to a single stem, they are apt to push too vigorously. M. Lepère had not a word to say in favour of the system.

FIG. 274.

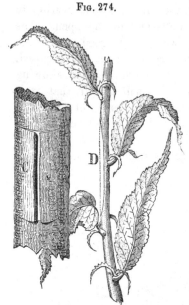

Details of the preceding Figures: C, incision of bare portion of stem; D, preparation of shoot intended to furnish it.

The U form is so pretty, successful, and generally adopted that the following on its formation by M. Lepère can hardly fail to be useful.

" This graceful form is very easy to establish, and I strongly recommend it to those amateurs who have but little wall space to devote to Peach-growing. Peach trees planted in this way afford the means of growing a number of varieties in a small compass, and of speedily obtaining a well trained tree in full bearing. After having chosen healthy trees eighteen months old full of buds at the base, they are cut down to within eight inches of the graft at the time of planting. When the first leaves begin to appear, two well placed shoots situated about six inches above the graft are chosen, one on each side of the stem. These are intended to form the two main branches that are afterwards to be trained in the U shape. The ends of these two branches are then turned directly upwards, care being

FIG. 275.

Nail Basket used instead of Nail Bag.

taken that the extremities are perfectly free, so that their development may not be interfered with.

"The space to be given in planting when the soil is of the best kind is about a yard to each tree, which will leave an interval of eighteen inches between each principal vertical branch, thus allowing sufficient room for nailing in the summer shoots.   When the soil is not so favourable for Peach-growing, the trees can be planted two yards from each other and trained in the form of the double U.   In this case, as in the other, the principal branches will be eighteen inches apart.   Three years ago I planted on a southern aspect some Peach trees in the form of the single U. They yield on an average one hundred Peaches each every year.   The wall against which they are trained is ten feet high, and they were in full bearing the third year.

"I give the preference to this form over the oblique cordon because, the principal branches being trained in a perfectly upright position, the sap is more equally divided amongst the smaller shoots, and if a tree or two happen to die in a fully formed plantation, the place they occupy on the wall which thus becomes empty is not shaded by the branches of the neighbouring trees.   The dead trees can therefore be easily replaced by young subjects from the nursery.   This is a great advantage for amateurs, who have not always full grown trees to fill up bare spaces. In the oblique form the inclined position which each tree is subjected to at the time of coming into leaf, causes

Fig. 276.

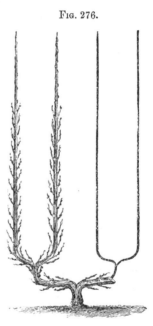

Peach Tree in the double U form. One side is left unfurnished to show the practice of marking on the walls the outline which the tree is to assume before beginning to train it.

a disturbance all along the upper edge of the branch when constant watchfulness of training is not pursued. Besides, if several trees happen to die, and the only trees available to replace them are those from the nursery, the place they will occupy on the wall will be shaded by the branches of the old trees, and the young ones will be injured for want of light and air. As I have already said, the U form is the most easy to train, the most graceful to the eye, and more prolific than the oblique." I have in many parts of France seen fine results obtained by trees grown on this simple principle. Occasionally the points of trees trained in the U and double U forms are united by grafting by approach. This does not in their case seem to be any advantage.

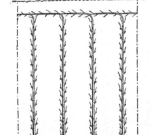

FIG. 277.

Peach trained in the Double U form, with the points of the branches united by grafting.

The reason why the Peach is so successfully cultivated at Montreuil is, that the cultivators pay thorough and constant attention to its wants, with which a life-long experience has made them familiar. The trees are at all times well attended to. I believe that quite as good and as certain results could be attained with the Peach in many of the southern parts of England and Ireland, particularly if its culture were made a speciality of, as it is in France. When cultivators devote themselves entirely to a subject, they soon learn all its wants, and moreover, attend to them at the right moment—a great point. But it is very different with private gardeners generally, whose hands are very full of other matters in spring and early summer, a time when the Peach requires much attention; the result being that it is too often neglected for a week or two at that season, with a consequent loss of health to the trees. There does not seem much help for this in private gardens, and the only hope is that, by the cropping of the borders as elsewhere suggested, gardeners generally may find it worth while to devote more attention to walls than they usually do.

I think it a matter for regret that public attention has been to some extent called away from the many uses and advantages of walls in our climate, and that we have made no progress in protecting or managing wall trees corresponding with our advances in other respects. Some persons have gone so far as to say that garden walls ought to be abolished altogether. One cannot believe that such people can ever have seen the excellent results produced by well-managed garden walls—results as beautiful as profitable. Why, even if we could erect glass-houses by the economical aid of a magic wand, the good fruit-grower would still find uses for a large extent of wall surface. As things are at present, all should aim at greater success in the protection and management of wall trees—a thoroughly practical and attainable aim. Our chief want of success now is due to not preserving the flowers and tender young leaves from the sleet, cold rains, and frost, during the cold and changeable spring common to northern France and the British Isles.

CHAPTER XX.

It has been frequently said that the minute division of property in land retards the improvement of agriculture in France. It may be so with farming, but it certainly does not hold good with market gardens. Those in and around Paris are very small, but they are the best and most thoroughly cultivated patches of ground I have ever seen. Every span of the earth is at work ; and cleanliness, rapid rotation, deep culture, abundant food and water to the crops—in a word, every virtue of good cultivation—are there to be seen. I doubt very much if such good results could be obtained by a larger system, and certainly in no part of Britain is the ground, whether garden or farm, so thoroughly cultivated, or rendered nearly as productive, as in these little family gardens, as they may be called, for they are usually no larger than admits of the owner's eye seeing the condition of every crop in the garden at once. The Paris market gardeners as a class keep to themselves, marry among themselves, and seem content with about as much ground as gives occupation to their family. They are as a rule a prosperous class. The gardens vary in size from one to two, and occasionally three acres, are usually walled in and furnished with a cottage, a few sheds, and a well.

In the neighbourhood of our English cities the price of ground is high—according to our scale, that around Paris is very high indeed. From information gathered on the spot, during September last, I may say the rent varies from 24*l*. to 33*l*. per acre. On entering a market garden the tenant has to pay in addition to his rent from 200*l*. to

600*l.* for stock, fixtures, &c. It is necessary to dig deep to get a return under these conditions!

Manure forms a very considerable item in the expenses of these gardens. One market gardener of the first

Fig. 278.

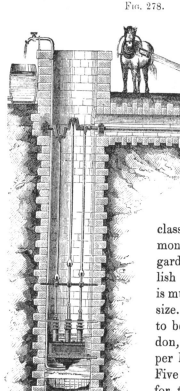

class paid 500 fr. (20*l.*) a month for manure. His garden was about three English acres in extent, which is much beyond the average size. Manure would appear to be dearer than in London, from three to four francs per horse being paid for it. Five francs a month are paid for that of each omnibus horse. These being in Paris all strong, large, well-nurtured stallions, their manure is the more valuable It is

Pump used in the Market Gardens of Paris.

usually piled in heaps near the entrance to the market gardens. Some of the crops are absolutely growing in nothing but decomposed manure; and it is used profusely for everything. From the beginning of May to the end of November the market gardeners have no use for hotbeds, and yet

every day there arrives one or more loads of stable manure,
all of which is piled into rick-like heaps, to be used chiefly in
winter and spring.   In November they commence to make
the hotbeds, and as hot dung arrives every day they mix
with it that which has been gathered during the summer—
thus insuring beds giving a moderate degree of heat.

But a more important and expensive item is the watering.
The Parisian market gardener, if not a scientific man, would
appear to be fully aware of the fact that by far the most
important constituent of vegetables is water.   As a rule, the
less of this they contain the worse they are.   It is owing to
the abundant watering of the market gardens, more especially than to anything else, that the Paris markets are in
all sorts of seasons and summers better supplied with
crisp, fresh, and delicious vege-
tables than those of any other capital
in Europe.   Every market garden
has its pump worked by the horse
of the establishment—the Naudin
system being generally preferred.
From   this the water is conducted into   old barrels nearly
plunged to their rims in the
ground at regular intervals over
the garden, and from   these
barrels distributed by watering pots.   These are always
of copper—the best I saw being flat-sided and oblong
instead of round-bodied.   The handle springs clean from
the top to the back of the vessel, so that when filled
and carried by the workman to the spot he wishes to water,
he merely has to pitch the pots (he always uses two at a
time) forward a little and let his hands fall back, so as to
hold the pot in the position which most favours the pouring
out of the water.   The pipe being very wide, and the rose
broad and freely perforated with large holes, the water is
discharged almost in an instant, and the workman again
proceeds to his barrel close at hand, and always kept filled
from the pump.   Thorough watering is thus effected, but
it involves a considerable expenditure for labour, one or

Fig. 279.

Watering-pot used by the Market
Gardeners of Paris.

two men being nearly always employed at it in each little garden during the sunny months.

The system of watering with the hose, generally adopted in the city of Paris, could not fail to attract the attention of the market gardeners: it is already used by several of them, the old system of pots and tubs being done away with. In these cases the pump is again employed to elevate the water to a cistern placed a few yards above the highest point of the garden, and near the manure heaps and sheds of the establishment. I examined a garden thus arranged, and found the system very satisfactory. Twenty-seven outlets for the water were established over the surface of a garden about two acres in extent. To these a hose of india-rubber is attached, with a few feet of copper tubing and a large profusely perforated copper distributor or rose at its other end. From this, when the water is put on, it flows in a gentle but dense shower; and the apparatus may be managed by a woman or a boy. The hose is not on little wheels as is the one used in the roads and parks, nor can it be by any means or in any shape dragged over the growing crops that occupy every inch of the ground except the very narrow alleys between the squares or large beds; therefore there is a little contrivance to facilitate its use without injuring any of the plants. The outlet, we will say, is on an alley crossing the garden, and the operator wishes his hose to play say thirty feet from the outlet, and up one of the narrow footways that leads from the alley. Three little wooden rollers held together on one piece of iron enable him to do this. The following simple diagram will explain it : H . There is a little wooden roller on the cross bar, and one short one with a margin on each of the upper limbs—the lower points form the teeth, and are stuck in the earth. Through this, placed at the mouth of the alley, the india-rubber hose glides as easily as a snake, without hurting a leaf. The mechanical arrangement of each outlet is such that a twist of the base of the hose which fits it is made to turn the water off or on in an instant. This very satisfactory apparatus cost its owner about 3000 francs. Having his own pump there is nothing to pay for the water.

Nobody could pass suddenly, as I have done, from our own markets and market gardens to those of Paris in the middle of any but a wet summer without being forcibly taught how advantageous it would be to be able to command water in our gardens. It is the custom, and a very frequent one, among the horticultural community, to grumble about our climate—the " dull," " cloudy," " changeable" climate of Britain ; to speak of that of other countries as paradisaical, and to attribute all our failures to " want of sun."

In 1868 we had sun enough to satisfy an Arab, and what was the result ? The worst ever remembered in the garden. It was natural to think the soft, green vegetables would suffer, but everybody hoped the heat would prove favourable to such useful members of the Solanum families as the Potato and Tomato ; whereas the Potatoes proved worse than if badly blighted, and even the heat-loving Tomato dropped its flowers before setting. Radishes disappeared with the dew of May.   The Cabbage tribe presented, everywhere that they had not completely perished, a sad spectacle—a mere bony framework of glaucous vegetation, with all the softer parts gnawed away by hungry tribes of vermin, the only things that flourished with the heat.   In this condition the Brassiceæ were sold, and—the fact speaks well for the appetite of the public—eaten. An extensive London market gardener showed me a field of Celery with not a single plant in it good or large enough to be culled for seasoning, and at nearly every root grubs gnawing away the plant. Those who are accustomed to realize hundreds of pounds for a crop, gathered barely as much of it as would make it worth while sending to the market ; while the private growers were quite as badly off.   During the month of July, and when Cauliflowers in British gardens had almost disappeared, I measured them in the market gardens of Paris a foot in diameter, of that pure creamy white and perfectly dense and firm texture which admirers of the Cauliflower like so much. Strange as it may appear to some, during the whole of the hot weather vegetables of the primest and the most delicate quality were to be had in the Paris markets, where even greater difficulties had to be met by the cultivator.

Be it observed that this is no contrast of 1868 ; it is the same thing, to some extent, every year. The year 1867, for example, was anything but a dry one ; yet, in passing through the central markets at Paris during the month of July, 1867, with an excellent cultivator who has every convenience and good ground in the prettiest and richest part of Surrey, he was more surprised with the Radishes and young Turnips than with anything else. "We cannot get anything like these at this season," said he. Of course not, and simply because we do not take the simplest precautions to secure them. We have them when the weather is dewy and favourable, and where the climate and soil are moist ; but a few weeks of dryness puts an end to all such luxuries, and should the drought continue, everything becomes worthless and uneatable, as during the past season. But are we in a position to boast of our horticulture while this is the case ? Does the routine work, which merely waits upon the seasons thus, deserve the name of skill ? I think not, and moreover that it is absurdly unsatisfactory to reflect that the very things which our watery and cloudy clime is supposed to be most favourable to, are to be found in greatest perfection with the French, in the drier and, for vegetables, less favourable climate of Paris ! The secret of it all is that the French market gardener, in addition to tilling and enriching his ground in the best manner, waters thoroughly and repeatedly every crop that requires water for its perfect development.

Our gardens are no more prepared to encounter a great, or even an unusual scarcity of water, than they are to meet a second deluge. The practice of dragging water considerable distances in pots and barrels is a very doubtful good ; watering is useless if not thoroughly done. I need not remind the reader of the many things to which water is almost the life. Extract the water from a juicy Lettuce, or any other appreciated vegetable, and how much remains ? Our soils are of course saturated with water in winter, when plants do not want it ; but it is often absent when they would absorb it as thirstily as the hart the cooling stream, and when the absence of it leaves them mere accumulations of tough fibre. I am not sanguine enough to hope that any words of mine can induce cultivators to adopt something like a system for watering gardens effectively; but there can be no doubt whatever that it would be a decided advantage to establish immediately in every large kitchen garden a small department near the best supply of water, to make it rich and light, and keep it thoroughly moist during the dry and warm months; so that a few crisp and delicate salads and vegetables may not during a dry season be as impossible with us as upon the Sahara. This small division might be established in most places without any but the most trifling cost, and the result would be very satisfactory indeed. Even a few very rich and light beds, closely cropped and looked after, and placed near a good supply of water, would repay the cultivator, and perhaps soon lead him to adopt the same plan of giving abundance of food and water on a larger scale. It need hardly be added that it would not be necessary to make any such arrangement in any very moist districts in the British Isles; but although theoretically our climate is very moist, there are many parts of the southern and midland counties where a modification of the Parisian plan would prove a decided advantage.

In addition to the abundant watering and rich manuring the Parisian market gardeners owe a great deal of their success to a close system of rotation, eight crops per year being frequently gathered from the ground. Were it not so the cultivators could not exist, so very limited is the ground each possesses. A considerable portion of the surface in one garden I visited was devoted to Cos lettuce, and very fine specimens of it; but beneath them there was a dark green carpet of leaves very close to the ground—the leaves of the Scarolle, which forms such an excellent salad, and is indeed one of the very best of all salads, and not yet sufficiently grown in England. The young plants have plenty of room to grow now amongst the closely-tied up Cos lettuce; but the moment the Cos is cut for market, the Scarolle has full liberty, and with abundance of water soon makes wide heads. Then perhaps some young plants of another vegetable are slipped in at regular intervals in the angles between four plants of Scarolle, which crop will be vigorous and halfway toward perfection when the great smooth Endive is ready for the market. As an illustration of the cropping, the cultivator described to me that of a portion of his ground for the past year. In the earliest spring the ground was occupied by Cos lettuce, and from between them a crop of Radishes was gathered. Cauliflowers were planted early among the Cos, and as they approached maturity the ground was of course wholly occupied by them, as one could not well put anything beneath a crop of perfectly grown Cauliflowers. When they were cut in May and June, an opportunity occurred of giving the ground that thorough culture and preparation which such a course of heavy cropping demands. Then a crop of Spinach was sown, and in the Spinach Cos lettuce. As soon as the Spinach was cleared off, a crop of Endive was placed alternately with the Cos. Then small Cauliflower plants were put in, yielding a fine crop in the autumn, and after them a small quick crop like Corn salad, and afterwards the ground was covered with frames.

Like everything else in Paris, and in France generally, the condition of these market gardeners has much improved during the past generation. Their houses are humble enough now, but I am told by M. Courtois Gérard, a capital authority on the subject, that they are palaces compared to what he remembers them to have been. Some of the crops, and particularly the forced crops, are now brought to invariable perfection in low narrow wooden frames. Eighty or ninety years ago, however, the market gardening of Paris was much less perfect; fewer crops were gathered during the year, the art of forcing early vegetables and salads was in its infancy, and the most advanced market gardeners had not gone beyond the use of the cloche to force their vegetables. It is not that frames were not known at the time, as they were known in Royal and other private places, but they had not entered the market garden. In 1780 a cultivator named Fournier first used frames, and with such success in forcing that a great number of his fellows soon imitated him. It was the same individual who first introduced the culture of the

Cantaloupe Melon, and he also first grew the Spanish or Sweet Potato. The first who forced white Asparagus was one Quentien, about the year 1792; the green Asparagus was also first forced by the same about 1800. One Basnard first forced the Cauliflower about the year 1811. The first forced Cos lettuces appeared about 1812, and the Endive about the same time by Baptiste Quentien. The Carrot was first forced in 1826 by M. Gros.

The workmen employed in these market gardens work, like their masters, very hard, but are pretty well paid. From inquiries made from different cultivators, the wages are from fifty to seventy-five francs per month with board and lodging. They have no fixed hours for work as with us, but in summer begin with the dawn and in winter hours before it. They often commence work at three in the morning, and continue it till eight in the evening in summer. In the dark winter mornings they cannot of course work in the gardens, but they can take the produce to market and go for the ever necessary manure. After a visit to one of these places I was invited by the proprietor to take a glass of wine with him. Hardly were we seated to this before he spoke some words from the door, and presently in came his two workmen, sunburnt, strong men, working barefooted in the soft moist soil of the garden, and they also had their glass of Bordeaux, touched our glasses, and again went out to work. I afterwards learnt that this was the rule with them—anything the master has the workmen partake of. Under these circumstances the gulf of distrust, and consequently other evils that exist where the workman is treated as a far inferior being, cannot be. M. Courtois Gérard says that to cultivate a garden of two and a half acres devoted to forcing in frames, and open air culture, it is necessary to have five or six persons—that is to say, the master and mistress, two men, a girl and a boy.

As to the masters, I was informed that many of them could not read or write; but noticed notwithstanding a good barometer in each house. They well know the value of this instrument, and I was told by one of the very best of them that it was of the greatest use to him in his cultivation, by helping him to take precautionary measures and adapt his labour to the weather. This individual had worked four-teen years at the business, and was desirous of disposing of his garden, feeling rich enough to retire and live on the fruits of his labour. These men have their vicissitudes not-withstanding the vigorous industry and excellent system of culture which is general with them. Some that I visited devote a considerable portion of space to a difficult crop—Cauliflower seed. This takes a long time—more than a year—to bring to perfection; one market gardener who had been in the habit of growing large and precious quan-tities of it for Messrs. Vilmorin, Andrieux, and Co., had scarcely gathered two pounds of it in consequence of the great heat of the season of 1868.

There is a " Société de Sécours Mutuel " among these market gardeners. To give an example of the way they work, I have merely to state that when a body of pro-vincial cultivators were almost ruined by inundations, the Paris society sent them more than 1200 lbs. weight of seeds to begin again with. Generally they seem independent, and are said to accumulate money; but their houses do not show the comfort that one could desire. However, few will doubt that it is better to have a large class of small proprietors in a thrifty and independent, if very humble condition, than one individual with his hundreds of acres, and every soul employed by him without a single thing in the world to call his own, except it be misery, poverty, and degradation. The Paris market gardener is very far from being mistaken for a " genteel " person, or putting in the smallest claim to the

> " Grand old name of gentleman,
> Defamed by every charlatan,
> And soiled by all ignoble use ;"

but he is a thousand degrees better than the poor wretch working in a London market garden, who is practically a slave, and a very wretched, badly-fed, badly-housed, and badly-clad slave too.

The cultivation of the Mushroom is of vast importance about Paris; and I will next deal with the doings of the Champignonnistes, a class of men who devote themselves entirely to its culture.

CHAPTER XXI.

MUSHROOM CULTURE.

MUSHROOM growing as carried on around, or rather beneath
Paris and its environs, is the most extraordinary example of
culture that I have ever seen either above or below ground,
under glass or in the open air.   To give the reader as good
an idea of it as I can we must visit one of the great
" Mushroom caves " at Montrouge, just outside the fortifications of Paris, on the southern side.   The surface of the
ground is mostly cropped with Wheat; but here and there
lie, ready to be transported to Paris, blocks of white stone,
which have recently been brought to the surface through
coalpit-like openings.   There is nothing like a " quarry,"
as we understand it, to be seen about; but the stone is
extracted as we extract coal, and with no interference
whatever with the surface of the ground.   We find a
" Champignonniste" after some trouble, and he accompanies
us across some fields to the entrance of his subterranean
garden.   It is a circular opening like the mouth of an old
well, but from it protrudes the head of a thick pole with
sticks thrust through it.   This pole, the base of which rests
in darkness sixty feet below, is the easiest and indeed the
only way by which human beings can get into the mine.
I had an idea that one might enter sideways and in a more
agreeable manner, but it was not so.   The artist who afterwards descended to take the sketches here engraved was in
such a state of trepidation when he got to the margin and
looked down, that my friend M. Durand of Bourg-la-Reine,
who was kind enough to get these two sketches taken for
me at very considerable trouble to himself, seriously medi-
tated having him slung in cords.   Down the shaky pole
our guide creeps, I follow, and soon reach the bottom, from

which little passages radiate. A few little lamps fixed on pointed sticks are placed below, and, arming ourselves with one each, we slowly commence exploring tortuous passages as dark as night and as still as death. I have heard that the first individual who commenced Mushroom growing in these catacomb-like burrowings was one who, at a particularly glorious epoch of the history of France, when a great many more brave garçons went to the fight than returned from the victory, preferred, strange to say, to stay at home and hide himself rather than form a unit in " battle's magnificently stern array." Industrious and discreet youth ! You deserve being held up as an example almost as much as the busy bee that improves each " shining hour."

The passages are narrow, and occasionally we have to stoop. On each hand there are little narrow beds of half - decomposed stable manure running along the wall. These have been made quite recently, and have not yet been spawned. Presently we arrive at others in which the

Fig. 280.

Mouth of Mushroom Cave at Montrouge.

spawn has been placed, and is " taking" freely. The spawn in these caves is introduced to the little beds by means of flakes taken from an old bed, or, still better, from a heap of stable manure in which it occurs naturally. Such spawn is preferred, and considered much more valuable than that taken from old beds. Of spawn in the form of bricks, as in England, there is none.

The Champignonniste pointed with pride to the way in which the flakes of spawn had begun to spread through the little beds, and passed on—sometimes stooping very low to avoid the pointed stones in the roof—to where the beds were in a more advanced state. Here we saw little, smooth, putty-coloured ridges running along the sides of the passages, and wherever the rocky subway became as wide as a small bedroom two or three little beds were placed parallel to each other. These beds were new, and dotted all over with Mushrooms no bigger than Sweet Pea seeds, and affording an excellent prospect of a crop. Be it observed that these beds contain a much smaller body of manure than is ever the case in our gardens. They are not more than twenty inches high, and about the same width at the base; while those against the sides of the passages are not so large as those shaped like little Potato pits, and placed in the open spaces. The soil with which they are covered to the depth of about an inch is nearly white, and is simply sifted from the rubbish of the stone-cutters above, giving the recently made bed the appearance of being covered with putty.

Although we are from seventy to eighty feet below the surface of the ground everything looks very neat—in fact, very much more so than could have been expected, not a particle of litter being met with. A certain length of bed is made every day in the year, and as they naturally finish one gallery or series of galleries at a time, the beds in each have a similar character. As we proceed to those in full bearing, creeping up and down narrow passages, winding always between the two little narrow beds against the wall on each side, and passing now and then through wider nooks filled with two or three little beds, daylight is again seen, this time coming through another well-like shaft, formerly used for getting up the stone, but now for throwing down the requisite materials into the cave. At the bottom lies a large heap of the white earth before alluded to, and a barrel of water—for gentle waterings are required in the quiet, cool, black stillness of these caves, as well as in Mushroom-houses on the upper crust.

Once more we plunge into a passage as dark as ink, and find ourselves between two lines of beds in full bearing, the beautiful white button-like Mushrooms appearing everywhere in profusion along the sides of the diminutive beds, something like the drills which farmers make for green crops. As the proprietor goes along he removes sundry bunches that are in perfection, and leaves them on the spot, so that they may be collected with the rest for to-morrow's market. He gathers largely every day, occasionally sending more than 400 lb. weight per day, the average being about 300 lb. A moment more and we are in an open space, a sort of chamber, say 20 feet by 12, and here the little beds are arranged in parallel lines, an alley of not more than four inches separating them, the sides of the beds being literally blistered all over with Mushrooms. There is one exception; on half of the bed and for about ten feet along, the little Mushrooms have appeared and are appearing, but they never get larger than a pea, but shrivel away, " bewitched" as it were. At least such was the inference to be drawn from the cultivator's expressions about it. He gravely attributed it to a ridiculously superstitious cause.

Fig. 281.

View in Mushroom Cave.

Frequently the Mushrooms grow in bunches or " rocks," as they are called, and in such cases those that compose the little mass are lifted all together.

The sides of one bed here had been almost stripped by the taking away of such bunches, and it is worthy of note that they are not only taken out, root and all, when being gathered, but the very spot in which they grew is scraped out, so as to get rid of every trace of the old bunch, and the space covered with a little earth from the bottom of the heap. It is the habit to do this in every case, and when

the gatherer leaves a small hole from which he has pulled
even a solitary Mushroom, he fills it with some of the white
earth from the base, no doubt intending to gather other
Mushrooms from the same spots before many weeks are over.
The Mushrooms look very white, and are apparently of
prime quality. The absence of all littery coverings and
dust, and the daily gatherings, secure them in what we may
term perfect condition. I visited this cave on the 6th of
July, 1868, and doubt very much if at that season a more
remarkable crop of Mushrooms could be anywhere found
than was presented in this subterranean chamber—a mere
speck in the space devoted to Mushroom culture by one
individual.

When I state that there are six or seven miles run of
Mushroom beds in the ramifications of this cave, and that
their owner is but one of a large class who devote themselves
to Mushroom culture, the reader will have some opportunity of judging of the extent to which it is carried on
about Paris. These caves not only supply the wants of the
city above them, but those of England and other countries
also, large quantities of preserved Mushrooms being exported, one house alone sending to our own country no less
than 14,000 boxes annually. There were some traces of
the teeth of rats on the produce, and it need not be said
that these enemies are not agreeable in such a place; but
they did not seem to have committed any serious ravages,
and are probably only casual visitors, who take the first
opportunity of obtaining more varied food than is afforded
them by these caves. To traverse the passages any further
is needless—there is nothing to be seen but a repetition of
the culture above described, every available inch of the cave
being occupied. We again find our way to the bottom of
the shaft, carefully mount the rather shaky pole one by
one, and again stand in the hot sun in the midst of the
ripe Wheat. In traversing the fields, two things relating
to Mushroom culture are to be observed—heaps of white
gritty earth, sifted from the débris of the white stone, and
large heaps of stable manure accumulated for Mushroom
growing, and undergoing preparation for it. That prepara-

PLATE XLV.

VIEW IN MUSHROOM CAVES UNDER MONTROUGE, WITH BEDS IN FULL BEARING.

tion is different from what we are accustomed to give it. It is ordinary stable manure, not droppings, or very short stuff, and it is thrown into heaps four or five feet high, and perhaps thirty feet wide. The men were employed turning this over, the mass being afterwards stamped down with their feet, a water-cart and pots being used to thoroughly water the manure where it is dry and white.

As many will feel an interest in the cave culture of the Mushroom, and perhaps wish to see it for themselves, I may state that it is difficult to obtain permission to visit the caves, and many persons would not like the look of the "ladder" which affords an entrance. Even with a well-known Parisian horticulturist I had some difficulty in entering them. We were informed that one Champignon-niste in the same neighbourhood demands the exorbitant price of twenty francs for a visit to his cave. As the visit is a work of some little time, no visitor should put the cultivators to this trouble without offering some slight re-compense—say five francs. The above cave is but a sample of many in the immediate neighbourhood of Paris.

We will next visit a Mushroom cave of another type at some little distance from that city. It is situated near Frépillon, Méry-sur-Oise—a place which may be reached in an hour or so by the Chemin de fer du Nord, passing by Enghien, the valley of Montmorency and Pontoise, and alighting at Auvers. There are vast quarries in the neigh-bourhood, both for building-stone and the plaster so largely used in Paris. The materials are not quarried in the ordi-nary way by opening up the ground, nor by the method employed at Montrouge and elsewhere in the suburbs of Paris, but so that the interior of the earth looks like a vast gloomy cathedral. In 1867 the culture was in full force at Méry, and as many as 3000 lbs. a day were sometimes sent from thence to the Paris market; but the Mushroom is a thing of peculiar taste, and these quarries are now empty—cleaned out and left to rest. After a time the great quarries seem to become tired of their occupants, or the Mushrooms become tired of the air; the quarries are then well cleaned out, the very soil where the beds rested being scraped away, and the place left to recruit itself for a year or two.   In 1867 M. Renaudot had the extraordinary length of over twenty-one miles of Mushroom beds in one great cave at Méry, last year there were sixteen miles in a cave at Frépillon.   This is a clean, lonely village, just touching on the gigantic cemetery which M. Haussmann has projected.

Fig. 282.

Entrance to large Subterranean Quarry.

The distant view of the entrance to the quarries has much the appearance of an English chalk pit.   But there is a great rude arch cut into the rock, and into this we enter, meeting presently a waggon coming forth with a load of stones, the waggoner with lamp in hand. To the visitor who has seen the low Mushroom caves near Paris, where it is sometimes necessary to stoop very low to avoid knocking one's head against the roof rocks, the surprise is great on getting a little way in. At least it is so as soon as one can see; the darkness is so profound that a few candles or lamps merely make it more visible. The tunnel we traverse is nearly regularly arched, masonry being used here and there, so as to render the support secure and symmetrical, the arches being flat at the top for six feet or so, and about twenty-five feet high; sometimes five feet higher.

Presently we turn to the right, and a scene like a vast subterranean rock temple presents itself.   At one end are several of us with lamps, admiring the young Mushrooms budding all over the rows of beds at our feet, which, serpent-like, are long and slim, and coil away into the darkness. At about 150 feet distance there is a group of three men and

Fig. 283.

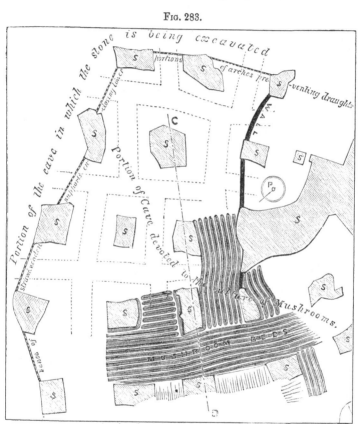

Plan of large Subterranean Quarry at Fortes Terres, Frépillon. S, S, S, represent the plan of the bases of the huge supporting pillars, and the dotted lines their union with the roof. D, C, shows the line of the section shown in the following cut, and P, place for preparing the plaster.

a boy, each with a lamp, again dispelling the darkness from the Mushroom beds, and occupied in placing small quantities of a sort of white clayey sand in the spots whence gatherings have been made a few hours previously. From both sides of this gloomy avenue the dark openings of others depart at short intervals, and the floor of all is covered with Mushroom beds, sometimes running along the passages, sometimes across them. These beds are about twenty-two inches high and as much in diameter, and are covered with silver sand and a sort of white putty-like clay in about equal proportions. In some parts of the cave the work of ripping out the stone by powder and simple machinery continually goes on. The arches follow the veining of the stone, so to speak ; their lower parts are of hard stone, the upper ones of soft, except the very top, which is again hard. There is but a slight crust of stone above the apex of each arch, and above that the earth and trees. Running along in parallel lines, and dis-

FIG. 284.

Section following the line C, D, in Fig. 283.

appearing from view in the darkness, one knows not what to compare them to, unless it be to barked Pine trees in the hold of a ship.

Everywhere on the surface of these little beds small Mushrooms were peering forth in quantity ; as the beds are regularly gathered from every day, no very large ones are seen. They are preferred at about the size of a chestnut, and are removed root and branch, a small portion of finely sifted earth being placed in each hole, so as to level the bed as in the caves at Montrouge. If the old superstition that a Mushroom never grows after being seen by human eyes were true, the trade of a Champignonniste would never answer here, as the little budding individuals come within view every day during the gathering and earthing operations. The most perfect cleanliness is observed everywhere in the neighbourhood of these beds, and the whole surface of each avenue is covered by them, leaving passages of ten inches or a foot between the beds. At the time of my visit (Sept. 29, 1868) the crops of the cultivator were reduced to their lowest ebb, and yet about 400 pounds per day were sent to market. The average daily quantity from this cave is about 880 pounds, and sometimes that is nearly doubled. It may be supposed that the profits from such an extensive culture are great; and so they are, but the expense is great also. The proprietor informed me that culture on a more limited scale than he pursued last year at Méry gave the best return in proportion to expense, the care and supervision required by so many miles of beds being too great.

All the manure employed is brought from Paris by rail, as the place is twenty-five miles from that city by road. In the first place, so much per horse per month is paid in Paris for the manure; then it has to be carted to the railway station and loaded in the waggons; next it is brought to the station of Auvers, and afterwards carted a couple of miles to the quarries, paying a toll for a bridge over the Oise on the way. That surely is difficulty enough for a cultivator to begin with! Then it is placed in great flat heaps a yard deep by about thirty long and ten wide, not far removed from the mouth of the cave, and here it is prepared, turned over and well mixed three times, and as a rule watered twice. About five or six weeks are occupied in the preparation, long manure requiring more time than short. The watering is not usually done regularly over the mass, but chiefly where it is dry and overheated. Every day manure is brought from Paris; every day new beds are made and old ones cleared out—the spent manure being used for garden purposes, particularly in surfacing or mulching, so as to prevent over-radiation from the ground in summer. The chief advantage the cultivator here has is the facility of taking his manure or anything else in or out in carts, as easily as if the beds were made in the open air. Near Paris, on the contrary, everything has to be sent up and down through shafts like those of old wells, and the men have to creep up and down a rough pole like mice. Many men are employed in the culture, the daily examination of sixteen miles of beds being a considerable item in itself.   Here and there a barrier in the form of straw nailed between laths may be seen blocking up the great arch to a height of six feet or so.   This is to prevent currents of air wandering about through the vast passages.

FIG. 285.

Extracting the Stone in Subterranean Quarries.

The mode of preparing the spawn here is entirely different to ours.   They prefer virgin spawn—that is to say, spawn found naturally in a heap of manure.   But as this material cannot be obtained in sufficient quantity to meet the wants of such extensive growers, they put a small portion of it into a Mushroom bed to spread, and instead of allowing this bed to produce Mushrooms it is all used as spawn, and is valued more than any other.   Of course abundance of spawn occurs in the old beds, but it is never used directly.   It is, however, frequently employed to spawn a small bed when virgin spawn cannot be obtained. In this case the small bed devoted to the propagation of spawn is placed in the open air, and covered with straw, and as soon as it is permeated with the spawn it is carried into the caves and used.   As the making and spawning of beds is a process continually going on, a bed of this sort must be ready at all times. It is never made into bricks as with us, but simply spread through short, partly-decomposed manure.

I was informed that coal mines are not adapted for growing Mushrooms, and the smallest particle of iron in the beds of manure is avoided by the spawn, a circle around it remaining inert. It is said to be the same with coal. If an evil-disposed workman wishes to injure his employer he has only to slip along by the beds with a pocketful of rusty old nails and insert one here and there.

FIG. 286.

View in old Subterranean Quarries devoted to Mushroom Culture, and in the occupation of M. Renaudot.

The beds remain in good bearing generally about two months, but sometimes last twice and three times as long. A useful contrivance for facilitating the watering of the beds has lately been invented; it consists of a portable water-cistern to be strapped to the back and fitted with a rose and tubing, so that a workman may carry a larger quantity of water, and apply it more regularly and gently than with the old-fashioned watering-pots—while one hand is left free to carry the lamp. An iron frame has also been invented, in which the bed is first compressed and shaped, the frame being then reversed and the bed placed in position. Another invention for earthing the beds over as soon as the spawn has taken will soon be in operation, if not already so.

As on an average 2500 yards of beds are made every month, simple mechanical contrivances to facilitate the operation will prove of the greatest advantage to the cultivator.

In addition to the caves in the localities above alluded to there are other places near Paris where the culture is carried on—notably at Moulin de la Roche, Sous Bicêtre, near St. Germain, and also at Bagneux. The equability of temperature in the caves renders the culture of the Mushroom possible at all seasons; but the best crops are gathered in winter, and consequently that is the best time to see them. I, however, saw abundant crops in the hottest part of the past very hot season. These Mushroom caves are under government supervision, and are regularly inspected like any other mines in which work is going on. As regards the depth at which this culture is carried on, it varies from twenty to one hundred feet, sometimes reaching one hundred and fifty and one hundred and sixty feet from the surface of the earth. They are so large that sometimes people are lost in them. In one instance the proprietor of a large cave went astray, and it was three days before he was discovered, although soldiers and volunteers in abundance were sent down. Is it possible that in a great mining and excavating country like ours we cannot establish the same kind of industry?

Culture in Cellars and in the Open Air.

Of course they are only professional Mushroom growers that carry on such extensive operations as those just described, but the Mushroom is

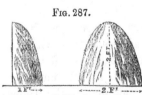
Fig. 287.

Newly-made Mushroom Beds.

also grown extensively in cellars and like places. As an analogous mode of culture is more likely to meet general wants, the following observations on the subject may prove acceptable. They are by M. Laizier, the President of the Mutual Aid Society formed among the Paris market gardeners.

" The manure to be used for this method of cultivation should be prepared in the same way as that for the open-air system described further on. Preference should be given to a cellar producing saline efflorescence; it should be as dark as possible, and exposed to no draughts. The warmth generated by the fermentation of the manure will subside, and the sowing of the spawn must not be commenced until the temperature of the bed

Mushroom Bed on rude Shelf against Wall of Cellar.

has fallen below 76° Fahr.; if it is above this the layers of spawn are liable to be burnt. Beds can be made in cellars in many ways. Those made in the middle should always be formed with two sides, while those against the walls should only be half as thick, on account of their having only one useful side. It is also possible to arrange them on shelves, one above the other. For this purpose strong bars of iron are driven into the walls, upon which are placed shelves of the proper size covered with earth, upon which is formed a bed, that is treated exactly as those made upon the ground. These beds are just as productive as any of the other kinds. They may even be made on the bottoms of casks, which should be at least two feet six in diameter; and they are built up in the shape of a

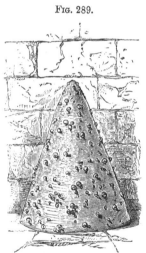

FIG. 289.

Pyramidal Mushroom Bed on Floor of Cellar.

sugarloaf, about three feet in height, and the pieces of spawn are placed an inch and a quarter deep, and sixteen inches apart. A barrel is sawn crossways into two pieces, each forming a tub. Holes are made in the bottom of each, and a thin layer of good soil is spread over them inside. They are then filled with good well prepared stable manure, just like that used in the case of ordinary Mushroom beds, the different layers of dung in each tub being well pressed down. When the tub is half full, six or seven good pieces of spawn are placed on the surface, and the remainder is piled up with manure, which is well pressed down, the operation being completed by giving to the heap the form of a dome. The tubs thus prepared are placed in a perfectly dark part of a cellar, and eight or ten days afterwards the dung is taken up until the spawn is visible, in order to see

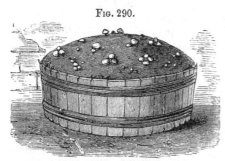

Fig. 290.

Mushrooms grown in bottom of old Cask.

whether it has commenced to vegetate and develope little filaments. If the spawn has struck, the surface must be covered with soil, care being taken to use only that which is fresh and properly prepared. The necessity of procuring good Mushroom spawn cannot be too strongly insisted upon, this being the indispensable condition for arriving at a good result. An excellent kind is sometimes met with that has been formed spontaneously in some old dungheap, which is called virgin spawn. When this cannot be procured, excellent spawn may be formed in the following manner:—A little bed of stable manure is prepared, either in a cellar or in the open air, and sown with good ordinary Mushroom spawn, the proper amount of care already described being bestowed upon it. As soon as the Mushrooms appear on the surface of the bed the spawn remaining must be speedily removed and placed in a dry, airy situation. The spawn thus prepared can be preserved for two or three or even four years. Specimens have been kept as long as fourteen years, from which excellent results have been obtained."

We will next turn to the culture of the Mushroom in the open air. In old times the market gardeners of Paris used to grow the Mushroom amongst their ordinary crops with great profit, but since the Champignonnistes cultivate it under no danger from cold in the caves, the market gardeners, who used to raise it to a great extent in the open air, do so now in a lesser degree. They begin with the preparation of the manure, and collect that of the horse for a month or six weeks before they make the beds; this they prepare in some firm spot of the market-garden, and take from it all rubbish, particles of wood, and miscellaneous matters; for, say they, the spawn is not fond of these bodies. After sorting it thus, they place it in beds two feet thick, or a little more, pressing it with the fork. When this is done the mass or bed is well stamped, then thoroughly watered, and finally again pressed down by stamping. It is left in this state for eight or ten days, by which time it has begun to ferment, after which the bed ought to be well turned over and re-made on the same place, care being taken to place the manure that was near the sides of the first-made bed towards the centre in the turning and re-making. The mass is now left for another ten days or so, at the end of which time the manure is about in proper condition for making the beds that are to bear the Mushrooms. Little ridge-shaped beds—about twenty-six inches wide and the same in height—are then formed in parallel lines at a distance of twenty inches one from the other.

In a market garden they may stretch over a considerable extent. Their length being determined by the wants of the grower. The beds once made of a firm, close-fitting texture, the manure soon begins to warm again, but does not become unwholesomely hot for the spread of the spawn. When the beds have been made some days, the cultivator spawns them, having of course ascertained beforehand that the heat is genial and suitable. Generally the spawn is inserted within a few inches of the base, and at about thirteen inches apart in the line. Some cultivators insert two lines, the second about seven inches above the first. In doing so, it would of course be well to make the holes for the spawn in an alternate manner. The spawn is inserted in flakes about the size of three fingers, and then the manure is closed in over, and pressed firmly around it. This done, the beds are covered with about six inches of clean litter. Ten or twelve days afterwards the growers visit the beds, to see if the spawn has taken well. When they see the white filaments spreading in the bed they know that the spawn has taken; if not, they take away the spawn they suppose to be bad and replace it with better. But, using good spawn, and being practised hands at the work, they rarely fail in this particular; and when the spawn is seen spreading well through the bed, then, and not before, they cover the beds with fresh sweet soil to the depth of about an inch or so. For cover, the little pathway between the beds is simply loosened up, and the rich soil of the market garden applied equally, firmly, and smoothly with a shovel. With these open-air beds they succeed in getting Mushrooms in winter. A covering of abundance of litter is put on immediately after the beds are earthed, and kept there as a protection. They have not long to wait till the beds are in full bearing, and when they are in that state it is thought better to examine and gather from the beds every second day, or even every day where there are many beds. And thus they grow excellent Mushrooms, and in great quantity, all the further attention required being to renew the covering when it gets rotten, and an occasional watering in a very dry season.

CHAPTER XXII.

THE CULTURE OF SALADS.

THE culture of salads for the Paris market is not merely good—it is perfection. Not only do the French gardeners supply their own markets with delicious salads all through the winter and early spring months, but also, to a considerable extent, those of some other countries, and send vast quantities to the English markets. Now it will probably occur to the reader that climate is the cause of the superiority of the French in this respect, and, indeed, some practical men repeatedly say so. Nothing can be more fallacious than this belief; and I have no hesitation in affirming that, by the adoption of the method to be presently described, as good salads as ever went to the Paris markets may be grown in England and Ireland during the coldest months of winter and spring. It is simply nonsense to say that it is the effect of climate; the winters in northern France are severer than our own, and I know many spots in England and Ireland which are preferable to the neighbourhood of Paris for this culture. Near that city I have often seen beautiful Cos and Cabbage lettuces looking as fresh under their coverings in the middle of winter, when the earth was frost-bound, as the budding Lilac in May: had they been treated as ours usually are, they would have presented a very different appearance. At all times of the year the gardens in which salads are grown round Paris are beautiful examples of cultivation. In the spring and summer, when they are grown in the open air, nothing can look more healthy; but it is their condition in the cold season, when little or nothing can be done with them out of doors, that demands most attention from us. As very ordinary cultivation suffices to grow them with us in the favourable parts of the year, and in the other our markets are supplied from France, it is obvious that it is as regards the winter and early spring supplies that we want improvement. That improvement is easily secured.

The first and the chief thing to do towards it is to procure some of the large bell glasses (cloches) used by the French for this purpose, which are more fully described in the chapter on horticultural implements. They are cheap—and they require no repairs, and are easily cleaned and stored when not required. The troublesome task of giving air is done away with in their case. Without air on "every possible occasion" the British gardener attempts nothing under glass. By adopting this simple article, he may forego that ceaseless trouble throughout the winter and early spring. In the hotter weeks of autumn, these glasses are tilted up on one side for an inch or so, with a bit of stone placed underneath; but when once winter comes in earnest, then down they go quite close, and are all through the winter in the same condition as what we call Wardian cases. By the way, the French recognised

Fig. 291.

Four plants of the Lettuce Petite Noire under the Cloche early in October.

this principle long before we did, and what is more, have made a far more practical use of it. For all sorts of winter salad-growing this huge bell glass of theirs is infinitely superior to anything that we use for like purposes. The plants get full light at all times, and, while perfectly preserved from the filth and splashing of the rains in winter, are not in the least "drawn" or injured by the confinement, the light coming in so freely at all points.

The glasses are nearly sixteen inches in diameter, and about as much in height. For the winter work they are sometimes placed on a sloping spot with a sunny aspect, or the ground is thrown into beds, each wide enough to accommodate three lines of glasses. In early autumn these beds are made and the plants placed upon them, so that they can be readily covered by the bell glasses when the time comes that growth is checked in the open air. It should be added that the ground chosen is thoroughly rich, light, and well and deeply stirred, and the Lettuces are sown at intervals of a fortnight or so, so as to secure a succession, and to provide for the wants of the various kinds. The plants put out in September for the early and mid-autumn supply may not require to be covered if the weather be fine; and if they do, the glasses are tilted up a little as before described. But when the sun begins to fail and the cold rains to check growth, about the end of October, then the crop to be cut in the following month must be covered; and when towards Christmas the frost begins to take hold of the ground, the glasses must be firmly pressed down, and a deposit of leaves and litter placed around them.

Fig. 292.

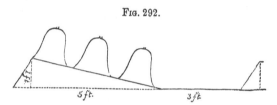

5 ft.   3 ft.

Sloping Bed for three rows of Cloches.

Thus, while everything else is at rest in the grip of ice, the plants will be kept perfectly free from frost, receiving abundant light from above, and growing as fresh as April leaves. Of course a deeper layer of this surrounding litter will be necessary in case of severe frost than in early winter. Covering them a little more than half-way up with a rather compact body of leaves and litter, effectually secures them from sharp frosts. When very severe frosts occur, mats made of straw are spread over the tops of the glasses; and should heavy falls of snow occur while these mats are on, they will enable the cultivator to carry it bodily away from the bed or beds; for it should not be allowed to melt on the beds or in the alleys between. In late spring the cloche is not required, nor is it for any except those crops that require artificial assistance. Thus the March and April supply is planted in October on a bed of light soil, with a surfacing of an inch or so of thoroughly rotten manure or leaf mould. These little plants are allowed to remain all through the winter unprotected; and when in spring the most forward Cloche Lettuces are cut, the glasses are immediately placed over the most advanced and promising of the little ones that have remained exposed. By that time they have begun to start up, encouraged by the early spring sun, and from the moment they receive the additional warmth and steady temperature of the cloche they commence to unfold the freshest and most juicy of leaves, and finish by becoming those great-hearted and tender products which one may see in such fine condition in the Paris markets in early spring. In the first instance three or five little plants may be put under each glass, and these thinned out and used as they grow, so that eventually but one is left, and that, without exaggeration, often grows nearly as big as the glass itself. Happily, no water is required, as the ground possesses sufficient moisture in winter and spring, and evaporation is prevented by the glasses and the protecting litter that covers the space between them. Thus a genial, agreeable moisture is kept up at all times, and the very conditions that suit Lettuces are preserved by the simplest means.

With the same glasses the various small saladings may be grown to perfection, or receive a desirable start. Thus, for instance, if Corn Salad be desired perfectly clean and fresh in mid-winter, it may be obtained by sowing it between the smaller Lettuces grown under these glasses; and so with any other small salad or seedling that may be gathered or removed without loss before or at the time the more important crop requires all the room. These bell glasses will be found of quite as much advantage in the British garden as they are in the French; they will render possible the production of as fine winter salads in our gardens as ever the French grew; they will enable us to supply our own markets with an important commodity, for which a good deal of money now goes out of the country; and, not least, their judicious use will make fresh and excellent salads possible in winter. At present the produce is so inferior and so dirty at that season, that it is generally avoided, and rightly so; for Lettuces when hard and wiry from alternations of frost, sleet, and rains—slug-eaten and half-covered with the splashings of the ground, above which they hardly rise—are not worth eating or buying. And though they may be grown well in frames and pits, the method herein described is better and simpler than that, and the Lettuces thus produced are far finer than those grown in English gardens in winter.

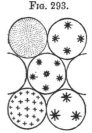

Fig. 293.

My first acquaintance with this mode of growing salad was made early in 1867. Since then I have had further opportunities of studying the subject, and it now appears to me that to discuss it in a general way is not sufficient. To understand the cloche and its use will not suffice; we must observe the culture of the varieties suited for each season, beginning with the Lettuce Petite Noire, a distinct winter kind, and requiring peculiar treatment.

Diagram showing the several stages of Lettuce Culture under the Cloche. The minute dots represent the seedlings, which are pricked off when very small, as shown in the circle with twenty-four asterisks. The central ring is the plan of a Cloche with one Paris Cos Lettuce in the middle, and five Cabbage Lettuces around it; above it, one with four plants of a Winter Cabbage Lettuce; and beneath it, one with three plants of the Cos.

CULTURE OF THE LETTUCE PETITE NOIRE.—This kind is grown to an enormous extent. Before leaving Paris in the first week of October last I saw beautiful crops of it growing, four plants under each cloche, each about five inches across, and without a speck of disease or dust. These plants, sown in August, are fit for cutting about the end of October, and prove very different to the rabbit food that serves us for salad as soon as the cold rains of autumn prevent its growing naturally to perfection. But this crop was an exceptionally early one; few sow it before the first days of September. It is sown on light, rich ground, well and deeply stirred, and covered with an inch, or a little more, of thoroughly decomposed and fine stable manure. The surface is made level and a little firm, and the impres-

sions of the number of cloches it is intended to sow made
upon it. One cloche will prove sufficient for a private
garden at one sowing; a few suffice for the wants of a
market garden. The Lettuce seed is then regularly sown
within the ring formed by the impression of the large bell
glass, and covered with a very slight coat of thoroughly
rotten manure—a substance that is always used in quantities in Paris market gardens. Then the cloche is put on,
the rim being gently pressed into the light, dark manure
mould. Shade is given when the sun shines strong in
early autumn, but air is never given. A peculiarity of this
excellent Lettuce is that it grows best without air.

As soon as the seedlings are strong enough to transplant,
they are pricked out, about thirty under each cloche. This
transplantation is done at a much earlier stage than is the
case with us. They do not let them wait to get crowded,
and then transplant them, tall and drawn, into their places,
but take them up gently and without disturbing the roots,
soon after the cotyledons are developed and when the first
leaves are beginning to appear. In transplanting, a surfacing of very fine and thoroughly rotten manure is placed
over the earth to the depth of an inch, and the earth is
often thrown into beds sloping to the south, so that the
cloches may enjoy the full sun. Whether the beds are flat
or sloping, they are wide enough for three ranges of cloches
placed in a quincunx manner, so that very little space is lost
between them. Before transplanting, the ground is marked
by the impression of a cloche, and the little seedlings are
inserted by the finger in the soft mould. Instead of getting
drawn, as would be the case if they were left together for a
longer period, the plants spread out into neat and firm little
rosettes, their leaves lying close to the ground, for the light
comes freely through the clean cloche, and there is not a
sign of ill-health or speck of dust to be seen on the leaves.
The first crop mentioned is not a general one, as there is
abundance of open-air salad to be had about the time these
will be cut; but the strongest batch of those that in
September were in the rosette stage, under the cloches,
are transplanted into their final places before the 15th of

October, planting four under each cloche, and they supply a great want, coming in in perfect condition towards the middle of winter.

Sometimes the crop is planted out in the narrow frames common to Paris market gardens, turning over the old dung beds before planting. The frames being narrow and shallow, the plants are near the glass, and as soon as they are planted the lights are put on, and, instead of giving air by the aid of these lights, the greatest care is taken to keep it out. No matter whether under cloche or frame, the Petite Noire must never get any air. Should severe frost occur, the glasses may be protected with straw mats. It must be observed that when the plants are transferred into the places where they are to attain their fullest size they are removed with good balls, and with some care to check their growth as little as possible. The plants pricked out in October are ready to cut at the end of November or commencement of December, which are the seasons when this Lettuce begins to come into general use. In addition to the crop put out during the first half of November, another is sown at that season, in the way before described, and in like manner. Should very severe weather render the Lettuces liable to suffer, dry litter is placed between the cloches, and higher up at their north side, so as to prevent the frost from entering the ground, and the contents of the cloches are daily exposed to the light and sun, except when they are frozen, when the covering is kept on or increased.

The forced culture in the same narrow frames usually begins about the end of November. At that period a hot-bed is prepared, sixteen inches thick, and about 55° F. in temperature, spreading on it some of the never-failing, crumbling, thoroughly rotten manure, and on it is planted seven lines of Petite Noire. This plantation requires greater care than those placed under the cloche, in consequence of the warmth and humidity; decaying or spotted leaves have to be picked off when they occur, and the plants must be protected with the mats more than those not excited by heat. The frames are surrounded by hot dung, and a dryer mass of it fills up the alleys between the

little frames to the very edge of each. At the end of
January or beginning of February the last forced crop of
Petite Noire is planted, under cloches placed on a gentle
hotbed a foot deep, and covered with about four inches of
the same mould-like manure, the bell glasses as usual being
placed in three ranks. The bed for them may be made
wide enough for six (three lines on each side of a narrow
alley), or for three only. In this January or February
planting, four plants of Petite Noire are planted under each
cloche, and one Cos in the middle. The tender Petites
Noires are good to gather in February and March; the
Cos remains a little later, nearly filling up the glass, and
forming one of those superb Lettuces to be seen in all our
great towns in early spring, and which are usually supposed
to come from some paradisiacal climate, instead of the hard,
cutting, and most ungenial winter climate of Paris. Certainly the climate that would produce them without gardening skill at the periods spoken of should be as mild and
smiling as that in which

"O'er the four rivers the first roses blew."

Laitue Verte Maraîchère.—This Lettuce is sown about
the first fortnight in October in the open air or on a sloping
bed under a cloche. It is pricked out, and twenty or thirty
are generally placed under one glass, which is taken off every
time that the weather permits. As it often happens that,
in spite of the care taken with it, this Lettuce will grow too
tall, it is generally taken up, and immediately transplanted
during the course of the month of November. For this purpose a new sloping bed is prepared and the plants are pricked
into it immediately, in which case only eighteen or twenty
are put under each glass. From this moment they receive
the same care as the other Lettuces sown at the same time.
Towards the end of December or the beginning of January
planting in frames and under cloches is begun. In the
former case eight rows are placed in each frame, each row
consisting of twenty-five plants, so arranged that there is
alternately a Petite Noire and a Cos lettuce in each row.
Under the cloches they are arranged so that there are four

of the former to one of the latter. The Cos lettuces thus grown may be gathered at the beginning of February. After they are gathered the beds are planted a second time, and towards the end of February or the beginning of March, that is to say, when the severe cold is no longer to be dreaded, a single plant is pricked out in the little spaces between the cloches. As soon as the Cos and the Petite Noire lettuces planted beneath the bell glass are gathered these latter are used for the second crop. By this method they may be gathered about three weeks after. At the same time warm borders on the south side of walls are also planted with Cos lettuces. Ten or twelve rows are drawn according to the size of the border, and planted with Cos lettuces about twelve or fourteen inches apart. After this crop has been planted out a few Radishes, Leeks, or Carrots are sown between the Lettuces. Generally speaking, these Cos lettuces are fit to gather towards the end of April or the beginning of May. Some market gardeners also sow Cos during the month of August, which, planted out in hotbeds under bell glasses, are generally fit for gathering in December and January.

LAITUE GOTTE.—This Lettuce cannot be raised in the same way as the Petite Noire, because it will not come to perfection without plenty of air. It is not so early as the Petite Noire, but is much esteemed, growing larger and more perfect than its congener. The Laitue Gotte is sown from the 20th to the 25th of October, on a sloping bed, and the same method of after-treatment is adopted as in forcing the Petite Noire, although it is less damaged by frost. Being a later kind, it may be left in the sloping bed until the Petite Noire is all gathered, when it may be used to fill the vacant places in the hotbeds. The manure of the hotbeds should be left undisturbed, but the soil in the frames should be well forked and made even. Towards the end of January or beginning of February the L. Gotte should be planted in the frames. It should have plenty of air, whenever the weather will allow it, by propping up the back of the light. If the Lettuce does not heart early the light should be removed as soon as the fine weather makes its appearance, so as to allow it to come to perfection in the open air. Instead of planting the Laitue Gotte in a frame, it may be planted on a hotbed under cloches arranged in three rows, three plants being placed under each, taking care to preserve them from frost in the usual way, and to give them air whenever the weather will allow of it. The Laitue Gotte may also be planted in the open ground under a cloche. The earth is well dug and raked, and an inch and a half of well decomposed stable manure thrown over it, smoothed and flattened. The cloches are then placed in alternate lines, with three plants under each. When the plants have struck, air should be given them whenever it is possible. The Laitue Gotte, when planted in frames at the end of January, arrives at perfection by the end of March ; those planted under cloches in February, at the beginning of April; and those planted under cloches or in frames at the end of February, towards the middle of April.

THE PASSION LETTUCE—Laitue de la Passion as it is called —is the only winter Lettuce grown in the neighbourhood of Paris in the open air. It is sown from the 15th of August to the 15th of September, according to the soil which is to receive it, and is pricked out rather thinly in October. It is generally left unprotected through the winter; however, it is prudent to defend it against severe frosts by covering it with long litter, which is taken off and put on again as it is required. This Lettuce is generally fit for gathering towards Passion Week, from which circumstance it is called Passion Lettuce by the Parisians.

CORN SALAD OR MÂCHE.—This plant is very much used in Paris, and is excellent as a salad. It is peculiarly agreeable when mixed with a sprinkling of Celery. The culture is of the simplest kind, the seeds being often sown amongst other crops, which must be placed somewhat thinly, and the Corn Salad is gathered before the other requires all the ground. They begin to sow the Ronde variety about the 15th of August, and continue at intervals till the end of October. That sown in August comes in for the autumn consumption ; that sown in September for winter use; and that sown in October is used in spring. During hard frosts the crops to be gathered during winter are covered with long litter. Italian Corn Salad or Mâche Régence is sown in October, and is sown more thinly than the preceding; it is considered the best variety. It may also be raised in the spare places between the plants, under cloche, in any open surface between plants in frames, or any cool light garden structure.

THE BARBE DE CAPUCIN is the most common of all salads in Paris in the winter and early spring, and for its culture the cloche is not required. It is perhaps too bitter for some tastes, but is sometimes used by English families, and is well worthy of culture in small gardens, being so very easily forced when other salads are scarce. This salad is of all others that which may be had with the least amount of trouble by any person in possession of a spot of rough ground, a cellar, or any dark place where a little heat might be used to start the blanched leaves of the Chicory in winter; it is therefore desirable that it be brought into common use. Should the taste be too bitter to those unaccustomed to it, or who do not like bitter salads, the addition of Corn Salad, Celery, or Beetroot, improves and modifies the flavour, and makes it a very distinct and agreeable salad. The gardeners of the commune of Montreuil sow every year a large quantity of wild Chicory for the purpose of forcing the Barbe de Capucin. For this purpose the crop is sown in April. It is sown both broadcast and in drills, which are traced at a distance of eight inches from each other. At least nine pounds of seed per acre must be used. In the course of the summer the ground should be turned up several times; when the frost sets in the roots are taken up with the fork, care being taken not to break them. They are then laid by the heels so as to have them always ready for use; and in the course of the month of October, the season when such work is usually commenced, a hotbed about sixteen inches deep is prepared, the heat of which is from 65° to 80° Fahr. The most favourable position for such a hotbed is in a cave or in a deep cellar without light or air.

When the heat of the bed has somewhat abated, the plants are tied up in bundles, having first carefully removed all the dead leaves and other portions liable to produce mouldiness, after which they are placed upright on the bed and watered frequently with a rose watering-pot; but, as usual, the waterings must be adapted to the heat of the bed. From the time the Barbe begins to grow, these waterings ought to be performed with great judgment, so as to prevent the interior of the bundles from rotting. At the end of fifteen or eighteen days, the salad is long enough to be gathered. From the time given above the Chicory can be blanched up to March and April; after every gathering however, the spent dung should be removed, and replaced by a fresh supply, so as always to keep the bed at the same degree of warmth.

In the market gardens of Viroflay large quantities of wild Chicory are cultivated. It is sown broadcast towards the end of the month of May or the beginning of June. The following February it is covered with an inch and a half of leaf mould, or, if that cannot be procured, mould from the pathways. Ten or twelve days afterwards it is cut just where the two kinds of earth join. Two or three gatherings are generally made, after which it is allowed to remain until the following year.

SCAROLLE (Broad-leaved Batavian Endive).—This fine salad forms a very considerable item in the culture of the Paris market gardens. It is deserving of being generally grown in England, being easy of cultivation, very large, and forming an excellent salad; indeed, it is on the whole perhaps the best we have. The mode of blanching is very simple: the leaves of the plant are gathered up, and then a single straw tied around them. This is only done five days before the Scarolle is ready for market. A crop of this was nearly ready to blanch in September, it was the second crop of the same plant that had been on the ground—the first and best having been gathered a few weeks. Some of the finer specimens of this second unblanched crop measured twenty and twenty-one inches in diameter on September 7th, 1868.

CHAPTER XXIII.

ASPARAGUS is grown much more abundantly and to a much larger size in France than it is in England. The country is half covered with it in some places near Paris; small and large farmers grow it abundantly, cottagers grow it—everybody grows it, and everybody eats it. Near Paris it is chiefly grown in the valley of Montmorency and at Argenteuil, and it is cultivated extensively for market in many other places. About Argenteuil 3000 persons are employed in the culture of Asparagus—at least so I was told by the son of the cultivator who took the best prizes for Asparagus at the Exhibition. His father not being at home, I traversed a considerable portion of what may be termed the region of Asparagus with this youth, who was of the intelligent type, and understood all about this dainty vegetable. I first saw it growing to a large extent among the vines. The vine under field culture, I need scarcely say, is simply cut down to near the old stool every year, and allowed to make a few growths, which are tied erect to a stake: they do not overtop the Asparagus in any way, but on the other hand the strong plants of that show well above the vines. It was not in distinct close lines among the vines, but widely and irregularly separated, say six or seven feet apart in the rows, and as much or more the other way. They simply put one plant in each open spot, and give it every chance of forming a capital specimen, and this it generally does. When the stems get large and a little top-heavy in early summer, a string is put round all, so as to hold them slightly together (the careful cultivator uses a stake), and the mutual support thus given prevents the plant from being cut off in its prime. We

all know how apt it is to be twisted off at the collar by
strong winds, especially in wet weather, when the drops on
every tiny leaf make the foliage heavy.   The growing of
Asparagus among the vines is a very usual mode, and a
vast space is thus covered with it about here.   But it is
grown in other and more special ways, though not one like
our way of growing it, which is decidedly much inferior to
the French method.

Perhaps the simplest and most worthy of adoption is to
grow it in shallow trenches.   I have seen extensive plantings that looked much as a Celery ground does soon after
being planted, the young Asparagus plants being in a shallow
trench, and a little ridge of soil being thrown up between
the lines of Asparagus.   These trenches are generally about
four feet apart, sometimes less.   The soil is rather a stiff
sandy loam with calcareous matter in some parts, but I do
not think the soil has all to do with the peculiar excellence
of the vegetable, and am certain that soils on which it
would flourish equally well are far from uncommon in
England.   It is the careful attention to the wants of the
plant that produces such a good result.   Here, for instance,
is a young plantation planted in March, and from the little
ridges of soil between the shallow trenches they have just
dug a crop of small early Potatoes.   Now, in England,
the Asparagus would be left to the free action of the breeze,
but the French cultivators—like the old Scotchwoman who
would not trust the stormy water and God's goodness as
long as there was a bridge in Stirling—never leave a young
plant of Asparagus to the wind's mercy whilst they can
get hold of a bit of oak about a yard long.   But when
staking these young plants they do not insert the support
close at the bottom, as we are too apt to do in other instances, but at a little distance off, so as to avoid the
possibility of injuring a fibre; each stake leans over its
plant at an angle of 45°, and when the sapling is big enough
to touch it or be caught by the wind, they tie it to the
stick as a matter of course.   The ground in which this
system is pursued being entirely devoted to Asparagus, the
stools are placed very much closer together than they are

when grown among the vines, say at a distance of about a yard apart. The little trenches are about a foot wide and eight inches below the level of the ground—looking deeper, however, from the soil being piled up.

The young plants are placed in these trenches very carefully. A little mound is made with the hand in each spot where a plant is to be placed so as to elevate the crown a little and permit of the spreading out of the roots in a perfectly safe manner. In fact they seem to be about as particular as regards depositing the young plants in the first instance, as a good grape-grower is about his young vines. They plant in March and April—using any kind of manure that can be had, but chiefly here, so far as I could see, the refuse of the town—the ashes, old vegetables, rags, and other matters, that the people throw before their doors, and which the dust-carts take away in the morning. They are very particular to destroy the weeds, and they also take good care to destroy all sorts of insect enemies in the mornings, especially during the early summer. Between the lines of Asparagus they plant small growing crops

Fig. 294.

This figure shows the depth of the successive annual earthings given to the Asparagus. After four or five years' growth the ridges disappear, and the highest points of the grounds are those over the crowns of the roots.

on the little ridges during the first years of the plantation, but are careful not to put the large vegetables there, which would shade and otherwise injure the plant. When they plant they spread a handful or so of well-rotted manure over each root, and they repeat this every year, removing the soil very carefully in the autumn down to the roots, putting on them a couple of handfuls of rotten manure, and spreading the earth over again, so that the rain is continually washing nutriment to the roots. When doing this they notice the state of the young roots, and any spot in which one has perished, or has done little good, they mark with a stick, to replace it the following March. Early every spring they pile up a little heap of fine earth over each crown. When

the plantation arrives at its third year they increase the size
of the little mound, or, in other words, a heap of finely
pulverized earth is placed over the stool, from which some,
but not much, Asparagus is cut the same year, taking care
to leave the weak plants and those which have replaced
others, to themselves for another year.

They would appear to cut the best of it when it is about
an inch and a half out of the ground—and here is the only
objectionable thing about their system. The top is very
good, but as a rule too short; but such a handle as they
give you to it! Now, it may be desirable to have something to take hold of, but to cut it as they do here, and as
we often do in England, is not wise, or conducive to the
thorough enjoyment of the vegetable. However, it is
simply a matter of the amount of covering given, or of the
depth at which it is cut, and therefore of the simplest
management. The care and culture may be applied as
described, and the Asparagus cut at pleasure. To procure
it in a thoroughly blanched condition, the French pile up
these little mounds of fine earth, which enables them of
course to get it much longer; besides, they can pull away
the soil conveniently, and get at the rising stems as low
down as they like. It is not, however, the fault of the cultivator that the Asparagus is so much blanched, for I have
been told by the first fruit and vegetable merchant in Paris
that his customers would not buy the finest Asparagus ever
grown if brought in a green state. This is why you see it
with a shaft like ivory and with the point of the shoot of a
red, rose, or violet tinge. Then again, some contend
that Asparagus blanched after the French fashion is far
more delicious than when it is eaten in the green state, while
others in England say it is worthless. From what I know
of their arguments, however, it is clear that those who say
French Asparagus is worthless, mostly know it from some old
bundles bought and eaten perhaps a fortnight after they
were cut in France. Let us hear the French side :—" In
certain localities they do not yet value the distinction
between blanched and green Asparagus, and occasionally
prefer the last. That is an error very prejudicial to the

consumer's interests. In the green Asparagus there is only the point edible; in the white it is often entirely so, and, moreover, it is infinitely more tender and delicate. All Asparagus cut when it is green is not fit to be eaten in the ordinary way, but may be used cut up small as an accompaniment to other dishes. To serve up green Asparagus is to dishonour the table! In the markets of Paris the green Asparagus is worth one franc a bunch, when the blanched is worth three francs; they do not eat it (the green Asparagus) —it serves for the manufacture of syrup of Asparagus.—V. F. LEBŒUF."

When the plantation reaches its fourth year the little mound of blanching earth is increased to fifteen inches in height, for then they expect to cut something worth while, and these mounds are made in the early part of March; and even after this, as they grow stronger the little mounds are increased; and they always keep a look-out for the feeble plants, with a view to replace them. To have Asparagus as it ought to be, they say you must cut every day, or every two days, according to temperature, so that it may be obtained at the right moment; indeed if they do not do this, the shoots become too high and too green. They place great importance on obtaining strong and healthy plants; and in the establishment which I visited they have three kinds, l'Ordinaire, La Hollande tardive, improved, and La Hâtive d'Argenteuil. The first is described as very fine, the second very strong, and the last as the earliest, most productive and best. Of course there are various modifications of the plan described herein, and in several instances I saw two rows placed in a rather wide trench in an alternate manner. As to the size and quality of the Asparagus produced by this method there can be but one opinion. Mr. Veitch and many other English horticulturists, who know what gardening is, as well as it is possible to know it, have been, with myself, surprised at it. The same difference holds good in the forced Asparagus—the slender pipe-shank productions of the English forcing-house being miserable compared to it.

Concisely: the French mode of cultivating this delicious

vegetable differs from our own diametrically in giving each plant abundant room to develope into a large healthy specimen, in paying thoughtful attention to the plants at all times, and in planting in a hollow instead of a raised bed, so that as the roots grow up they may have annual dressings of enriching manure. They do not, as we do, go to great expense in forming a mass of the richest soil far beneath the roots, but rather give it at the surface, which is consistent with the nature of the root. And in this way they beat us with Asparagus as thoroughly as Messrs. Meredith, Henderson, or Miller, beat them with hothouse grapes. A man who knows how to spend two and a half francs for his dinner in Paris enjoys Asparagus for a longer time and of much better quality than many a nobleman in England with a bevy of gardeners. In the first-class restaurants you usually pay high for Asparagus, as you do for all other vegetables, but it is served very cheaply in many respectable ones—so much so, indeed, that it is partaken of by all classes.

As the culture of this vegetable is so important, and the French manage it so well, I venture to go further into detail by giving the following account, written by a well-known and very successful cultivator of Argenteuil, and first published in the *Gardener's Chronicle*. I have made some few alterations, with a view to rendering the meaning simpler and clearer to the reader :—

"PREPARATION OF THE GROUND.—When a convenient piece of ground has been selected, it is first of all to be mellowed by spreading on its surface a good dressing of horse or sheep manure. The ground is to be dug up to a depth of sixteen inches in fine weather at the beginning of winter, during which season it is to be left at rest. In the month of February following—at least, as soon as severe frost is no longer to be expected—the ground is to be laid out in furrows and ridges, in order to shape shelving beds, and the excavations which are to receive the plantations. For this purpose the following operations are to be performed. First, there are to be drawn the whole length of the ground, and by preference from north to south, two lines, leaving between them a space of fourteen inches, intended for the site of the first half-shelving bed. Reckoning from the interior base of this half-shelving bed, a distance of twenty-four inches is to be measured for the first trench. The earth taken from it will serve to form the shelving bed. The second shelving bed, which will be a large one, is to measure twenty-eight inches in width at its base, and fourteen inches in height. Next comes the second trench, then the third entire shelving bed, and so on, until the whole piece of ground has been occupied. Thus, the first half-shelving bed will measure in width fourteen inches, and in height eight inches; the first trench in width twenty four inches, the second entire shelving bed in width twenty-eight inches, and in height fourteen inches, &c., as shown in the accompanying figure. The earth of the shelving beds being intended to cover over the plants from time to time, these beds will gradually diminish in height, and the whole piece of ground will become nearly level at the end of five years, when the Asparagus plantation will be in full productiveness." [In justice to the extensive market grower and successful prizetaker who thus describes his culture, we are bound to respect his diagram; but a readier and less precise method is more generally pursued, such as that indicated by Fig. 296, roughly drawn from memory.]

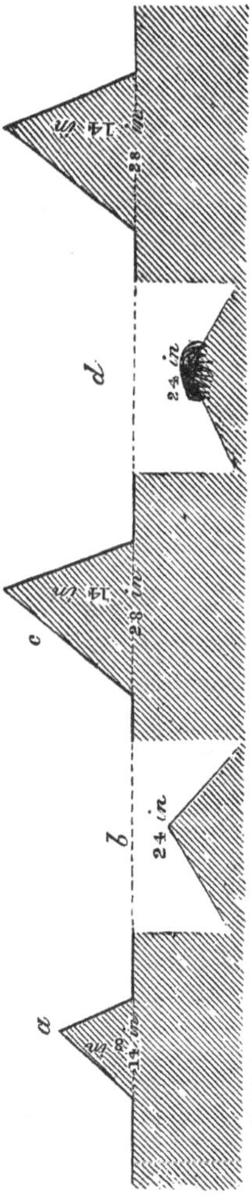

FIG. 295.

d. Ground or trench, with plant set on the hillock.

c. One of the shelving beds.

b. Ground or trench.

a. A half-shelving bed.

" FIRST YEAR.—The first plantation is to take place during

the months of March or April, and should be performed in the following manner :—In each trench, through its entire length, small holes, eight inches in diameter and about four inches deep, must be formed about thirty-six inches distant from each other. In the centre of each of them a small hillock of earth about two inches high is to be raised, upon which the Asparagus plant is to be laid down, care being taken to divide the roots equally in every direction; the roots are then to be covered over with half an inch of earth; and one or two handfuls of very good manure are to be added, and covered over with about an inch and a half of earth, at the same time forming a small hollow of about an inch deep over each plant, to indicate its position. In order exactly to know the position of the plants, and to shelter them and their shoots from accidents, a small stake is to be set to each, inclining it at an angle of 45°, in order not to injure the roots, and placing it a little away from the plant.

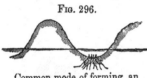

Fig. 296.

Common mode of forming an Asparagus plantation.

" Every morning, towards the months of April and May, slugs and snails are to be carefully looked for and destroyed. Beetles are also much to be feared in the Asparagus plantations. Twice every day during a fortnight it will be well to pursue these insects with rods, so as to hinder them from depositing their eggs on the stalks of the Asparagus ; these eggs develop at the end of three weeks into black maggots or worms, which prey upon the Asparagus stems and dry them up. Yet these insects are not the only ones which are to be dreaded. The white worms (or maggots of tree beetles) are very dangerous, and it will be well constantly to put in use the most proper means to get rid of them, for they eat the roots and destroy the plants. It will be useful also to set mole traps, for while tracing their underground roads the moles cut the roots. Frequently during the season the plantations should be thoroughly cleaned, taking care to never bruise or in any way injure the young plants, for any accident to these is of course directly prejudicial.

" Common vegetables, such as late Potatoes, Cabbages, &c., ought not to be planted on the ridges of beds, which, however, may be made useful (but only during the first years) by growing on them early Potatoes, Lentils, Kidney Beans, Salads, and such other vegetables as are of little inconvenience from their dimensions. In the month of October, during fine and dry weather, the small stalks of the Asparagus are to be cut off at six inches above the ground. The ground is to be lightly cleaned, and the shelving beds must be dug up to a depth of twelve inches, maintaining their conical shape. The Asparagus is to be lightly covered with manure, the plants being laid bare with a flat hoe, for a diameter of eight inches, and up to the crowns. Proper care ought to be taken not to injure the roots with the implement. On each plant lay one or two handfuls of good manure, free from all noxious substances. While spreading the manure, mark out with a small stick the site of the plants which have failed during the course of the year ; these must be renewed in the month of March following. The manure is at once to be covered over with about three inches of the best mellow earth at hand, and over the plants is to be made a small conical hillock about two inches high. This operation is the last to be performed for the year.

" SECOND YEAR.—In March or April begin by replacing the plants which have failed in the preceding year, selecting vigorous plants a year old, and setting them in the same manner as recommended for the first year. Stakes are to be placed near the foot of each plant, always at an angle of 45°. In the beginning of April a cleaning is to be made on the shelving beds and on the grounds ; it will be well to perform this operation the day after a sprinkling of rain, in order the more easily to break the clods. As soon as the Asparagus stems become firm, fasten them to the stakes, in order to protect them against the wind, which might break them. In the month of October the dry stalks are to be cut off at eight inches above the ground ; the shelving beds are to be turned up, always lightly hollowing out the trenches. Manure is to be spread on the shelving beds, which are then to be dug up. The stakes, having become useless, are to be taken away. Lastly, the laying bare of the roots is to be done by taking away the earth, as already directed, the dressing of decomposed manure placed over them, and lastly, the manure is in its turn covered with a couple of inches of the finely pulverized soil.

"THIRD YEAR.—In the middle of the month of March, during fine weather, small knolls, from six to eight inches high, are to be made over each plant, taking nevertheless as a basis the comparative strength of the crowns, more or less large, or of a more or less determined development; those which may be too feeble, or having served the preceding year to supply the bad ones, or those which had failed, are to be covered over with a hillock of only four inches high, and should then be left to themselves. From the other plants, three, or at most four Asparagus heads may be gathered; but they are not to be cut off with an Asparagus knife, but removed with the fingers. However, there is a particular sort of knife, square-shaped at the end, and having teeth on one side, forming a saw, which will be useful to take away the earth about the stalk, and will make it easy for the fingers to reach the subterranean stock, which care must be taken not to injure. With regard to the gathering, one finger must be got behind the Asparagus stem at its base, and by bending it, it will easily come off the stock. In this manner all injury to its neighbours, which may easily happen with an Asparagus knife, will be avoided; and there will not be left any wounded ends, from which the sap will flow and spread around, occasioning rapid decay. Care should be taken to close up the hole made for the gathering of the Asparagus, and the knoll is at once to be formed anew. In the month of April, the stakes are to be again used, and the stems fastened to them in due time. After having, in fine weather, done all that is necessary in the way of cleaning, in the month of October the dry stalks are to be cut off about ten inches above the ground, and the dead rubbish thrown out of the Asparagus plantation. From the whole surface of the trenches, and to a depth of four inches, the earth is to be taken away and thrown upon the

ridges; this earth is to be substituted by a layer of very
good manure, which layer is to be of a thickness of about
an inch and a half, if night soil is made use of, or of about
two inches if it is only common manure. At the same
time a portion of the end of the old stalks is to be taken
away, preserving that nearest to the crown, so as to indicate
the exact site of the plants for the fourth year. After
having spread the manure, the ridges must be dug up, and
the manure covered with an inch or two of earth from
them, a small hillock being left over the crown of each plant.

" FOURTH YEAR.—About the middle of March, in dry
weather, or the day after a sprinkling of rain, knolls of the
height of from ten to twelve inches must be formed over
each plant with the fine earth from the sides of the ridges.
The feeble plants marked with a small stick at the preceding laying bare, are to be covered over with hillocks of
a thickness of from four to six inches only. While earthing
up the Asparagus the ends of the dry stalks are to be taken
away. The gathering is to take place from the largest specimens during one month at the most. Then they are to be
left to run to seed. The most feeble ones are to be spared
in order to strengthen them. At the second dressing in the
month of May, earth is taken from the shelving beds, in order
to cover over, to an extent of an inch or two, the whole surface
of the grounds, so as to protect the Asparagus plantation from
the dryness of the summer. The stakes should be five feet
high. In the month of October the stalks of the Asparagus
are to be cut off at fourteen inches above the ground, and
the plantation is to be cleared of the rubbish; manure is to
be spread on the ridges, which are to be made up from the
knolls in the trenches; and are then to be dug up to a
depth of sixteen inches. Notwithstanding the manure laid
upon the shelving beds, the roots of the Asparagus are to
be laid carefully bare in the manner already described.
Upon the crowns are to be put a few handfuls of good
manure, which is to be covered over with two inches of
good mellow earth; the little knolls which are to be formed
over the centre of the plants, are to be over three inches in
height.

" FIFTH YEAR.—The making of hillocks on the Asparagus is to begin in the month of March; they are to be fourteen inches high, and their diameter is to be in conformity with the diameter of each specimen or ' stool.' The gathering is to consist of the heads on all the large plants, and of some only on the feeble ones; the gathering may last two months at most. In order to get fine Asparagus, the heads are to be gathered once every day, or every other day, or every third day at farthest, according to the degree of temperature. This is the way to obtain rosy, red, or violet Asparagus. In order to get it green it will be sufficient to let the heads grow during four or five days more; they will lengthen and become green. The second dressing is to be made as in the preceding years. The stakes are to be put in as soon as the necessity is felt, and the stems, having regard to the increase of their height and weight, must be firmly tied, so that the wind may not disturb them and that they may not be broken. In the month of October following, the dry stalks are to be cut off at fourteen inches above the ground. The plantation is to be cleared, and the ridges are to be replenished by adding to them the earth of the knolls which have been raised on the plants for the gathering. Then the manure is to be spread in the manner already indicated; and the digging up of the ridges is next to take place.

"SIXTH YEAR.—When the Asparagus plantation shall have reached its sixth year, it will then be in full productiveness. The forming of knolls is to take place in March during fine and dry weather; the knolls must always be fourteen inches high, reckoning from the subterranean stock. The care to be taken is to be the same as in the preceding years, particularly with regard to cleanliness and staking. As for insects, they will be less to be feared than during the first years of the establishment of the plantation. The beetles can no longer lay their eggs on the stalks, since they are cut during two months, and when allowed to start up the time of the laying of eggs is past. In the month of October the shelving beds are to be turned up in conformity with the manner shown for the preceding year; the shelving beds and the plants are to be manured, as has been explained for the fourth year. As the Asparagus plantation may last fifteen or twenty years, the operations and the care to be taken are to be repeated from year to year in the manner above indicated. Generally, in a well established Asparagus plantation, the gathering, reckoning from its beginning, is to take place during two months, whatever may be the climatic circumstances under which the plantation is placed. It must have been seen that the expense is not very great; the chief object is the care which must be taken. The main point is to get good plants, in order to obtain good produce. By properly following the rules laid down here, satisfactory results will be obtained."

The mode of forcing Asparagus chiefly consists in digging deep trenches between beds planted for the purpose, covering the beds with the soil and with frames, filling in the trenches between the beds with stable manure, and pro-

Fig. 297.

Preparation for forcing Asparagus. The trenches are dug out and filled with stable manure, the earth being heaped on the beds. These are covered with rough frames, up to the edge of which the heating material is piled.

tecting the frames with straw mats and litter to keep in the heat. In the beginning of November the pathways between the beds of Asparagus are dug up about two feet deep, and as much wide. Divide the soil coming from the pathway very carefully, and put about eight inches thick of it on the surface of the bed. Fill up the trench with good new horse-dung, and place frames on the bed. The manure should rise as high as the top of the frames, and the lights be entirely covered with mats and litter to prevent the heat accumulated in the frame from escaping. About a fortnight or three weeks after, the Asparagus begins to show itself on the surface of the bed. Many market gardeners cover the whole of the bed inside of the frame to a thickness of three or four inches with dung to force more

L L

quickly the vegetation, but in this case the manure must be removed when the Asparagus begins to shoot. When the shoots are about three inches out of the ground they may be cut. The mats must be taken off in the day-time, but the heat must be well kept up or the roots and buds will fail to push. The beds are forced every second year only. The gathering of the Asparagus may continue for about two months, but no longer, or the plantation would be injured. When the gathering of the Asparagus is over, the frames and dung linings are taken away, and the soil which has been dug up from the alleys is put back again.

The preceding note applies to the forcing of the better qualities of Asparagus chiefly. I visited last September a place at Clichy in which quite a specialty is made of forcing the smaller sized Asparagus. It is the garden of M. Caucannier, Place de l'Eglise, and contains a number of iron houses, just on the same plan as those in the Jardin Fleuriste, already described. Indeed, if I mis-take not, those in the Fleuriste are copied from them. There are frames within each house, just as in many propagating houses in England, and beneath them the Asparagus is forced for the markets, and in incredible quantities. The houses are heated by hot water, and the culture in other respects resembles that which is practised in forcing gardens in England—that is, when the plants are taken up to be forced indoors or in pits. The disturbance weakens the roots a good deal, and by this méthod the large table Asparagus is never forced. M. Caucannier and other growers produce it specially in a small state for cookery.

CHAPTER XXIV.

OBSERVATIONS ON SOME OF THE VEGETABLES OF THE PARIS
MARKET — CULTURE OF THE SMALL CARROT — THE CAR-
DOON — FORCING THE CAULIFLOWER — THE SWEET POTATO
—EARLY POTATOES — OLEANDER CULTURE — CULTURE OF
THE ORANGE—SHOWING ROSES IN FRANCE—FORCING THE
WHITE LILAC.

A VISIT to the markets of Paris is sufficient to interest
many in the vegetable culture of that capital. There is
so much difference in the supplies to that market and the
London one that there is certainly much to be learnt
on both sides. That so great a difference should exist
in the supplies of cities so near each other is somewhat
remarkable. The Parisians make as much use of that
delicate, wholesome, and excellent vegetable Seakale as
we in England do of the Bread Fruit Tree; and the
Briton who leaves London in a hot and dry July,
having failed to get a tender vegetable or salad at dinner,
arrives in Paris next morning, and finds the streets in the
neighbourhood thickly strewn with every variety as tender
as if the climate were a perpetual May.

But, although abundant intercourse has long existed
between the two countries, the fact that the observers are
rarely practical men, and therefore not capable of seeing
differences and their value and causes, and the difficulty of
getting information about the subjects, noticeable improvements have not been exchanged from side to side. Therefore, in addition to dealing with the subjects in which
the French are far ahead of us—Salads, Asparagus, and
Mushrooms, for example—I have thought it well to speak
of any varieties of vegetables with which we should be
better acquainted, or which are likely to prove useful, and

L L 2

to add a few remarks about the culture of any of them where desirable. In this I simply do the best I can for the time, and believe the subject is far from being exhausted. My acquaintance with it only began in 1867. To save trouble in inquiries, I add that, should any reader find difficulties in getting seeds of any subject mentioned in this book true to name, he may be certain of getting them of the best quality from MM. Vilmorin, Andrieux, and Co., of the Quai de la Mégisserie, or of MM. Courtois-Gérard and Pavard, Rue du Pont Neuf, Paris.

OBSERVATIONS ON SOME OF THE VEGETABLES OF THE. PARIS MARKET.—The Cardoon is much more grown and eaten in France than in England, and its culture is well understood. The variety most grown and usually considered the best is the Cardon de Tours (a spiny var.). A spineless variety, Cardon Plein Inerme, is not sufficiently known. It is as good as the former, and preferable on account of not being fiercely armed with spines. The Artichoke (Cynara scolymus) is grown to a very much larger extent in France than in England, and its culture is said to be attended with much profit. It is used in every Parisian restaurant. The variety considered best is the Gros Vert de Laon. Camus de Bretagne is the kind that is often used raw. Of Asparagus most people agree in considering that of Argenteuil the best, though of the distinction between the several varieties there is little certainty. Of Beetroots, there is nothing to surpass our English varieties; the best French one is the Longue. It is cultivated to a large extent at Les Vertus, near St. Denis, and brought to market cooked, so that the smallest portions may be sold with salad. It is used much more than with us by the poorer classes, especially with Barbe de Capucin in the winter.

The little Carrot which is grown to such unvarying perfection is the Rouge Courte à Chassis. This and the so-called choice varieties of Carrots are far from being always obtained true. Cultivated as it is in Paris it is infinitely preferable to the larger and coarser sorts grown with us, but the difference is chiefly owing to the mode of growing it. The best salads known are grown in the vicinity of Paris, and among them the various Endives assume a great importance. Chicorée Fine d'Eté and C. Rouennaise are the best summer kinds ; C. de Meaux is the large one, used in a cooked state as we use Spinach ; and C. de la Passion is a very large variety, passing through the winter well without protection. The Scarolle for winter or autumn salad is a really noble kind of Endive, with smooth leaves, a vigorous constitution, fine flavour, and every good quality that such a plant should possess ; and yet it is not at all sufficiently known or grown with us. The best kind is the Ronde or Verte, but the Blonde is also good. Of the wild Chicory there is an improved variety, Chicorée améliorée, which forms little heads four inches or so in diameter in early spring, and is then very acceptable in a salad-loving country. By putting a cloche over stools of this variety, these little heads, may be had all the winter. To blanch them slightly is an improvement, but this variety must on no account be employed to yield the Barbe, that popular Parisian winter salad. That is simply the common Chicory. It is grown in vast quantities near Paris, and prepared for use chiefly in caves at Montreuil—Montreuil of the Peaches.

Of Lettuces, as of Endives, the best known are found in and sent from the Paris market. The earliest is the Crepe, or Petite Noire and the Gotte or Gau ; a fine variety for summer use is the Blonde d'Eté; the Palatine or Laitue Rousse is also a most tender and delicate variety, keeps long, and is worthy of general cultivation; and the Laitue de la Passion is an excellent winter kind, that may be grown without a cloche through the winter. In summer and autumn the Grosse grise or Brune paresseuse is also an excellent variety, forming a good heart. Of the Cos lettuces, the Verte Maraîchère is the one so largely grown and exported for spring use, and the Blonde Maraîchère is the summer kind preferred and grown by the market gardeners. Radishes, we need not say, are found better in Paris than anywhere else. The Radish of the Paris markets has lately been sent out by English houses under the name of French breakfast Radish. The French name of the best

variety of this is the Rose Demi-long à bout blanc.   The
earliest Potato is the Marjolin, the source of an important
culture on the slopes of the hills above the Bois de Boulogne.   It is kept all the winter in the light, and yet free
from frost, so that when planted in spring marketable tubers
are quickly produced, and the ground when cleared of early
Potatoes is fit for Haricots in May.   Good King Henry (Che-
nopodium Bonus Henricus), a really good and delicate herb,
is used to some extent in private gardens, but does not form
a product of the markets.   Arroche or Orach, both of the
red and white varieties, is much grown in private gardens.

Of Chervil there are quantities grown which to us seem
incredible.   It is much used in salad.   One seed house
alone sells about 1000 lb. weight of seed of it per annum.
Bulbous Chervil is an excellent vegetable not found in the
markets, but which ought to be grown in all private gardens.
Such leguminous plants as have curious snail-like seed-
vessels are occasionally grown under the name of Chenil-
lettes, to decorate salads and form imitation snails.   They
are of no importance, and with us are rarely seen out of
botanic gardens, and not often in them.

The Ciboule, or Welsh Onion, is grown in quantity for
salad.   Tarragon is grown in great quantity for use in
vinegar, and also in a lesser degree for salads.   The Girau-
mon Turban Gourd (or Potiron Bonnet Turc) is much used
where the Potiron would be too large.   Salsify and Scor-
zonera are both grown in much larger quantities than with
us.   Of Corn Salad, so very important an article in the
winter salad consumption of Paris, the Mâche Ronde or
Doucette is the best variety for autumn and winter, and
the Mâche Régence d'Italie for spring use.   With Cucumbers
we are far ahead ; with Melons we go in quite opposite
grooves.   The English Melons would not find buyers in the
Paris market, nor probably would the French in ours.
They are eaten in quite different ways in the two countries
—in France with pepper and salt ; and some people, for
whom the rich flavour of the English Melons is too much,
can enjoy those of the Paris restaurants.   The large kind
grown by all the Paris market gardeners is M. Cantaloup,

Prescott Fond Blanc. Of the long Turnips, or Navets, the long Hâtive des Vertus and de Croissy are the best varieties. Small Onions are largely used with Peas, the kind preferred being the Blanc Hâtive, sown in August—this is a good kind. Of their keeping Onions, Jaune des Vertus is considered the best. Sorrel is of importance in the Paris markets, being largely used somewhat as we use Spinach. The variety preferred is the Large de Belleville. Of the Dandelion there is an improved variety, good for winter use, like the improved Chicory ; the common kind is very largely used.

The Potiron Gros Jaune is the enormous gourd of which the finest specimen is annually crowned in the market, and is the source of some amusement. It is sometimes grown about 200 lb. in weight, and last year a specimen was in the market which reached 250 lb. It is largely used by the poorer classes for making soup in winter.

In Peas and in Cabbages we are in advance of the French. It may, however, be worth noting that a superior and very hardy variety of the Choux de Milan—the Gros des Vertus —is grown to a vast extent in the neighbourhood of Pontoise, and sent to the Paris market in March and April.

Brussels Sprouts are grown to a vast extent near Paris, especially about Rosny and Noisy. The variety is the ordinary one. They are used in much greater quantities than with us.

The Cauliflower is cultivated to great perfection around Paris, the varieties used being the Petit Hâtif or Petit Salomon for earliest use ; the Demi-dur or Gros Salomon for summer ; and the Lenormand for autumn. Brocoli is not grown by the Paris market gardeners, the market being supplied with Cauliflowers from Brittany in spring.

It need scarcely be said that Haricots are grown and used in France to a degree of which we can have but a poor conception. They are used every day in winter, in the smallest as well as the grandest restaurants in Paris ; the earliest is the Nain Hâtive de Hollande. The one which supplies the quantities of ordinary Haricot is the Flageolet Ordinaire, or de Laon. The Bagnolet, or Suisse Gris, is excellent for using green, and for making conserves, and is largely grown for these purposes; and the Beurré, or Haricot d'Alger Noir, is the excellent mange-tout, which is not at all known with us. The pod is quite tender, of a yellowish white, and it is allowed to become larger than those of fully grown Scarlet Runners, and then cooked entire. This vegetable is both distinct and good, and deserves universal cultivation in the British Isles. A new variety, called H. Cosse violette, with violet pods, is also very tender and of good flavour.

CULTURE OF THE SMALL CARROT OF THE PARIS MARKET.— Every visitor to the Halles of Paris or the streets near them during the earlier hours of the day, must have noticed vast quantities of pretty, dwarf, tender little Carrots. They are always fresh, always to be had, and never contain a particle of the tissue which makes the coarser Carrots so much less valuable. Even when we do grow the best varieties of dwarf Carrots in this country, they never present the cleanly appearance of those of the Paris market gardens, nor are they so tender and good; the following article, therefore, on cultivating them both out of doors and in frames, by M. Courtois-Gérard, of Paris, may prove useful to admirers of this vegetable in its most perfect condition. Practically, and in a few words, the success of the Paris gardeners with the small Carrot results from sowing it, both in frames and in the open air, on very rich friable ground— the surface for a couple of inches being purely decomposed stable manure, and from giving it abundance of water whenever it requires it—thus securing quick unchecked growth and tenderness of texture. However, we will let this experienced cultivator speak for himself:—

" The common Carrot has produced several varieties, but the early or Dutch red, introduced into France about 1800, the Demi-long, and the Rouge Courte à Chassis, are the chief kinds grown in the market gardens of Paris.

" CULTURE IN FRAMES.—At the beginning of December, a hotbed is prepared of fifteen or sixteen inches in thickness, the heat being allowed to rise to from 65° to 80° Fahr. The frames are next placed in position, and filled with manure rotted into the state of mould, mixed with earth to the depth of six inches. By using this soil we obtain Carrots of a brighter red and better quality than when grown in garden mould only. When the heat has reached a genial point the seed is sown, and seven rows of the Petite Noire Lettuce are generally pricked into each frame. But although by this method we get two crops from the same frame, we do not think that there is much advantage to be gained from it, for it is not certain whether the Lettuces produced compensate for the harm that they do to the Carrots. These Lettuces are fit to cut in January. After they have been gathered, a little mould is spread over the place they occupied, and if the weather is dry the Carrots are given a slight watering. In the course of January, when the crop has grown up, the linings of the frames are turned over and raised as high as the top of the frame, so as to increase the heat of the bed. At the beginning of January, a second crop of Carrots is generally sown, but in this case a less amount of heat is required, and a sowing of Radishes is substituted for the Lettuces. When proper pains have been taken, the first Carrots may be gathered in the first fortnight of April. If the weather is fine during the latter half of the month of March, and the frames that cover the Carrots are required for other subjects, they may be taken off, in which case the Carrots may be gathered later. In February and March Carrots are again sown on heat, but in the open air. After this period straw mats are sufficient to preserve the sowings from the frost. These Carrots succeed those which were sown in December and January, and prepare for those sown in the open air. After the Carrots sown in February and March are gathered, Radishes are sown, and after they are gathered, turnip-rooted Celery.

" SOWING IN THE OPEN AIR.—The first sowings in the open air are made in September. In the eastern districts they sow large quantities at this period. From the commencement of the earliest frosts care is taken to cover the sowings with litter, which is taken up whenever the weather is fine enough. When this sowing is successful, the Carrots may

be gathered towards the month of May. Other sowings
are made in February and March, from which time they
may be continued regularly until July. But at whatever
time the sowing takes place, the ground ought to be well
prepared, and the seed sown broadcast, in the proportion of
about nine pounds to every acre. After the seed is sown
the ground is slightly covered, and then trodden down with
the feet, after which a layer of fine and thoroughly rotted
manure is spread over the whole; the ground is then raked
lightly, and watered whenever it is necessary. As soon as
the young plants make their appearance, the crop, which is
generally too thick, is carefully thinned out. Three months
after the time of sowing, the more forward Carrots may be
gathered, the results of the latter sowings being left until
November. When the Carrots are gathered, the neck of
each is cut, and the roots are prepared, after which they are
covered with long litter, or else placed in a house for storing,
so as to have a ready supply during the winter. In the case
of light and fertile soils they need not be pulled up, as it will
be only necessary to cover up the Carrot beds, so as to be
able to gather them when wanted. The market gardeners
of Meaux preserve their early Carrots by digging trenches in
the autumn three feet wide, two feet six inches in depth, in
which they place their Carrots, and cover them with straw
during the frosty weather. In this way they are able to
keep them until the end of February or beginning of March,
which is the time at which they begin to sell."

THE CARDOON.—The Cardoon, being a plant of very vigorous habit, must be grown in the best and richest soil of the
garden, and well watered frequently. If it is sown in April
and not watered abundantly many of the plants will go to
seed during the summer, for which reason it is better to
defer the sowing of it until May, when it may be performed
either in the open ground or in a seed bed. It is better
to adopt the former method, as the Cardoon having a very
smooth, fibreless, conical root is ill adapted for transplanting.
Those, however, who prefer the latter method may sow it
in a seed bed and plant it out when old enough. In a
well-dug bed about seven feet wide, two furrows are traced

at a distance of about four feet from each other. Marks are made along these furrows three feet apart, and three or four seeds planted at each spot thus indicated. If the soil renders it necessary a spitful of earth may be removed and replaced by well rotted manure, and the seeds sown about an inch deep. The seeds should be sown in quincunx fashion. If the weather is dry and warm, the seeds should be well watered, and they will strike in a few days. As soon as the little plants are above the ground, the weakest should be carefully removed.

Those who prefer sowing in a seed bed should wait till the plants are four or five inches in height and then transplant them into the open ground with great care, the little root being already pretty long. The earth round them should be well pressed down and watered, and the plants shaded until they have again rooted. As it is not until the month of August that the Cardoon begins to be vigorous, crops of salads may be sown and gathered in the meantime. It cannot be repeated too often that the finest Cardoons can only be obtained by frequent and copious watering, the dose being increased as they grow larger. If the weather is warm and dry, at least a wateringpotful of water should be given to each plant every other day. In the month of September the blanching process is commenced, and this is done in quite a different way to that practised in this country. The plants are simply tied up rather closely, and then a lot of long litter placed round each in a close tidy way, the straw or long litter being tied by small bands of the same material. The longest leaves of the head are left free above this blanching material. But the Cardoon is so fiercely armed that it requires a little care to get at the great plants to tie them up, &c., without being severely pricked. To obviate this three sticks are used—one of them short, and connected with the other two by strong

Fig. 298.

Mode of Tying-up the Cardoon for Blanching.

twine. The engraving will show this simple contrivance
and the mode of using it at a glance.   The workman
standing at a safe distance pushes the two handles under
the plant, and then going to the other side and seizing them,
soon gathers up the fiercely armed leaves.   Another workman then ties it up in three places, and then the straw is
placed around so as to quite exclude the light, and also tied
up like the Cardoon itself. In three weeks the vegetable is as
well blanched and as tender as could be desired.   To blanch
the Cardoon properly and render the leaves perfectly tender,
it should be deprived of light and air for at least three
weeks.   It is then cut just below the surface of the earth,
and divested of its straw covering; the withered leaves are
sliced off and the root trimmed up neatly.   If it is desirable
to preserve the Cardoon for winter use it should be simply
tied up, as before directed, in the month of November, and
uprooted carefully with a ball of earth attached to it, and
plunged in fine rotten manure or leaf-mould in a dark
cellar. The decayed leaves should be removed every week
or so.   Under this treatment they become sufficiently
blanched in a fortnight, and may be preserved in a good
condition for at least two months.

FORCING THE CAULIFLOWER.—The best Cauliflower forced
around Paris is the Petit Salomon.   It is sown in the
open air during the first ten days of September on very rich,
light, and fine earth.   When the young plants are well up
—that is to say, commence to show the first two leaves—
they are pricked out into shallow frames, surfaced with a
couple of inches of thoroughly rotten manure.   They are
very particular about transplanting them when very young,
and before they are drawn, watering before moving the
young plants, so that they may be removed with the least
possible mutilation of the roots, and they are pricked in
with the finger at about three inches one from the other.
At the end of November the plants are strong and hardy,
but they must not be allowed to grow too quick, and
therefore they are again transplanted, leaving a little more
space between them.   This second transplanting is to prevent the too rapid growth of the plant, and to enable it

better to resist the cold. So long as it does not freeze, it is better to leave the plants exposed to the air. When it does freeze they are protected as much as need be, opening the frames, so that the plants may enjoy the sun, and taking care to protect them carefully with straw mats at night, sometimes surrounding the sides with litter to prevent the entrance of cold in that direction. In February these Cauliflowers are planted on gentle hotbeds from which Lettuces have been cut. Between the Cauliflower plants are placed the Lettuce known as the Gotte and the Petit, and Gros Salomon Cauliflowers are planted alternately, so as to insure a succession. Other kinds of vegetables are placed between by some, but the Laitue Gotte is considered the best and most profitable for this purpose. Gradually as the season advances more air is given to the plants, and when they get too near the glass the frames are elevated by placing stiff wads of straw under their corners. About the beginning of April, if the weather be fine, the frames are removed, that they may be used in the culture of Melons. In case of late frosts, an arrangement is made to give some protection by means of straw mats. About the 10th or 12th of April the hearts are seen forming in the Petit Salomon, and eight days afterwards in the Gros Salomon. Thenceforward the Cauliflowers are visited every two days, and when the heart of one is seen formed as big as a hen's egg, some leaves of the lower part of the plant are broken and laid upon it, so that it may be deprived of light and thus kept perfectly white. When these leaves wither or shrivel they are taken off, a fresh one put over the heart, and then the old ones laid on top. They are thus regularly watched, blanched, and cut when at perfection.

THE SWEET POTATO.—Louis XV., it is said, was exceedingly fond of this vegetable, and had it grown for his table in the gardens of the Trianon and Choisy-le-Roi. From his day until about the year 1800 the Sweet Potato was relegated to hothouses and botanic gardens, but about the last named period M. le Comte Lelieur, who was appointed manager of the royal gardens, had some grown at·St. Cloud.

The Sweet Potato soon became fashionable once more, and the many market gardeners of the day grew the vegetable largely. Later its cultivation was again abandoned for the sake of more profitable plants, and at the present time MM. Découfflé and Gontier are the only persons who pay any attention to it.

Instead of stopping to inquire into the modifications that the cultivation of the Sweet Potato has undergone, we will confine ourselves to saying that at present three varieties are cultivated—the red, the yellow, and the New Orleans violet. They are all grown in hotbeds, and they are propagated in the following manner :—At the beginning of January a few tubercles are selected from those which appear to be the best preserved, and planted in a hotbed, the frame of which must be covered with mats during the night. In the course of a short time they begin to grow, and the young buds must be taken off when they reach the height of from two and a half inches to three and a half inches; they are then pricked into pots of about two and a half inches in diameter, which are plunged in heat and covered with a bell-glass, after which they may be watered as they require. As soon as the young plants strike, an event that soon takes place, the bell-glass must be lifted gradually until they are strong enough to dispense with it altogether without drooping.

Such readers as care about this root—which, by the way, is of agreeable flavour when well cooked—may grow it most readily and effectively by placing it in a frame or pit after the spring crop has been taken out; or, indeed, on a ridge like the ridge Cucumber; but the pit or frame is the safest way generally—the lights being taken off. As pits and frames are frequently empty from about the 1st of June till autumn, room might be readily spared for it without loss, and a useful vegetable added to our stock, which, fine as it is, is yet in want of variety. The roots may be bought in Covent-garden. The red variety is the best. The way to treat them is to pot them about the end of April; start them in a gentle heat, and have them fresh and stubby for planting out in the pit or frame about the 1st of June.

They would be the better for the lights for a few days. In this way they will be found to do better than when grown in a stove, and probably prove a more grateful vegetable than the Chinese yam in its best state.

EARLY POTATOES.—The supply of early Potatoes for the market is an important branch of industry about Paris, a considerable portion of the slopes of the hill sides to the north of St. Cloud being devoted to it. I only speak of the subject here to point out that the cultivators commonly allow the Potatoes to sprout vigorously indoors before planting them out, and thus secure crops so early as to have them out of the ground in time to put in summer crops. Some of the houses of the cultivators are stored with Potatoes freely exposed to the light in winter. It matters little where they are placed, provided they enjoy plenty of light and are kept perfectly free from frost. The usual plan is to have a room fitted up with rough shelves, and placing the Potatoes in the old oyster and fruit baskets of the markets, store them on the shelves till ready to plant out.

FIG. 299.

Early Potatoes arranged for " Sprouting " indoors.

Or the shelves may be dispensed with, and rough wooden trays, with feet like those shown in the accompanying figure, used instead. These may be piled one above another, and may be quickly made out of old boards by the commonest workmen.

Only one layer of tubers should be placed in each basket or box. The variety used is a Kidney Potato, the Marjolin : the roots are in preparation as described so early as November, and are planted out in February, the crop being gathered in May and June. There is no need to warm the place they are in, except indeed to keep the frost from getting in, as the tubercles get on very well without it. In case they show signs of weakness, the windows of the house should be opened whenever the state of the weather allows

it to be done without danger.   Planted with all the pains
necessary for the preservation of the young shoots, Potatoes
treated in this way come to perfection much sooner than
those which are planted without any previous preparation ;
indeed all the gardeners who supply the Paris markets with
early Potatoes prepare their seed in the way we have described above.

OLEANDER CULTURE.—Visitors to the Continent in the
summer months can hardly fail to be struck with the employment of certain plants for decorative purposes, of which
we in this country make comparatively little use.   Here, if
a few Orange trees or Portugal Laurels, perchance a
Pomegranate, are grown in tubs and put on the terrace in
summer time, it seems to be considered that enough has
been done in that way.   There is no reason, however, why
many other plants should not be used in the same manner.
Some may remember the beautiful effect produced on a quay
fronting the lake of Lucerne by a number of standards of
this kind, including not only the plants mentioned, but
Pittosporums, Yellow Jasmines, Evergreen Oaks, Euonymus,
Aucubas, and Figs.   At Vienna a similar assortment may
be seen in front of some of the principal cafés, where one
may sit in the open street under the shadow of the Pomegranate and the Oleander.

This latter plant, too, is an immense favourite with the
Parisians.   In fact, the Oleander forms, with the Myrtle and
the Pomegranate, one of the most important articles of
Parisian commercial horticulture.   The reasons for this are
obvious—the elegant habit, glossy foliage, profusion of
bright rosy or white flowers, endowed, moreover, with an
agreeable almond-like perfume, offer recommendations
hardly to be exceeded by those of other plants.   The
culture, moreover, is easy.   Indifferent as to the treatment
it receives in winter, it may be kept in cellars or garrets—
almost anywhere, in fact; hence its frequency abroad in the
windows of the artisan and at the doors of the merchant's
office.   The shrub may be propagated either by layers or by
cuttings; but of late years, in France, the former method
has been abandoned, as it is found that cuttings produce

plants of better habit, and in greater numbers. In this country the Oleander is rarely seen in perfection, and most probably because it is generally grown indoors. The treatment given it on the Continent insures the plant a perfect rest in winter: as it cannot grow in the cellars, caves, and dark orangeries in which it is placed. Therefore, when put in the open air, the accumulated growing power of the plant pushes forth equably and immediately: the shoots, being produced in the open air, are perfectly indifferent to any changes they may have to undergo therein, and the plants enjoy the full sun and uninterrupted light.

It may be noticed in two different conditions about Paris—in the large specimen form in tubs of various sizes, and as small neat plants in six-inch pots. These last are sold in great numbers in the markets, and flower as abundantly as the best managed of the large specimens. The finest examples of large specimens I have ever seen are those in the garden of the Luxembourg Palace, and I have much pleasure in giving the following account of their cultivation by Monsieur Rivière fils, son of the talented and excellent superintendent of the Luxembourg Gardens. Judging by the habit of the Oleander, as generally seen with us, it might be supposed that it would not make an ornamental tree for a terrace, but nothing can be finer than the immense specimens seen in the Luxembourg Gardens, the heads being as round and dense as a Pelargonium grown by Mr. Turner, and sometimes as much as ten feet through; and as for the little plants grown in six-inch pots, nothing can be prettier. They are certainly far handsomer objects than Orange trees, grow equally well or better in tubs, and are more worthy of culture in this way.

" This beautiful shrub is a native of Algeria and the south of Europe. In a state of nature, it prefers damp and fresh soil; it is consequently found in abundance on the banks of rivers and the edges of marshes. In the wild state it rarely reaches the height of more than from three to five feet, but under cultivation it may grow even to nine or ten feet. Its flowers are of a delicate rose colour, and from seed horticulturists have succeeded in obtaining yellow,

white, and double-flowering varieties, which form some of
the most beautiful ornaments of our gardens.   This plant
contains abundance of sap, which is very poisonous, and
consequently very dangerous; it is therefore advisable never
to put any of the flowers in the mouth, and to take care
that no children should be allowed near the plants.   The
hotter the district in which the plant is grown, the more
poisonous is the sap.

"The Oleander puts forth its flower-bearing branches a
year in advance, and then blossoms for two consecutive
years, so it is as well not to cut them down in the autumn
after the first time of flowering.   The beautiful specimens
so much admired in the Gardens of the Luxembourg during
the fine weather are from sixty to one hundred years old.
They are grown in tubs three or four feet square, and in a
compost made in the following proportions : half soil and cow-
dung, a quarter rotten stable manure, a quarter turfy heath
mould; the whole being well mixed at the time the tubs
are filled.   The operation of re-potting should be performed
every five years, about the month of May.   The sides of
the tubs being moveable, the earth is taken away from the
roots of the tree, which is itself lifted up about three inches,
so as to remove the soil all round it.   This being done,
broken flower-pots, or similar substances, are thrown into
the bottoms of the tubs for the purpose of drainage, as is
usually done with large shrubs planted in this manner.
The shrub is then lowered into its former place, and covered
up with the mixture just described.

"The Oleander is generally placed out of doors about the
10th of May, and as this plant grows naturally under a
burning sky, it is advisable to give it as much sun as possible.   A few days after it is put out, the surface of the
soil in the tubs should be covered with cow-dung, and
during the whole of the summer season they should be
copiously watered at least three times a week.   As soon as
October comes, the waterings are diminished, and all the
dung that is not entirely decomposed is taken away, the
surface of the soil being stirred up with a pointed stick to
make it more permeable.   The Oleander being extremely

sensitive to cold, the plants should be taken under cover
once more about the 15th of October, where they must
remain until the 10th of May, during which time they
ought not to be watered more than three or four times
every month.   In France the Oleander tree is attacked by
a parasite called the Chermes nerii, which does it a great
deal of injury.   While in the greenhouse no pains should
be spared to deliver it from its enemy by means of a stiff
dry brush.   The mischief caused by this insect will often
kill the tree; prompt means must therefore be taken to free
the trees from this pest as soon as it makes its appearance.
If, in spite of all your care, the Chermes still keeps up its
depredations, you must not hesitate to prune out all the old
wood that is attacked.   By this means the evil may be
entirely remedied, a new set of shoots appearing and bearing
flowers the following year."

The preceding details refer exclusively to the treatment
of the larger specimens.   The pretty little free-blooming
Oleanders are grown about Paris in pots, five or six inches
in diameter, in sandy soil, and these pots they very soon
fill with roots.   They are plunged all the summer in the
open ground, and grown at all other seasons near the glass
in those low houses so much in vogue in Parisian nurseries
and gardens.   They flower profusely, and receive the
same treatment as Orange trees, as regards housing in
winter.   They are allowed to rise with an undivided stem
for about four inches, and then break off into several
branches.   There should be no difficulty in growing them
wherever there is a sunny shelf in the greenhouse, by
securing a clean, while discouraging a soft or luxuriant
growth, giving them a rather dry rest in winter, and
abundant water and light in summer.   In winter any cool
house will do to store them, or even a shed.

CULTURE OF THE ORANGE.—In the following account of
the cultivation of the Orange by Mr. H. Jamin fils, the
son of the most successful cultivator of it in Paris, it will
be clearly seen why and how we fail, and why a person
with an old coach-house or any other rough structure with
a few sashes or windows on its north side may grow hand-

somer Orange trees than those with the fairest of conservatories. It should be understood that it refers to the culture of Orange trees for placing in the open air in summer, and not with a view of growing them for the sake of their fruit. Where fruit is required from Orange trees in this country an entirely different system must be pursued, and there are signs that before long all the finer Oranges will be abundantly grown under glass with us.

"The Orange is propagated by grafting on the stock raised from seeds of Citrus Medica (the Common Lemon), or from those raised from seeds of the Common Bitter Orange. For the trade, plants grafted on the Lemon stock are the most suitable, the Lemon growing more vigorously than the wild Orange tree; but to secure the plant long life, the latter is the most preferable. The reason of this will be easily understood; the difference between the Lemon and the Orange is much the same as between the Quince and the wild Pear: like the Quince, the Lemon makes all its roots at the surface of the soil, the wild Orange goes deeper, and consequently the tree is better able to resist the wind and the vicissitudes of the season; naturally there is more analogy between the two woods, and the result of experiments is that the plants live much longer. An Orange tree grafted on the Lemon may live about a hundred years; after that time it decays and perishes; an Orange grafted on its wild congener may live over 300 years—witness the Grand Bourbon in the Orangery at Versailles, near Paris, which tree is now more than 400 years old, and is grafted on the wild Orange.

Sow the seeds early in the spring in a light but not too sandy soil, and in pots (twenty-five to thirty per pot); put the pots upon a dung-bed (lukewarm), and keep the soil fresh, but do not have any steam in the frame, and to prevent this give a little air. When the seeds have come up, encourage them to grow to three or four inches high. Afterwards put them in a warmer bed, and keep a damp warm atmosphere in the frame; shade them against the burning rays of the sun; and when they are seven or eight inches high, give them a little air, increasing it as they get

stronger. Let them pass through the winter in a greenhouse, where the temperature must not descend lower than
40° Fahrenheit, and in early summer put them on another
hotbed in the open air plunged in leaf mould or cocoa fibre.
Leave them plunged on this hotbed through the summer,
and give them plenty of water, and from time to time a
little liquid manure.

About the end of August in the same year graft them by
the same method as that practised for Roses in the winter,
and put them on a hotbed, keeping as much damp vapour
about them as possible. Shade them during the sunshine,
cover at night, and keep them close as long as the grafts
are not well united together; they will be safe long before
the early frost. Keep them in the frame during the winter,
and the next spring divide and pot them in rich light soil
mixed with a very little silver sand to prevent the soil becoming hard : put the pots on a hotbed in a frame, and
after they are rooted give them plenty of air. In the
middle of June, make a hotbed in the garden and put them
on it without any covering whatever, giving plenty of
water during the hot weather, and three or four waterings
of liquid manure to encourage active growth. Before the
first frost they must be housed, and they will do through
the winter in a greenhouse where the temperature is kept
three or four degrees over the freezing point.

During the spring of the following year pot the plants
afresh, and place them on a hotbed covered with a frame ;
keep it close until the roots begin to shoot, and give
air carefully ; shading the frame against the burning rays
of the sun, and when frosts are no longer to be feared,
taking the lights off entirely. When they have done their
growth, and the wood is sufficiently ripened, pot them afresh,
and leave them in a greenhouse for a week or two. In
June make a hotbed in the open air, covered five or six
inches with dung-mould or cocoa refuse, and put them in
it. This is the last season during which the Orange need
be grown upon a hot dung-bed. The greatest obstacle to
the success of the Orange as a terrace-plant is the persis-
tence of the gardeners and nurserymen in treating it as a

greenhouse subject. I do not mean to say the Orange should be treated like common shrubs, but it is possible, with very little care, to grow them in England almost as well as in northern France.

Many writers on this subject give the south exposure as the best for an orangery, and therein is the mistake. To insure the success of Oranges grown in boxes or in pots, they must not in any case be allowed to grow in the houses; all their growth must be made out of doors. It is a matter of fact, that if the orangery is to the south, no matter what the trouble you take to prevent their starting, the plants will begin to shoot a long time before the weather is mild enough to permit of their being placed in the garden. A good orangery should have a northern exposure, with plenty of windows to admit the light, and every convenience to give full air when it is not frosty. It will be very easy to heat the orangery in such a position, as the temperature required is only two or three degrees over the freezing point. It must be remembered that Oranges are grown out of doors all the year round in parts of France and Spain where it freezes every winter. If the plants, after all the care taken to prevent their growth in the houses, begin to vegetate, and if the young shoots are more than an inch in length, it would be far preferable to cut them back than to let them retain a growth which is sure to be disfigured and spoiled in the open air.

The watering must be very carefully done, as too much water would be more pernicious than too little, and especially for the large plants, where the soil is in greater quantity; one or two injudicious waterings are enough to kill the best established plants. Good drainage in the bottom of the box or pot will prevent many accidents. In the winter they want very little water. Before watering them the grower should feel the leaves of the tree, and if flabby, as though on the point of flagging, it is time to give them water. This applies only to the large plants, the large quantity of soil employed for them keeping its moisture for a long time. The small plants must be watered more frequently, but still with great moderation in winter. During

the summer water must be given freely, but not in excess.
The best time to give it is in the morning; and at night
the plants will require a little syringing on the leaves, but
only in the hottest time of the year.   Liquid manure given
with great moderation will do them good and quicken their
vegetation.   The small plants which have passed beyond the
hotbed stage should be potted in a very rich light soil, and
not too sandy, say nine parts of soil divided as follows :—
Three of maiden loam, two and a half of yellow loam, one
and a half of old dung mould, one of peat, and one of sand.
In potting plants of a larger size, the soil should be a little
stronger, and be composed as follows :—Three and a half of
maiden loam, three of yellow loam, one of thoroughly rotten
dung, a quarter of peat, and one part of sand."

SHOWING ROSES IN FRANCE.—A Rose-growing friend has
suggested to me that it might be well to mention any
novelties in arrangement adopted by the French in showing
Roses, but I know of little worthy of recommendation.
The great exhibition of Roses at Brie Comte Robert—surprising accounts of which appeared in the daily papers at
the time—was, in some respects, a very different affair to
what might have been expected from the reports of it
spread abroad.   Brie Comte Robert is situated in a very
pleasant country, twenty miles or so from Paris—a country
without hedges or ditches, yet picturesque and pretty from
the number of fruit trees dotted over the land, and with (at
the time of my visit) the ears of ripening wheat bending
into the straight well-made roads—a country with rich sandy
loam and gentle hills, like parts of Kent, and for the main
part covered with wide level spreads of wheat and vines.
Brie Comte Robert is an ordinary and rather straggling
little French town, with an interesting old church traced
with the beautiful art of the olden time, and grey with the
lichens of a thousand years ; and finally, Brie Comte Robert
has a fête and a great Rose-show, as all the world has been
informed.

The Rose-show, although pretty and remarkable of its
kind, is not quite a marvel, but simply an adjunct to the
village fair.   Now, the fête of a small place like this is not

at first sight, or when examined in detail, a thing to be enraptured with. Imagine a grassy yard or small field, in the centre of which are a few tables, and the little hut of a person who divines the future; and all round the margin, a lot of small, meagre, dirty, canvas tents occupied with various things, from temporary restaurants and gingerbread stalls down to diminutive billiards and little games in which the yokels of the district invest a sou a time, and now and then win a trifling work of art worth about a centime. Imagine, in short, the mildest and smallest corner of Donnybrook fair, with every drop of " divilment" squeezed out of it, and you have a pretty good idea of the sight that greeted my eyes as I entered the show-yard of Brie Comte Robert. But at one end there was a very large oblong tent, and on entering that a very different sight presented itself. Here all was fragrance and beautiful colour. All the Roses were placed on the ground—no stages of any kind being used. First of all, there ran right round the great oblong tent a sloping bed of sandy earth, about five feet wide, covered with young Barley, the seed of which had been sown eight or ten days before. On this were thickly placed the Roses —eight rows deep, or thereabouts. They were for the greater part shown in small earthenware bottles, about five inches high, with long narrow necks and wide globose bases; and, placed amongst the Barley-grass, they looked very well indeed. Generally three or more Roses were placed in each bottle, which was made of ordinary garden-pot stuff, and of the same colour; and they looked so much better than those of glass used by some exhibitors, that their use should be made compulsory. Thus the most conspicuous thing in the tent was a dense bed of Roses around its sides. In the central parts of the tent there were beds of various shapes in which the Roses were plunged in moss, and mostly arranged in masses; for example, a bed of 700 blooms of Général Jacqueminot, edged with a line of Aimée Vibert; a bed of Madame Boll, edged with white and red Roses, all the flowers plunged singly in dark green moss, and so on. The competitors vied rather in quantity than in quality, and one exhibitor showed as many as 600 varieties or supposed varieties—certainly he had that number of bottles. Others showed large numbers also, but in most cases the Roses were inferior to those seen at an English show. As for the varieties, they were chiefly such as abound in England. There were quantities of that fine Rose, Maréchal Niel, to be seen, one bed of it being ten feet in diameter, the blooms plunged singly in moss. The largest exhibitor grouped his flowers very prettily by arranging wavy lines of yellow and white varieties through the long mass of rose and dark-coloured ones.

FORCING THE WHITE LILAC.—The production of the white Lilac seen so abundantly in Paris during the winter and spring is often a source of curiosity. To meet with a mass of it in October, quite white and deliciously sweet, is a pleasant surprise to the English visitor. You may see large bunches of it in every little flower-shop in the month of January, and it is always associated with the early Violet and the forced Rose. This Lilac is the common kind, and yet it is perfectly white. French florists have tried the white variety, but they do not like it—it pushes weakly and then does not look of so pure a colour as the ordinary Lilac one. They force the common form in great quantities in pots, and to a greater extent planted out, as close as they can stand, in pits for cutting.

The plants that are intended for forcing are cut around with a spade in September, to induce them to form flower-buds freely, and they are at first judiciously introduced to a cool house, but after a little while given plenty of heat, in fact, from 25° to nearly 40° C. = 77° to 104° F. At the same time abundant humidity is supplied, both at the root and by means of the syringe, but the chief point is, that from the day the plants are placed under glass they are not allowed to receive a gleam of light, the glass being completely covered with the paillassons, or neat straw mats, such as are much used for covering frames, pits, and all sorts of garden structures in winter. Thus they get the Lilac to push freely, and gather its white blooms before the leaves have had time to show themselves. The great degree of heat—a degree which we never think of giving to anything of the kind in England, and the total shade to which they are subjected, effect the bleaching. The French commence to cut the white Lilac at the end of October, and continue the operation till it comes in flower in the open ground. In the same establishments enormous quantities of Roses are forced, small, pretty, and unopened rose-buds being in great demand in Paris.

CHAPTER XXV.

" Went out at early morning, when the air
Is delicate with some last starry touch,
To wander through the Market-place of Flowers
(The prettiest haunt in Paris), and make sure
At worst that there were roses in the world."
E. B. BROWNING.

FLOWER, FRUIT, AND VEGETABLE MARKETS—LIST OF PLACES
IN WHICH THE MORE INSTRUCTIVE FEATURES OF PRACTICAL
HORTICULTURE MAY BE SEEN—THE CLIMATES OF PARIS AND
LONDON COMPARED.

SOMETHING about the markets is surely not out of place in
a book on the gardens of Paris, for all places where the produce of gardens is to be seen in its fullest perfection ought
assuredly to be as interesting as any garden, and so they
are when orderly and spacious enough to be seen by others
than the porters and small tradesmen who force their way
in daily. No garden in existence possesses half the interest
of the flower, fruit, and vegetable departments of the fine
Halles Centrales, and it is an interest that is perpetual, for
every day brings its fresh materials, every week its changes
of supply. About twelve o'clock at night, before Paris has
gone to bed, the growers have already arrived on the spot
and begin to expose their freshly gathered produce in the
market or on the wide footways of the streets around, and
for eighteen hours after that time the whole scene is one of
animation and bustle.

There is no market where wholesale business is better
arranged or more expeditiously done than here; but
what interests us most are the provisions made for the
retail trade — for the purchases of the general public.
In Paris far more than in London it is the custom to
go or send to the market daily in every class of house,
rich or poor. Thus they are not dependent on the

greengrocer, whose stock is often yellow on his hands, but go where numbers of competitors are placed side by side, and where from the nature of the arrangement the majority of vegetables exposed must be fresh.   To secure them in that state is our chief want.   As regards the quality of the products when delivered by the grower, there is rarely anything to complain of, for the market gardener is usually an excellent cultivator; but the bruising and filth and delay they encounter before reaching the customer in London often render them barely edible, while the very poor, in buying the cheapest, often get that which is less fitted for human food than the garden refuse thrown to the pigs in many country places.   Everybody knows the utility and even the necessity of abundance of fresh vegetables to keep man in perfect health.   I believe that the proportion of vegetables eaten by the humbler classes of Paris and London is as seven to one, while all the advantages of as perfect freshness and wholesomeness as can be secured in a great city are with the former.   The arrangement of the markets has much to do with this difference.

Our people are great consumers of the universal Potato, which suffers little from carriage or keeping; but it is almost impossible for them to use any other vegetable as a regular article of food, while the French workmen have a daily variety.   There is no country in the world where vegetables can be grown more abundantly and cheaply than in the country round London, which can pour its produce into the great centre in an hour or two by rail, and yet for the want of a sufficiency of markets, space, and order, the public is to a great extent deprived of a benefit second to none other.   As for our chief fruit and vegetable market, our famous Covent Garden, it is a disgrace to civilization.   So long as the largest and richest city in the world depends upon Covent Garden as at present arranged, for its fruits and vegetables, so long must it find them very deficient.   Why, the want of room alone is sufficient to frequently make important differences in the prices, not to speak of the treatment the produce gets at all times, and especially in wet weather, the piles of baskets that convey it to market being

PLATE XLVI.

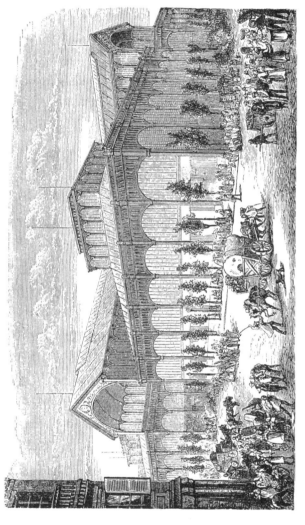

THE HALLES CENTRALES.

invariably heaped up over loads of dung.   What a contrast
between the central market in Paris and this famous spot !

Can we not secure a good wide market accessible to river,
rail, and streets somewhere on the new Embankment, and
leave Covent Garden to some one branch of the trade?
Can we do nothing to remedy a state of things which is not
only discreditable to our system of managing such matters,
but must have a positively bad effect on the supplies of
almost every family that invests in a Cauliflower?   The
new cattle market at Islington and the new meat market
in the City are things to be proud of—they, like the
Thames Embankment, are really worthy of London and
the Victorian age ; but as yet we do not seem to have
moved a step towards the establishment of a fruit, vegetable, and flower market.   Were this done with as broad
and excellent an aim as has been shown in the two markets
just named, we should have a feature added to London
which from its nature would assuredly be of the greatest
utility and benefit to the public at large.   We should also
have a grand exhibition of all that is fresh and lovely,
indicative of the fecundity and beauty of nature and the
industry of man throughout the year, and presenting new
objects of interest every day.

In the Paris market, in addition to every provision for
wholesale trade, there are streets of stalls containing everything the purchaser requires, classified so that the neat and
well-to-do market women who vend the same sorts of produce
are brought into close proximity and competition with one
another.   The advantages gained by the public are obvious—
the thrifty housewife has not only the opportunity of purchasing everything good and at a moderate price ; she also has
an immense variety to choose from, and can compare prices.
But it is needless to enumerate all the blessings that a good
retail and wholesale market confers upon its neighbourhood.
One of them, however, we do not often think of, and might
omit were it not that Mr. Sala admiringly alluded to it
recently in London.   Above all advantages, said he, is that
so plainly written in large letters—NO CREDIT !   In those
little streets of neatly arranged stalls in all the Paris markets

the name and number of the occupant is plainly printed; there is usually a free passage between each two rows, along which the purchaser can leisurely walk and survey the produce, and in fact there is every convenience for both purchaser and seller. The adoption of the same system of stalls in our grand new fruit and vegetable market, which we may, I trust, look forward to, would be a great improvement; but London is now so vast in extent that nothing less than a good series of well-managed markets will ever supply its population with a sufficiency of fresh vegetable food, which is the most wholesome and necessary of all.

The history of the Halles Centrales illustrates to some extent the essentially practical turn changes and improvements have taken in Paris of recent years. At one time the site was occupied by a vast graveyard, where the greater portion of the dead of Paris were gathered for centuries. At one time it lay outside the walls, but Paris gradually surrounded it with its narrow old streets, and eventually the place became a horrible nuisance. Then the government caused the vast accumulation of human remains to be removed by night in covered carts, escorted by chanting, torch-bearing priests, to the subterranean quarries that lie under Paris, and which, now filled with the piled bones of millions of men, are known as the Catacombs.

Some of the pavilions are not yet complete, but they will be on the same plan as those already in existence. The most noticeable and admirable features of this great covered market are the neat stalls for retail dealers before alluded to, lightness of design and good ventilation, and the roomy, airy character of the whole. It is constructed so as to be a protection against extremes of weather at all seasons; it is cool and shady in summer, the system of cellars underneath roomy and good, and with many useful arrangements for storing away the provisions, both live and dead. The roof is of zinc, the flooring partly asphalte, partly flags, and, like every new building, or avenue, or wide street in Paris, trees adorn the margin of the wide footways around it, shading the scene of almost ceaseless animation beneath.

PLATE XLVII.

INTERIOR OF THE HALLES CENTRALES.

There are many other markets in Paris, but all of them are smaller than the Halles, which offer most interest to the English visitor.   A good deal of the choicer produce is, however, taken to the Marché St. Honoré, after having been sold wholesale in the central market.   When finished, the Halles will cover about five acres.

There are thousands of Parisians whose garden is the window-sill, or a basket mossed over in the sitting-room, or

Fig. 300.

The Flower Market at the Madeleine.

a glazed case, and to most of them the flower-market is a nursery; and an excellent nursery too, for they can get numerous pretty plants in them in the best of health for a trifling sum.   Considering that a few miles of sea have for ages separated many marked customs of both peoples, for good and bad, and that 14,000 miles of sea have not pre-vented English habits, that have never crossed the Channel, from spreading to the Antipodes, it is vain to hope for the adoption of such a feature as the flower-markets of Paris in

our great towns; yet few could be more agreeable or useful.
They are in themselves, as Mrs. Browning remarked, the
" sweetest spots in Paris," and certainly do good by enabling
the poorer classes to freely enjoy things that are generally
admitted to have an ameliorating influence.   In Paris the
larger flower-markets are not in permanent buildings, but
occupy spaces which may be compared to that in Trafalgar-
square—the plants being placed in groups on the gravel or
flags, and the flowers and choicer plants under temporary
tents.   The market once over, the space is cleared.   In the
great central market and in the minor markets there are
also rows of stalls for flowers; shops vending them are
numerous, and occasionally a solitary stand with abundance
of them is seen here and there in the streets.   The regular
flower-markets are held at the Place de la Madeleine, the
Château d'Eau, the Quai aux Fleurs, and in the Place St.
Sulpice—twice weekly in each place.

They usually show in abundance all popular flowers—from
spring flowers to Chrysanthemums; but Palms and fine-leaved
plants generally—Cactuses, Mesembryanthemums, &c., in variety, are also to be seen; as well as young vegetable plants, pot-
herbs, Shrubs, Roses, Oleanders, and Pomegranates.   Oranges
are also sold in quantity.   Flowers ready cut for bouquets and
room decoration are particularly well done and very abundant.
The distinctive feature of the whole of these markets of cut
flowers consists of flower-buds—these are sold in quantities,
and arranged in a way that is unknown to us.   Bunches of
Roses may be seen all of one kind and colour, and all young
unopened buds.   Of some of the very dwarf Roses they pick
buds little bigger than a half-developed Fuchsia flower.   All
are very pretty, from little pink and white ones to the large
golden cones of such Roses as Maréchal Niel.   Sometimes
the white Roses are surrounded by a band of Forget-me-not
flowers; oftener the Myosotis is sold in bunches alone, and
so is nearly every other pretty garden flower; sometimes
they are mingled with grasses and the spray of such small
profuse flowers as Gypsophila, and common, but none the
less pretty flowers, such as Forget-me-not, Lily of the
Valley, and Woodruff.

PLATE XLVIII.

THE PLACE ST. SULPICE. ONE OF THE OPEN SPACES USED AS BI-WEEKLY FLOWER MARKETS.

List of places in which the more instructive features of practical horticulture may be seen.

It is so much better to illustrate a subject by eye-proof than by any other, that I venture to give a list of gardens in which visitors to Paris may see for themselves some of the things spoken of in this book. We may breakfast in London in the morning and dine without inconvenience in Paris the same evening, so that those wishing to examine for themselves any of the subjects discussed will have little difficulty in doing so.

FRUIT CULTURE.—M. Nallet, Brunoy, Seine-et-Marne: This station is passed on the way to Fontainebleau and Lyon, and is also that at which you alight to go to the great Rose-show occasionally held at Brie Comte Robert, so that numbers of English travellers have an opportunity of visiting it. Nobody taking an interest in fruit-culture should pass the station without seeing it. It is within an hour or so of Paris, should a special visit be paid. The garden is within a few minutes of the station, and M. Nallet is a most obliging and amiable amateur.—The Imperial kitchen and fruit garden, or Potagerie, at Versailles (for Pear culture on trellises in open quarters, winter Pear culture against walls, and horizontal cordons).—M. Chardon-Chatillon, Fontenay aux Roses, a little to the south of Paris.—Rothschild, Ferrières, Seine-et-Marne: This noble place has a fruit garden which though not very extensive, is exceedingly well managed. The fruit room is the best structure of the kind yet made, and usually stored with a fine stock. The Grape-room, in which the Grapes are preserved in bottles on the system previously described, is also worth seeing.

M. Rose-Charmeux, Thomery, near Fontainebleau: This garden is entirely devoted to Grape culture against walls and also in houses. The village is celebrated for furnishing Paris with a great quantity of its favourite Chasselas de Fontainebleau; indeed popularly it is supposed to supply all; but this is not the case, the next-mentioned place supplying a good deal, though hardly known. Nearly all the ground around the village is netted over with walls devoted to Grape culture. In fact it

N N

is as much devoted to the production of Grapes as Montreuil is to Peaches.—MM. Crapote et Cirjean, Conflans, St. Honorine : A very interesting establishment for vine-culture in the open air.—M. Lepère, Montreuil : This is the well known Peach-grower, and it need scarcely be remarked his garden is worth seeing at all seasons.—M. Chevallier, Boulevard de l'Hôtel de Ville, at Montreuil : This place is by no means so well known as M. Lepère's, but very well worth seeing. It is not nearly so large or so long established as the garden of M. Lepère, but it would be difficult to find more perfectly beautiful wall trees than are here to be seen. The ground all round this village is devoted to Peach culture.—M. Bac-Ivry, near Vitry : This is the garden of an amateur containing very nice examples of the horizontal cordons, pyramids, and trellised trees. It is but a few miles from Paris, the omnibus passing the door.—Of amateurs successfully cultivating fruit trees in small gardens in Paris, M. Laclaverie, Avenue du Roule, and M. Mattifat of Neuilly, may be mentioned.

Of nurseries near Paris where fruit culture is practised to any extent, the following are the most worthy of being seen— Jamin et Durand, Bourg-la-Reine. The partnership existing between MM. Jamin et Durand will be dissolved during the present year, and in future two separate nurseries may be looked for at Bourg-la-Reine. Attached to each will be found an interesting "school" of fruit culture—i.e., a garden devoted to fully grown specimens of fruit trees both against walls, trellises, and in the open.—M. Croux, nurseryman, Vallée Daulnay, Sceaux : He has also a very interesting fruit garden.—M. Cochet Suisnes, Brie Comte Robert, Seine-et-Marne : Horizontal cordons and Pears in the columnar form, and young trained trees are here in very good condition.—M. Deseine Bougival, Seine-et-Oise : Here there are large nurseries and also a school of fruit culture. There are in France many other large nurseries very interesting to the fruit-grower, such as Baltet's at Troy, Leroy's at Angers, Oudin's at Lisieux, Leconte's at Dijon, but only those within easy reach of Paris are named.

The Fig is grown best in the neighbourhood of Argenteuil,

by various cultivators. Its culture is well worth seeing. The Apricot is very extensively cultivated in the neighbourhood of Frieul, Vaux and Meulan, Seine-et-Oise. It is, however, needless to mention instances of culture away from walls and which do not afford us any practical lessons. Of governmental schools for fruit culture that nearest to Paris is in the Bois de Vincennes. It is quite new, and is described elsewhere.

I have been informed that the agricultural school of Grignon has a very good fruit garden, but I have not visited it. The nearest example to Paris of the planting of railway embankments with fruit trees is on the line from Gretz to Colommiers, Chemin de Fer de l'Est.

VEGETABLE CULTURE.—The finest examples of the culture of vegetables are found in the market gardens round Paris. The best are near Asnières, and also near Grenelle and Vaugirard. A ready way to get a general idea of their state, is to take a seat in the upper story of one of the trains of the railway that runs round Paris.

M. Courtois-Gérard, the well-known seedsman and writer on market gardening, has been kind enough to furnish me with the following list of representative market gardens:—

M. Pinson, 88, Rue de Charonne, Paris.
M. Julienne-Hichelle, 105, Rue de Beuilly, Paris.
M. Ledru, 16, Rue Mongallet, Paris.
M. Dagorno, 30, Rue de Picpus, Paris.
M. Dulac, 18, Sentier St. Antoine, Paris.
M. Hebrard, 70, Rue du Pot au Lait, Paris.
M. Marie, 4, Rue des Plantes, Paris.
M. Conard Louis, 5, Rue Volontaire, Paris.
M. J. Lecomte Impasse Maconnais, Boulevard de Poissonniers, Paris.
M. Gros, Rue de Paris, Charonnes.
M. Langlois, Rue Croix Nivert, Vaugirard.
M. Dumier, 16, Rue de Beuilly, Charenton.
M. Stiswille aîné, 12, Rue de Beuilly, Charenton.
M. Noblet, 18, Rue Bouery, La Chapelle.

M. Ponce, 53, Route de la Révolte, Clichy.
M. Dupont, 50, Rue de Hartre, Clichy.
M. Crosnier, 30, Route de Chatillon, Montrouge.
M. Leger Claude, 237, Avenue de Paris, St. Denis.
M. Chevalier, 10, Route de St. Denis, St. Denis.
M. Chevet père, 1, Rue Valentine, Bobigny.
M. Chevalier, Rue Montpensier, Vincennes.
M. l'ivert, Rue des Vignerons, Vincennes.
M. Jolleaume, 23, Rue des Marais de Villers, Montreuil St. Cors.
M. Duloc, 2, Rue de Montempoivre, St. Mandé.
M. Houdart, 2, Rue de la Grange, St. Mandé.

Asparagus is grown to the greatest perfection at Argenteuil,

and in the valley of Montmorency. Among the best cultivators in the former town are M. l'Hérault, of the Rue de
Calais, and M. Lerot-Salbœuf, Rue de Sannois. Asparagus
is forced both at Argenteuil and in the market gardens
within the fortifications of Paris. M. Caucannier, Place
de l'Eglise at Clichy-la-Garonne, has a curious and interesting establishment for forcing Asparagus on a large
scale in houses and by means of hot-water pipes.

THE CLIMATES OF PARIS AND LONDON COMPARED.—Most
people who have visited Paris are under the impression
that for clearness, salubrity, dryness, and heat, the climate
of the fairest of European cities is incomparably superior to
that of London. The idea has no doubt arisen from the
fact that most visitors to the French capital choose either
summer or autumn for their trip. At these periods even
our own smoky metropolis is at its best; but the hard-
working citizen, who for the first time finds himself walking
down the boulevards or the Rue Royale upon a lovely June
or August afternoon, sees the Paris climate in its fullest
perfection. The air is free from smoke, the buildings and
houses are either dazzlingly white or of a delicate cream
colour, and even the mud itself is of a clearer and brighter
hue than the greasy, metallic-looking paste with which the
Londoner is so familiar. Let him, however, choose November or December for his excursion, and he will soon
discover that Paris can be as cold and cloudy, and even as
foggy, as our own city. A few figures from various unim-
peachable sources, both French and English, will, it is
hoped, do much to dispel the prevailing notion of the great
superiority of the climate of Paris over that of London.

The climate of Paris may be taken as being typical of
that of the whole of the north-west of France, its change-
ableness, however, being somewhat less than that of the
districts bordering on the sea. In general characteristics
it may be said to stand midway between the climate of the
north-east portions of the Continent and that of the shores
of the Channel. It is less cold in winter than the former,
being warmed by the breezes from the Atlantic Ocean, but
colder than the south and west. In summer it is more

temperate than the south and east, but hotter than the extreme west. The mean temperature of Paris, taken from a series of official and private observations running over thirty-six years, may be taken at 51·55° F. The lowest temperature observed during fifty-two years was 2° below zero F.; the highest during the same time was within a fraction of 99° F.

These figures are worthy of a little consideration. For a similar period the averages of the observations taken in London by the officers of the Royal Society are as follows: Mean temperature 50·50° F.; highest temperature, 97° F.; the lowest, 5° below zero F. The mean temperature of Paris is therefore a fraction over 1° F. higher than our own, while the highest temperature only exceeds ours by something less than 2° F.

It will be also instructive to compare the mean temperature of the four seasons in both places with each other.

		Paris. Fahr.		London. Fahr.
Mean Temperature,	Spring	50·0	. . .	49·0
„	Summer	64·8	. . .	62·5
„	Autumn	52·0	. . .	51·0
„	Winter	39·5	. . .	39·0

It must, however, be borne in mind that in the suburbs of London the mean temperature is 2° F. below that of the city, and that on winter nights, when Jack Frost is striving his hardest to destroy all the vegetation within his reach, there is often as much as 4° F. difference between the thermometers of the city and the suburbs. The cause of this variation is twofold. In the summer a large quantity of heat is radiated by the masses of brickwork everywhere to be found about the city, to say nothing of the amount absorbed and given off again during the night; while in the winter the city is obviously warmer during both day and night, on account of the extra heat caused by the numerous fires, both industrial and domestic, that are constantly burning within its walls. Paris, as a city, being under precisely similar conditions, we may feel safe in assuming that the same difference exists between the mean temperature of the Observatory and Montreuil as there does between that of Somerset House and Tottenham for instance. Luke Howard, one of our first and most acute British meteorologists, on the strength of many thousands of observations made at Plaistow, Stratford, and Tottenham, gives the difference between the mean temperature of London and the country at 2° F. exactly, and a careful examination of his data has proved his figures to be correct within a fraction. This difference sinks to less than half a degree in spring; it increases in summer and autumn, and often rises on winter nights to as much as $4\frac{1}{2}°$ F. It is a singular fact that towards the end of spring, when the fires are being discontinued, and the sun has not yet reached his full power, it sometimes happens that the day temperature is somewhat greater in the country. This is doubtless to be attributed to the veil of smoke and cloud that is hanging over the metropolis. The effects of the higher mean winter temperature in the city are singularly apparent in the earlier budding and blooming of the trees, which frequently begin their spring life several days before their suburban cousins—a fact which may be easily verified by a walk from Haverstock-hill to Tottenham-court-road, just as the Elms are beginning to bud, or when the Pear trees are putting on their early spring livery.

The amount of annual rainfall in London only slightly exceeds that of Paris, although any unprejudiced person would feel inclined to give it as his opinion that the number of rainy days in London greatly exceeded those in Paris. The French authorities that have been consulted differ somewhat in their calculations, owing possibly to having collected the rain with dissimilar instruments. The English figures are from Luke Howard, the French from Gasparin and Bouvard.

	Gasparin. Inches.	Bouvard. Inches.	Howard. Inches.
Rainfall in Spring	5·6	4·0	5·0
„ Summer	6·8	6·0	6·5
„ Autumn	5·3	6·4	7·5
„ Winter	4·6	4·8	6·0
	22·3	21·2	25·0

CHAPTER XXVI.

HORTICULTURAL MACHINES, IMPLEMENTS, APPLIANCES, ETC.

TRANSPLANTING LARGE TREES.—Not the least remarkable feature of the public gardening of Paris is the excellent system of removing trees there practised. For the following article on this subject I am indebted to my friend M. Edouard André, the talented designer of Sefton Park at Liverpool :—

"The city of Paris, prior to having formed the large parks and public gardens which she now possesses, had no regular system of transplanting large trees, with the exception of the old-fashioned carts which had been used at Versailles and the other royal parks, and at M. de Rothschild's châteaux at Boulogne and Ferrières, principally for the purpose of removing large Orange trees in tubs, and occasionally for transplanting old and valuable trees.

" These carts were designed and constructed in the time of Louis XIV., and it may be well imagined that they were extremely cumbersome and inconvenient. In recent days, however, when the chief gardeners and the city architects were often called upon to extemporize shady avenues in a few days, it became absolutely necessary for them to put their heads together to invent some new machine which would work more easily and with less damage to the lives of the trees. The first apparatus built consisted of a frame bearing two moveable wooden rollers, one on the fore-carriage and the other at the back, each provided with holes in which to place the ends of the levers when hoisting up the tree. A round case made of sheet iron was hung in the centre suspended from the rollers by chains, which, when the tree was raised up by the levers, held the earth-ball and roots.

" We do not intend reviewing all the improved means

successively employed prior to the actual model now
in use (Fig. 301) being adopted; but confining ourselves to the apparatus figured here, we have only to state

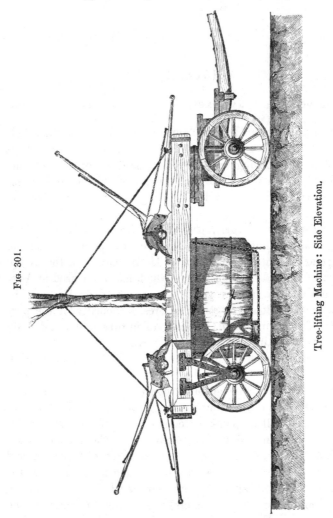

Fig. 301.

Tree-lifting Machine: Side Elevation.

the way in which the removal of large trees is managed in
Paris.

"We take, for example, a specimen tree, thirty years old,

thirty feet in height, the trunk of which has a circumference of three feet at a height of three feet from the ground, its total weight with the earth-ball being nearly two tons. The operation is commenced by staking out, round the stem, the circumference of the earth-ball, which will be on an average about four feet in diameter for most species, and larger according to the size of the trees to be removed. A second concentric circle is then made about two feet outside the first, the space between which will be the place for the trench to be dug for pre-paring the tree. The soil is then re-moved from this trench to the depth of three feet, and the small and delicate roots are drawn out of the earth, left hang-ing, and carefully pre-served. The earth-ball is then under-mined to prevent the roots from adhering to the subsoil; two thick planks, a foot wide, and a little longer

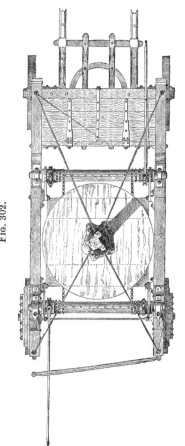

Fig. 302.

Plan of Tree-lifting Machine : the Tree in position.

than the ball, are placed underneath parallel with the width of the cart, so that they sustain the weight of the earth when the tree is lifted. Privet stems are now placed vertically, close together, all round the earth-ball, tied at the top and bottom with ropes, so as to prevent the earth from crumbling away, and also to protect the small roots from the inclemencies of the weather.

" The removal of the tree is then commenced in the following manner :—Two stout thick planks, strong enough to support the cart with the tree slung in it, and a little longer than the entire excavation, and having iron plates about two inches higher than the surface bolted on each side so as to prevent the wheels from slipping off, are placed parallel to each other across the excavation with the exact width existing between the wheels. The moveable bars at the back of the cart are then removed, and the cart is backed into the ways until the trunk of the tree is exactly in the centre of the frame. The moveable bars are then put in their place again so as to strengthen

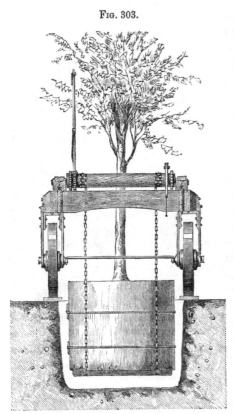

Fig. 303.

Tree-lifting Machine : back view.

the back of the wheels, which do not run on an axletree, but are fitted in wrought-iron frames hanging from the upper part of the cart, as shown in the woodcut. The chains attached to the rollers on each side of the cart are now lowered and passed under the planks before described, which are placed under the earth-ball. When all is fast, four workmen begin simultaneously to turn the handles attached to the cast-iron cog-wheels, by which great power is obtained on the rollers. The tree is raised slowly and steadily until it just swings clear of the ground, and then nothing is left to be done but to steady the tree before it is hoisted up to its proper height. For this purpose there is at each corner of the cart a strong wrought-iron hook, to which is attached a block, through which runs a strong rope fixed at the other end to a leather collar. These four ropes are then raised up together and the collar firmly fastened on the stem of the tree about seven or eight feet from the top of the earth-ball. The tree can be now easily removed without fear of its falling over.

Fig. 304.

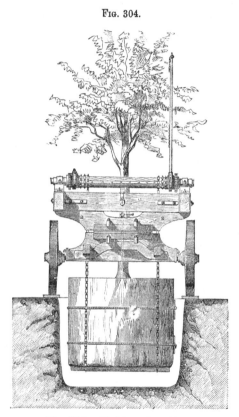

Tree-lifting Machine : front view.

" The horses are then attached to the cart, which is drawn slowly off the ways, and the tree can be removed with safety to its future resting-place. If the tree be vigorous and healthy, a hole a little wider than the one from which it has been removed should be dug beforehand, the earth being placed carefully on one side if it should be of a kind to suit the tree about to be planted, and if not, it should be replaced by suitable soil. The average dimensions for the hole, for an earth-ball of four feet in diameter, should be about seven feet, so that eighteen inches are preserved all round the tree to be filled up with good vegetable soil. The depth should be equal to the height of the earth-ball, or a little more if the tree be of a species with tap roots. The bottom of the excavation should be filled in with a little good soil, which will allow the top of the earth-ball to be a little higher than the surrounding ground, in accordance with an instinctive notion, which almost invariably induces us to place trees used as isolated specimens in lawns on small hillocks.

" When this is done the planks or ways are placed in position as before described, and the cart is very carefully drawn on them until the earth-ball is exactly in the centre of the hole. The tree is then slowly lowered, and when it touches the ground the guy-ropes from the corners of the cart are pulled tight, so as to have the tree perfectly upright and steady; the chains are unfastened and hoisted up round rollers; the two planks beneath the earth-ball are undermined and removed, and the privet shoots taken off. They then proceed to fill up the hole, particular attention being paid to the small roots, which are each separately covered in. When this is finished and the tree is considered sufficiently steady, the ropes are removed; the bars are taken out of the back of the cart which is drawn away, and the bars having been refixed all is ready for another removal.

" An abundant watering, if the removal has been made in the growing season, will be the end of the operation. The tree must be now protected against the wind, being as yet merely dependent upon its own gravity, as the roots take time to get hold of the ground. This result is obtained by placing at about half-way up the stem of the tree a padding of straw, round which three or four long pieces of wire-rope are attached; these are carried out on all sides of the tree and firmly fastened to strong stakes driven in the ground. We may then bid defiance to the strongest winds that blow.

" If drought is to be feared, the stem and main branches of the tree can be surrounded with plaited straw watered from time to time, or by a coating of clay mixed with cow-dung and covered with rough canvas, which is much about the same colour as the bark. Sometimes in the Boulevards of Paris they water trees surrounded in this manner by pouring water through a funnel from the top, between the clay and the trunk of the tree. These auxiliary means for keeping the tree alive may be supplemented by many others, such as covering it entirely on the south side with canvas, to preserve it from the sun and drought if it is of a rare kind; by watering the ground well if it is dry, or by draining the hole with rubbish or drainpipes if the soil be too damp, &c.

Fig. 305.

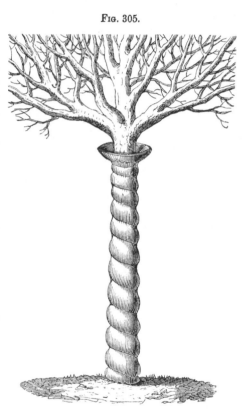

Trunk of large Tree recently planted enveloped in Moss and Canvas, to preserve the Stem from the action of the Sun.

" The ordinary season for transplanting large deciduous trees is from October to April, and from March to April or August for evergreens. But with sufficient care it is quite possible to transplant trees all the year round, provided the weather be suitable, the roots uninjured, the soil good, and that they be kept well sheltered, and watered.

" In choosing the tree to be transplanted, its age and species must be duly considered. For instance, it is useless to remove a tree that is sixty or eighty years of age, as it will never produce as fine foliage as it did before its removal, nor will it make any remarkable progress in size. It is better only to remove those not more than fifteen years old and under without any earth-ball at all, taking especial care to preserve all the roots intact. The best age for transplanting larger trees is from twenty to thirty years. The number of species ordinarily removed is limited, as only the more common kinds of trees are subjected to the process, no one caring to run the risk of losing a rare and valuable tree. In Paris, experiments made on various species have given the following results :—

" Success nearly always certain : Elms, Planes, white and red, Horse-chestnuts, Limes, Ailantus, Catalpa, Paulownia, Celtis, Planera, Sophora, and Willows.

" Success uncertain but sometimes satisfactory : Poplars, Sycamores, Maples, Alders, Mulberries, Beech, Ash, Magnolias, American Walnuts, Cercis, Diospyros, and several other exotic trees not yet sufficiently experimented upon.

" Success very rare : Robinias, Cratægus, Hawthorns, and nearly all the Rosaceæ, Birch, Laburnum, and many Leguminosæ, Oaks (European and American), Pavias, Elms, and Gleditschias."

With respect to the value of this machine as compared with any in use in England, there can be no doubt that the Paris machine is the best. Trees are there removed daily without the least difficulty or fuss, that, if removed in this country, would probably be honoured with a notice in the local papers. The best of our English machines must be taken to pieces for the removal of every tree : the beams have to be taken off in order to bring the wheels in position, then they have to be replaced in position, as well as the lifting apparatus. Besides, the machines are unwieldy and awkward. The advantage of the French machine is, that by removing the iron rod which connects the hind wheels and the hind cross-beam, the machine is put to the tree without trouble or awkwardness. The lifting power is by means of racks, pinions, and levers.

Besides the above-described excellent method for the removal of large trees, there is a very good method employed for the transplantation of small trees, specimen conifers, evergreens, and like subjects. Round each tree a circular trench is opened large enough for a man to move about in it at his ease. The depth should be equal to that of the deepest large roots, and a ball of earth large enough to insure the safe removal of the tree should be left. All the smaller roots found in the trench should be carefully preserved.

Fig. 306.

Small Machine for Lifting Specimen Shrubs and Conifers.

The ball is shaped into the form of a truncated cone, with its smallest portion below. It is next surrounded with light deal boards, separated from each other by the distance of three-quarters of an inch or so, like the staves of a barrel. They are next secured temporarily by a suitable rope. A man then descends into the hole and fixes the rope by means of the screw apparatus shown in Fig. 307, so as to press the planks firmly against the soil of the ball. The press is then removed and the same thing done higher up, within say four inches of the top, an ordinary cask hoop being first nailed round the planks before the screw is unfixed. The ball being firmly fixed in its proper position, it is hove over so as to get to its underneath part. The bottom of a cask having its boards fastened together with a circular piece of sheet iron rather larger than itself is passed under, the iron being pierced with two or three holes and turned up so that it

Fig. 307.

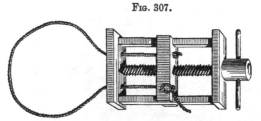

Screw used in preparing specimens for removal, as shown in the preceding Figure.

may be nailed against the planks. In some cases the stem of the tree should be fixed by iron wire to the sides of the improvised cask.

When it reaches its destination it is gently inclined to one side and the bottom boards removed. The hoops are next unfastened, the boards removed, and the roots carefully arranged in their natural position, some good earth being spread over them. The amount of success capable of being attained by this method may be seen throughout the squares of Paris, hardly a single tree having been killed during the plantation of the myriads now growing so luxuriantly in that city. Some at Vincennes have died, it is true, but after having been transplanted in the rough and ready way usually resorted to.

The apparatus costs a mere trifle, as will be seen from the following estimate. A press made of oak and beech, with the rope included, only costs eighteen francs; if it were made of iron it would possibly cost less. For a ball six or seven feet in circumference and eighteen to twenty inches high, the boards, hoops, cask bottom, sheet of iron, and nails would cost less than a couple of francs. If still greater economy is desirable, what are known as Yankee flour barrels may be used, if they are cut in two and taken to pieces. With these simple appliances two men can prepare five trees a day ready for hoisting on to the cart intended to receive them.

CARRIAGE FOR TRANSPORTING ORANGE TREES.—The fashion of growing large Orange trees in tubs is so general in France that some efficient means of moving them from place to place becomes necessary. Many contrivances have been tried, and several are in use, but the best and handiest is that employed for the carriage of the large specimens in the gardens of the Tuileries. For the following notice of it I am indebted to my friend Mr. John Gibson, the able and deservedly popular superintendent of Battersea Park, who has long taken a deep interest in the public gardening of Paris :—

" The machine used in the gardens of the Tuileries for removing large Orange trees in tubs, of which a longitudinal representation is given on the next page, is the most useful contrivance I have seen in use for this purpose. Its simplicity and the facility with which the tubs are lifted for transit are its chief recommendations; no taking to pieces or removal of the side beams, prior to loading, is necessary, beyond the removal of the hind axle, which consists of a strong wrought-iron bar with a hook at each end, the hooks fitting into an eye fixed on the inside of the stock of the hind wheels. They are made fast with a pin through each hook; when this bar is removed the machine is backed to the tub, one of the hind wheels passing it on each side until the tub is midway between the fore and hind wheels where the lifting apparatus is fixed. This being done the axle bar is fixed and the machine is ready for loading.

FIG. 308.

" The stirrups attached to the lower end of the upright

lifting rods are now lowered to the bottom of the tub by
means of the rack and pinion machinery until the two iron
bars, which are previously pushed under the tub, can be
placed in the four stirrups ; this being done, all is ready for

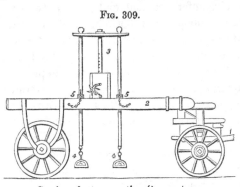

Fig. 309.

lifting the tub
in an upright
position by the
rack and pinions,
which are worked
by a man on each
side.   When the
tub is high enough
for travelling it is
secured by means
of a pin through
the four upright
lifting rods in-
serted at 5, the

Carriage for transporting Orange trees.

1. Fore carriage. 2. Side-beams. 3. Lifting screw.
4. Stirrups for carrying tub. 5. Pins.

tub being lowered on to the pins for travelling.   The
whole operation does not occupy the three men required to
work it more than two or three minutes.   The machine is
drawn by one horse, and it will be seen how easily and
quickly the magnificent Orange trees alluded to are brought
from their hibernatory in the spring to their summer

Fig. 310.

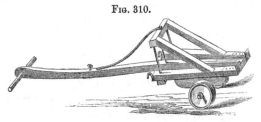

Truck for moving Plants in tubs and large pots.

quarters, and as easily taken back in the autumn.   The fore
part of the machine is made to 'lock' so that it turns in
little more than its own length.   It is in every respect a
most complete apparatus for this and for like purposes."

TRUCK FOR TUBS OR VERY LARGE POTS.—This very handy
little truck, Fig. 310, is what the French use for moving large
plants in tubs and large pots.   It would be difficult to find

anything more useful in its way. Large specimen plants
are quickly and easily moved by this means. The pot or
tub is caught by the little iron feet, then thrown on its
side and tied firmly if a long distance has to be traversed.

TUBS FOR ORANGE TREES, &c.
—Oranges, Oleanders, &c., are
so much grown on the Continent
that good kinds of tubs are of
high importance. There can be
little doubt that the square tubs
now employed in the public gardens of Paris are the best and
most durable. I mean those with
the hollow cast-iron frame and
bottom, and wooden sides. In
their case renewing the sides from time to time is not a

FIG. 311.

Tub for Orange Trees.

matter of much expense. The tub here figured is a well-made
wooden one, with a wide ornamental margin of metal. The
effect with good specimens is superior to that of the square
ones in common use, but it is very expensive.

GARDEN CHAIRS AND SEATS.—The kind of chair shown
in Fig. 312 is seen in quantities in all public places in Paris.

FIG. 312.

It has a convex seat made of flexible strips
of metal springing from the sides and
joined together in a little central piece.
These chairs stand any weather, and are
nevertheless as elastic as a drawing-room
one. A very neat, elegant, and comfortable conservatory, pleasure-ground,
or summer-house chair is composed of
three of these seats united in one, the
larger framework of the back and sides
being made of rustic iron about as thick as
the thumb, the smaller spray being tied to the larger by imitation osier twigs. This is made by M. Carré, the maker of
the greater number of chairs in this way.

There are many modifications of this kind of chair. One
on much the same principle, but with the elastic bands crossing from side to side instead of all ending in the centre, is made

by Tronchon, of the Avenue d'Eylau, who has a large collection of such articles. His modification

Fig. 313.

Seat with box for climbing plants.

of the elastic chair is certainly stronger than that of Carré, but for durability and general good qualities the chairs made in imitation of split cane work are the best of all.

Fig. 313 shows a combination of moveable seat and shallow bower, with a box at the back for planting climbing plants wherewith to cover the trellis work. The best plan would be to train some graceful and rapid growing annual creeping plant on this. So shaded and decorated, it might prove very acceptable in some positions.

The next illustration shows a form of seat seen at the Paris Exhibition of 1867. It consists of a not uncommon form of garden seat with a tent-like shade supported as shown in Fig. 314. This shade can be rolled up in a moment by means

Fig. 314.

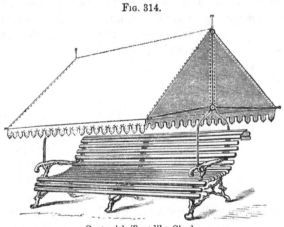

Seat with Tent-like Shade.

of the chain at the ends, and let down with equal facility. This seat would seem to be a want out of doors in summer, and also in conservatories and like structures in winter and spring; that is, where people sit and read in them. There is a modification of it in which the back of the seat is reversible.

GRAFTING MASTIC.—The thorough knowledge of grafting possessed by the French has long ago led them to invent various kinds of grafting wax or mastic, which greatly facilitate grafting. These, while distinct improvements for propagators and practical gardeners having much grafting to do, render grafting on a small scale and in the gardens of amateurs a pleasant and interesting operation. The mixture of clay, dung, &c., commonly employed for grafting in this country is not such as many amateurs care to make, and it is scarcely worth while doing it for the sake of a graft or two. The best of the French compositions for grafting is that called *Mastic l'Homme Lefort*—an awkward name for an excellent article now sold in this country by Messrs. Hooper & Co., the seedsmen, of Central Row, Covent Garden, W.C. One of the most able fruit growers and horticulturists in the country has recently given his opinion on this article in the *Gardener's Chronicle* :—

It is a substance of about the consistency of common white lead, somewhat resembling half-melted gutta-percha, and having a very pleasant and agreeable perfume. It is quite easy of application, being readily spread over the parts with the blade of a knife or a flat piece of wood, like butter on bread. Although in the box, away from the air, it will keep pliable and moist for many years, it very soon hardens on the outside after being exposed thinly on the graft, and, as it were, hermetically seals up the point of junction, and thus prevents all access of air to the cuts. It is at the same time quite elastic, and easily removed when required. It was largely tried in various ways in the Royal Horticultural Gardens, Chiswick, by Mr. Thompson, who reported favourably on its merits. I have myself used it in grafting all sorts of hardy fruit trees, and approve of it very much indeed. In grafting tall standards it is better than clay, which it is difficult to fix at all times. This, on the contrary, can be applied with the greatest ease in any position, and a very little of it suffices spread thinly round about the junction of the scion with the stock. I wish particularly to recommend its use in grafting Vines. For this purpose it is far superior to clay, or any other paste that I have used—and I have grafted some hundreds. The objection to the use of clay or moss is that in general when the Vines are growing a moist atmosphere is maintained in the house ; in fact, to induce the scion to break strongly it is regularly syringed. The clay, &c., is thus kept continually moist, and roots are emitted into it from the stock, and frequently from the scion also. When this takes place, and I have seen it many times, there is but little chance of the graft succeeding. With the Mastic, on the contrary, no roots are possible, however much the moisture and heat applied externally to the graft and stock, and, as a consequence, success in Vine grafting becomes almost a certainty. I have also found the Mastic useful in placing over wounds or bruises on plants generally, thereby enabling them to heal quickly. For this purpose, for Vine grafting, and for all the more delicate operations of grafting, I strongly recommend it : further, it is very handy, always ready for use, and so easily applied. With a sharp knife, a bit of matting, and a little sixpenny box of Mastic l'Homme Lefort (at which price sufficient for 100 grafts can be purchased) any gentleman or even lady can go grafting trees, at any moment, with the greatest facility, and finish the operation without soiling the fingers. B.—It has a distinct advantage over other kinds of grafting mastic, inasmuch as it may be used cold, whereas other kinds have to be heated before being used.

IMPROVED FRUIT SHELVES.—In the Pear-room at Baron Rothschild's at Ferrières there is a new and excellent plan

FIG. 315.   FIG. 316.

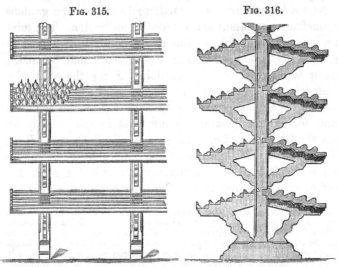

Portion of Pear stand at Ferrières.   End view of Pear stand.

for arranging the fruit—the successive shelves of splendid Pears being so formed that every individual one can be examined without touching any. I need scarcely say that in the case of a fruit requiring so much nicety of judgment and attention as the Pear does, in the gardener who makes the most of his collection, and has each kind, or even each perfect fruit, eaten at the right time, this is an improvement. The Pear-room in the Imperial garden at Versailles usually presents a fine sight. There the old flat form of bench is in use, and all the shelves are closed in by wooden doors, so as to exclude the light from the fruit.

FIG. 317.

Position of each line of Pears in the Fruit room at Ferrières.

DRYING FRUIT ROOMS.—Chloride of calcium is sometimes

used by the French for drying the air of their fruit rooms.
Mr. Thompson recommends in his book *chloride of lime* for
drying the atmosphere in a fruit room, but he no doubt
means chloride of calcium, which is a much more powerful
absorbent of moisture. The fumes of chlorine given off by
the former substance, which is simply bleaching powder,
would be injurious to the
colour and flavour of the
fruit. Chloride of calcium
is a cheap salt, costing
only a few pence per
pound. It may be ob-
tained at any large opera-
tive chemist's or drysal-
ter's, and should be pre-
served in well-corked jars.
For use, a pound or so
may be spread on plates
about the room, and should
be renewed as soon as it
shows a tendency to run
into a liquid. A few
pounds of this material will

Fig. 318.

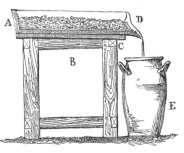

Arrangement for the use of chloride of calcium in the Fruit room. A, Tray or box
about twenty inches square, and lined
with lead; B, Support; C, Slope on one
side; D, Outlet; E, Jar to receive liquid.

be sufficient for a large fruit
room, for the whole fruit-preserving
season. Its damp-absorbing power may
be renewed by heating the wet salt to
redness on a fire-shovel or old frying-pan;
but it is so cheap, and so readily obtainable, that the process of renewal is
hardly worth going through. This substance gives off no fumes of any kind,
and cannot be in any way injurious to
the Grapes, like the salt recommended by
Mr. Thompson.

Fig. 319.

The Panier.

THE PANIER.—This is the article alluded to in the description of the garden
in the Bois de Boulogne. It is much
used for carrying vegetables, and also
frequently for conveying manure amongst
close rows of vines, and has many similar uses. At first it

appeared a ridiculously antiquated thing to me, but afterwards I often saw it in efficient use. Where materials have to be carried through houses, and in positions where barrows of any kind could not be employed, it might be useful, and there is no way by which one man can carry so many vegetables as by using it.

PLOUGH-HOE.—This is used for cleaning the numerous long straight avenues in the imperial demesnes. A few men, each guiding one of these, clean the weeds from an avenue almost as quickly as they can walk along it, but the texture of many walks would not permit of its use at all. At St. Cloud and other places where it is used, the surface is quite sandy, and wherever this is the case it may be used with advantage, particularly in places where many wood-walks and drives have to be kept in order. They could not be used on such firm walks as we have about London.

FIG. 320.

The Plough-hoe.

THE BINETTE.—This is a handy implement that I think would prove more useful for stirring the earth between crops than anything we employ. It serves as a draw hoe, and the forked portion is very efficient in loosening hard ground. There are various slight modifications of the one here figured. The handle is usually about as long as that of the common draw hoe.

FIG. 321.

The Binette.

FRAMES FOR FORCING.—The French market gardeners use an immense quantity of frames, and it is by their aid they procure most of the tender and excellent forced vegetables sent to the markets in early spring. These frames are made

of very rough wood; are narrow—not exceeding four feet
in width; and arranged in close lines completely immersed in
the heating material.   They are usually about twenty inches
high at the back and fourteen in front.   Undoubtedly the
principle is better and cheaper than our own.   We employ
large and well-made frames in private gardens, and for the
most part place them so that all but the base is exposed to
the influence of the weather, and the plants therein are
more liable to changes of temperature and cold.   By having
the frames narrow, all the sidework rough and cheap, and
the frames placed in close lines, we get the greatest amount
of heat at the smallest cost.   By having nothing but the
surface of the glass
exposed, little heat is
lost, and when the
frames are covered by
the neat, warm, and
flexible straw mats,
they are as snug as
could be desired.
When it is simply
desired to preserve bedding plants through the winter,
the spaces between the rough-sided frames are merely
filled up with leaves and slightly heating materials.   About
two feet of space is left between each frame, or just
enough for the convenience of the workmen.   Generally
they are put together by the workmen of the market
gardens: two stout posts being driven firmly in at one
end, and an end-board nailed to them.   Then at every four
feet or so minor posts are driven down, and the rough front
and back boards nailed to them.   Numbers are also made
on a plan by which they can be readily taken to pieces and
stored in a small space while not in use.   By this means
the ground covered by forcing frames in winter is cleared
for ordinary open-air crops in summer.

MATS FOR COVERING PITS AND FRAMES.—In our cold
and variable climate, the winter covering for many minor
glass structures is of the greatest importance.   It is a thing
at present managed in a very expensive and by no means

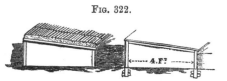

FIG. 322.

Narrow frames used for forcing by the market
gardeners of Paris.

satisfactory way. The French mode of doing it is much cheaper, neater, and more effective; and in passing through their market gardens and forcing-grounds in winter, it is one of the first things that seems to the English horticulturist as worthy of imitation. The covering used consists of straw mats about an inch thick, the sides as neat as if cut in a machine, the mat knit together by twine, and its texture such that it may be rolled up closely. One of these mats, which is much better as a protection than a bass-mat, costs about one-third the present price of that, while in point of appearance and amount of protection given the advantage is all in favour of the French paillasson. The figure given

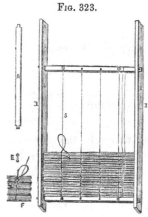

Fig. 323.

Frame for making Straw Mats.

represents a simple frame for making these mats in the nurseries of M. Jamain, the celebrated cultivator of Orange trees, and I append his description of it. There are several frames for this purpose; and there is also a machine for making these mats, which are indispensable to the French gardener; but the one here described is the best and simplest for private use. " Get two pieces of timber (1) about three inches thick, four inches wide, and as long as required. Pierce these timbers, as shown in Fig. 323, and introduce A in the holes to maintain the same width between the sides, and support the nails or screw, as shown in the cut. These nails are to keep the string tight (5). The board may be shifted from hole to hole so as to make mats of any desired length. The length of the string must be about three times as long as the straw mat, and rolled round a little reel, shown at E. The straw must be placed on the machine so as to have all its cut or lower ends close against the sides, the tops meeting in the middle, and so thick as not to have the mat thicker than three-quarters of an inch when finished. The stitches must not be wider than three-quarters of an inch, and be worked as follows (see F of the figure). Take a little of the straw with the left hand, and work the reel with the right, first over the straw, then over the bended string, coming back underneath, and swiftly passing it between the two strings, pulling tightly and pressing the straw, so as to have a flat stitch, and not thicker than three-quarters of an inch at the most. The same operation is repeated until the mat is finished. The machine described has been at work for the last twenty years in our nursery, at Paris, and is still as good as new. An ordinary workman may make daily from thirty to forty yards run of these straw mats with it."

Fig. 324.

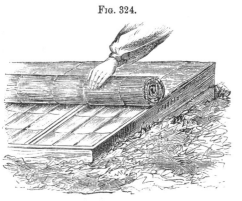

All new or strange things of this sort are adopted slowly by horticulturists ; but that they would immediately use this, if they had an opportunity of seeing it in working order, I have no doubt; and I hope yet to see it

Straw mat (paillasson) used for covering frames.

in general use in British gardens. In France these mats are found so useful that they are employed for many purposes besides that of covering frames, and they even form a very effective temporary coping for walls in some cases. I doubt very much if anything I can say for them will give a full idea of their utility. In all gardens where men are regularly employed they may be made during bad weather in winter; and as there is often a difficulty about procuring enough of useful indoor work for men at such times, the making of these mats will be a gain from that point of view alone. In country places, where straw is abundant,

their cost would be a mere trifle.   Around Paris so great is
the demand for these mats that, in addition to being made
abundantly by hand as described above, they are also made
by machinery.   There is indeed an establishment for manufacturing them thus belonging to M. Dorleans, 37, Rue du
Landy, Clichy.   The nurseries of the city are supplied by
him, and many people find the machine mats cheaper than
those they make by hand.

THE NUMÉROTEUR.—Numbering instead of labelling is
now adopted in so many gardens and nurseries, that
this instrument cannot fail to be useful.   The following
description of it originally appeared in the *Gardener's
Chronicle*:—" Horticulture is a science so vast, and em-
braces subjects so different, that however good a man's
memory may be it is insufficient, and hence it becomes
necessary to give it mechanical aid.   Among the means
employed are tickets or labels written upon parchment or
paper, or small pieces of wood or zinc; but these are soon
effaced, and are very liable to get lost or displaced.   A very
good plan frequently adopted consists in the use of small
bands of lead, which are rolled round the stems or branches
of the plants.   Upon this lead a number is marked, corresponding with a catalogue, in which the name and any particular remarks are entered.   This method is sure; but to
carry it out several things are necessary.   First, there is
wanted a series of numbers from 1 to 10, or rather from
1 to 9, the zero, combined with other figures, making the
numbers 10, 20, 100, &c.   Then this series of numbers
must be fixed upon a block of wood, and the figures have
to be impressed upon the leads by means of a small hammer.
So that to mark the leads we want—1st, a pair of scissors
to cut the metal; 2nd, a set of numbers; 3rd, a block to
receive them; and 4th, a hammer to strike and indent the
figures in the leads.   This apparatus therefore becomes
troublesome, especially when it is necessary to change its
place, as is the case when it has to be used in different
parts of a large garden, or in a field.   Besides, it suffices
for one of the little figures to be lost to render the whole
series useless.

" A consideration of these inconveniences induced an in-
genious cutler, M. Hardiville, of the Rue St. Jacques, in
Paris, to invent the Numéroteur, or Numbering Pincers.
This instrument in its ge-
neral form resembles a large
pair of scissors, in which
the blades instead of being
cutting are flat and blunt,
with the upper extremity
prolonged. On the inner
side of the upper of these
blades is fixed a series of
ten figures arranged in
order, from 1 to 9, followed
by 0. These figures are
placed at the end of small
steel shanks screwed into
the blade, and upon the
opposite blade, which is flat,
the figures are marked in
hollows, so that, without
grouping, one is able to
effect with certainty any
necessary numerical com-
binations. A pressure of
the blades suffices to indent
the figure in the piece of
lead that has been placed
between them, and the lead
is then withdrawn and
placed in the same way beneath whatever other figure
or figures may make up the
number required. The blades
of these numbering pincers
work upon a movement

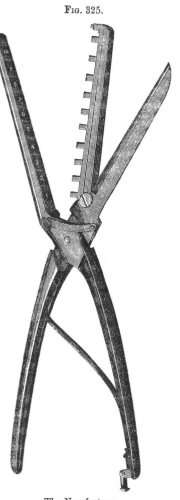

FIG. 325.

The Numéroteur.

similar to that of a pair of scissors, the alternate opening and
shutting of the curved portion or handle also opening and
shutting the two opposite blades, so that it is only necessary to

put the plate of lead straight with the figure which is wanted, and then to make a pressure, to have this figure indented on it.   At the end of one blade, in a line with the figures, is a small punch, with which, if necessary, to pierce the lead, in order to admit of passing through it a wire thread, by which it may be suspended.   To complete this instrument M. Hardiville has added, on the side of one of the branches, a small blade, which, by means of a spring adapted to the other branch, forms a pair of scissors with which to cut the leads.   When the scissors are not needed, the spring is unfastened, being made to do so easily and quickly, and the blade then tightens itself against the branch of the pincer without any trouble.   At the base is a moveable spring which serves to open the branches.   Thus we see that this instrument is very complete, but its value is augmented by its not being complicated, and especially by its being of a reasonable price—ten francs."

THE SÉCATEUR.—Of garden cutlery I will only mention the sécateur, and this is an instrument that every gardener should possess himself of at once.   I know well the prejudice that exists in England among horticulturists against things of this kind, and their almost superstitious regard for a good knife.   I also believed in the knife, but when I saw how useful is the sécateur to the fruit growers of France, and how easily and effectively they cut with it exactly as desired, I became at once converted.   A sécateur is seen in the hands of every French fruit grower, and by its means he cuts as clean as the best knife-man with the best knife ever whetted.   They cut stakes with them almost as fast as one could count them; they have recently made some large ones for cutting stronger plants—such as the strong awkward roots of the briars collected by the Rose growers.   Of these sécateurs there are many forms, several of the best being figured here.

First we have the Sécateur Vauthier (Fig. 326), a strong and handy instrument.   Its sloping semi-cylindrical handles have their outer side rough, which gives a firm hold; the springs, though strong, resist the action of the hand gently; the curvature of the blade and the adjustment are perfect;

and lastly, the principal thing, the action is so easy as never to hurt the hand. " During the many years of my experience," observes M. Lachaume, a fruit grower who describes this implement in the *Revue Horticole*, " I have used tools of all kinds, and the tools have also used me a little; but I have never met with anything which gave me so much satisfaction as the Sécateur Vauthier. Every desirable quality is combined in it, and I recommend it with perfect confidence. The strongest branch will not resist its cutting, nor a single branch, however well concealed, be inaccessible to it. Moreover, the double notch on the back of the blade and hook (in which a wire is shown in the figure) will enable the

Fig. 326.

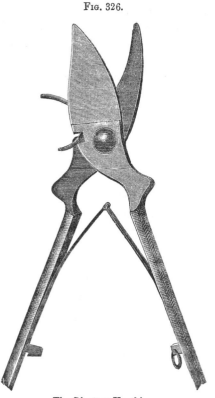

The Sécateur Vauthier.

operator when employed at his trellises to cut every wire without using the pincers."

The Sécateur Lecointe (Fig. 327) is another variety recommended by the leading French horticultural journal. The inventor was led to devise this kind of spring in order to avoid the annoyance arising from the frequent breakage of the form usually employed. It is said that this form of spring secures an easy and gentle action of the instrument,

and has the advantage of lasting longer than others, from
not being so liable to break, while it secures a firmness and
evenness in working which is not otherwise attained. A
further improvement is pointed out in the fastening, which
consists of a stop which catches when the two handles are
drawn together, a projecting portion on the outside acting
as a spring which is to be pressed when the instrument is
required to be opened. M. Lecointe of Laigle is the
inventor.

Fig. 328 represents the sécateur of older date than the
preceding, and one more generally used. It is much

FIG. 327.   FIG. 328.

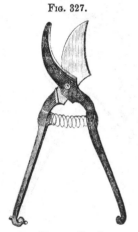   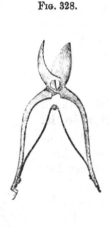

The Sécateur Lecointe.   The Common Sécateur.

employed at Montreuil. There can be no doubt that
where much pruning of any kind is done, and particularly
pruning of a rather rough nature, the sécateur is a valuable
implement. For pruning in which great nicety of cutting
is required a good and properly shaped knife is best. The
sécateur was first invented by M. Bertrand of Molleville.

THE RAIDISSEUR.—This is the name for the little wire-
straining implement which plays such a very important part in
the wiring of garden walls, or erecting of trellises for fruit
growing in France. It is an implement which, though
insignificant in itself, is calculated to make a vast improve-

ment in our gardens and on our walls. It will save labour,
time, expense, and make walls, and permanent trellises for
fruit growing infinitely more agreeable to the eye and useful
to the cultivator than ever they were before.

There are various forms which I need hardly describe, as
they are so well shown
in the accompanying
cuts. The first (Fig. 329)
is a reduced figure of
one about three inches
long, and of which I
brought some specimens

FIG. 329.

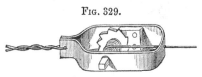

The Raidisseur.

from Paris. The engraver has placed it in the best position
to show its structure. The wire that passes in through one
end is slipped through a hole in the axle; the other end is
attached to the tongue, as shown in the engraving, and
then by the aid of a key, Fig. 330, placed on the square

FIG. 330.

Key of Raidisseur.

end of the axle, the whole is wound much as a guitar
string is wound round its peg. The first form figured is
very much used in the best gardens, and always seemed to
me to do its work effectively.

The next figure is that of the Raidisseur invented by
Collignon and re-
commended by
Du Breuil. It
does not differ
much from the
preceding. D
shows the point
of insertion of

FIG. 331.

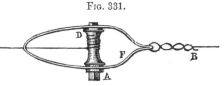

Collignon's Raidisseur.

the wire that has to be tightened; B the fastening of

P P

the other end of the wire; and A the head on which the
key is placed.   Fig. 332 is a side view of the same implement.   The fore-
going kinds are

Fig. 332.

Side View of Collignon's Raidisseur.

galvanized, just
like the wire.
That shown by
Fig. 333 is a
very simple one,
not galvanized, which was much used in the fruit
garden of the Paris Exhibition.   This last form is surely
such as can be readily and cheaply produced in any
manufacturing town.   The best of these tighteners cost but
a few pence; and if it were not so, it would still be profit-

Fig. 333.

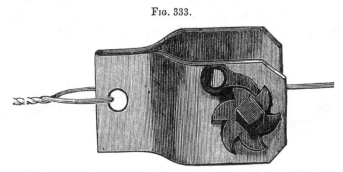

Raidisseur used in the garden of the Exhibition.

able to employ them, in consequence of the great saving
they effect, by enabling us to use a very thin wire, which is
quite as efficient and infinitely neater than the ponderous
ones now generally employed by us, where the nail and
shred have given way to some costly system of wiring.

   Since writing the foregoing I have found a much improved and very simple raidisseur in use at Thomery.   Fig.
334 represents its actual size.   It is simply a little piece of
cast-iron costing little more than a garden nail—so small that
its presence on wall or trellis does not look awkward, as in
the case of some of the larger kinds, and very effective.
I never met with it except in the garden of M. Rose-

Charmeux at Thomery. The walls there are very neatly wired by its help, and it is equally useful for espaliers. I have indeed never visited a garden in which the walls and trellises were so neatly done, and all by means of this simple strainer and the galvanized wire. Fig. 334 shows the wire strained tight, and is a little more than half the size I recommend. Messrs. J. B. Brown and Co., of 90, Cannon-street, have at my request cast a great number of these, and can supply them in any quantity and at a very low rate. They are made of malleable cast-iron, and are galvanized. The edges of the division in the head of this little implement being sharp, those of the specimen I brought from France were filed to prevent them cutting the wire in the straining; but any danger from this source

FIG. 334.

The simplest and best form of Raidisseur.

is quite obviated by allowing the wire to be loose enough to permit of one coil being wound round the neck of the raidisseur before the real strain is applied. It is almost needless to add that the wire is simply placed in the groove in the head of the raidisseur, which is then turned, and finally tightened with a key like that for the other forms.

MATERIAL FOR TYING PLANTS.—The drop wears away the stone in a far larger sense than is usually accepted with this trite saying. Petty cares often help to wear away the soul, and petty details occupy much of our life. Small indeed, then, must be that which we can call beneath our notice. The tying of plants, of fruit trees, of anything and everything in a garden, is not often a conspicuous effort; but it occupies on the whole a great deal of time, even in small places. In larger ones operations of this kind often occupy several men for weeks at a time. The material usually employed with us is bass matting, and in most large gardens a number of bass mats are annually cut up and used for this purpose. Of late years they have trebled in price. There is the labour of cutting them into shreds, and of selecting the best strings for tying, but after all the trouble a perfect and a cheap material is not the result.

This expense may be done away with, and a much better material secured, by simply planting a few tufts of the common glaucous Rush (Juncus glaucus) in some moist spot, or, where much tying is to be done, a few dozen tufts. The stems of this plant are smooth and ready for use at any moment, and are suited for tying everything except the strong or " mother branches " of fruit trees (for which twigs of the yellow osier are best fitted) and the finest and youngest shoots of hothouse plants. The Rush may be cut green and used out of hand, or it may be cut soon after flowering for winter use in a dried state. When wanted in winter it is desirable to steep it in water a couple of hours before it is used, so as to insure the requisite flexibility. It forms a neat and lasting tie, and is not knotted like the matting, but simply twisted, then pinched off with the nail or cut with the knife, and one of the ends turned back a little. For tying the young shoots of fruit trees to an espalier it is admirable, as it is for most other purposes of training. When men are accustomed to it, they work with greater facility with it than with anything else. When green it is a matter of no trouble for a horny hand to pinch it instead of cutting it off; thus the workman has not the trouble of employing a knife, and has both hands free.

The dried grass of Lygeum Spartum is also used in France to a great extent for gardening purposes. It is a Spanish grass which I have grown pretty freely on cold soils in England, and which will do well on warm ones everywhere with us. It is suitable for very strong and durable tying. Thus the two best materials for this purpose may be grown in any garden without cost. If the expensive matting were as cheaply got as these, the fact that the Rush and Grass are ready rolled in twine-like fashion, should make us prefer them.   In tying carefully it is necessary to twist the matting, and thus a good deal of time is lost.

In addition to the above hardy plants, the " grass " of which may be directly used for tying, the leaves of the New Zealand Flax, Phormium tenax, are very largely employed for that purpose about Paris.   This plant is grown everywhere in greenhouses for room decoration.   The long leaves being produced in great abundance, the old leaves that are cut away are preserved, thus securing a strong and excellent material for tying.

FIG. 335.

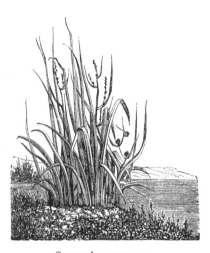

Sparganium ramosum.

MATERIAL FOR TYING GRAFTS AND BUDS.— The dried stems of Sparganium ramosum, the Bur reed, have replaced woollen thread for budding purposes in France. In texture they are peculiarly suited to this purpose, being soft, dense, elastic, and tough, so as to enable the operator to secure an effective tie.   The plant is a common waterweed, growing everywhere in Britain along the margins of ponds, streams, and ditches, and about three feet high.   It is therefore a very cheap material, and may be cut and stored in any quantity for budding and grafting purposes.

I have seen it in extensive use in some of the largest and best nurseries in France, and have no doubt that it is an economical and real improvement.   The stems of the common Bullrush (Typha latifolia) are used for like purposes, but not so extensively.   By means of these, many French grafters have been enabled to do away

with all expense for woollen and cotton thread. The Sparganium is gathered in summer when fully grown ; the leaves,
which are united at the base, separated, and placed to dry
in a shed or barn hung up in bundles. When required for
use they are cut into the necessary length, from fourteen to
twenty inches, steeped in water for a few hours, and then
slightly dried by pressure or wringing. In large field
nurseries, where there is no water, bundles of the Bur reed
are simply kept moist and flexible by being buried in the
earth, and they may also be kept so by placing them in a
cellar. It must not be used very wet, and if too dry it
is more liable to crack. It is found to bend best when ap-

Fig. 336.

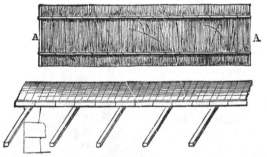

Mode of Protecting Walls. A, Paillasson or neat straw mat, two feet wide,
held between laths, for placing under the permanent copings while there
is danger of frost.

plied edgeways to the body which it is to envelope, and
slightly twisted. For all kinds of budding and grafting,
except large cleft-grafting and the like, it is as good a material as can be found.

PROTECTION FOR WALL AND ESPALIER TREES.—Having
several times spoken of the deep temporary copings the careful French cultivator uses for his fruit wall, I here give a
rough figure showing a section of the tile-coped wall, and
projecting from beneath it the supports for the temporary
protection. The French take a good deal of trouble with
temporary copings, and find them of the greatest value in
getting regular crops; for the frosts are severe in the

northern parts and all around Paris, and, in fact, over nearly all the region north of the river Loire—the most important of France. The best appliance of this kind I have ever seen consisted of narrow lengths of bituminized felt nailed on cheap frames from six feet to eight feet long, and about eighteen inches wide. The use of these on walls devoted to the culture of choice Pears, Peaches, &c., would result in a marked improvement. The temporary coping has a great advantage in being removable, so that the trees may get the full benefit of the summer rains when all danger is past, and not suffer from want of light near the top of the wall, as they would if such a wide protection were perma- nent. I believe that similar copings would be much more effective than any of the netting and canvas pro- tections now in use in Eng- lish gardens.

Fig. 337.

Wall with permanent coping of tiles, temporary one of straw mats, and canvas in front.

The commonest temporary coping seen in France is made of straw nailed between laths ; it seems to answer its purpose very well, but is not so neat and satisfactory as that made of bituminized felt. Whatever kind of protec- tion be employed, care is taken to throw the wet well off the wall ; the slightest experience of the effects of frost on vegetation will show the wisdom of this course. Of what does it avail to place a net or a few branchlets of trees before a fruit wall, if we allow the cold rains and sleet to dash on to every tender little brush of pollen-bearing stamens on the wall ? Even when the French do employ canvas in front of a wall they usually use the wide temporary coping too, thus keeping the wall dry and preventing radiation.

Whatever imperfect efforts we make to protect our wall trees, nobody in England ever thinks of protect- ing espaliers, but the French sometimes do it with success. Fig. 338 shows a mode of arranging two

rows of trees in a manner different to that already shown
on the double trellis.   The main supports are strong posts.
French gardens are usually surrounded by walls, and

Fig. 338.

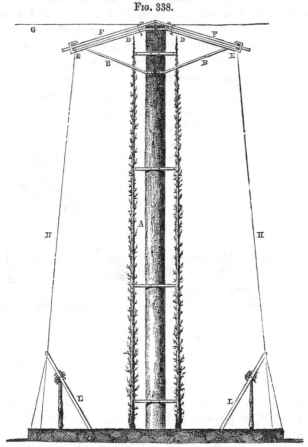

Double Espalier with a row of Cordons on each side, showing Mode of Protecting
the whole in Spring. A, Wooden support; B, B, Supports for protection;
F, F, Wide temporary copings of neat straw mats, held by iron brackets;
H, H, Galvanized wires fixed at E, E, descending at intervals and fixed
in the iron posts, L, L, and to stones in the ground; G, One of the lines running across the Espaliers from the walls of the garden.

in establishing a system of trellising for growing the choicer
pears, it is considered wise to stretch an occasional wire from
the trellis to the adjoining walls, or from one trellis to

another.   Thus if a whole square is devoted to galvanized
trellises for Pears—at say nine feet high, and at from fifteen
to twenty feet apart, the intermediate space being cropped,

Fig. 339.

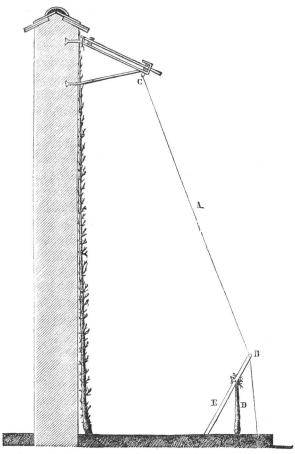

Wall protected with wide temporary coping and canvas curtains.   E, Iron sup-
    port; D, Horizontal Cordon Apple on Paradise Stock; B and C, eyes in the
    iron to permit of galvanized wires being passed through. These wires support
    the canvas which is stretched from C to B.

the trellises, in addition to being individually well supported,
afford each other a mutual support by means of strong
wires running across all the lines of trellising, say at thirty

feet apart. At the bottom run rows of horizontal cordon Apples of the most important kinds. The posts are placed closer together in erecting the trellises than when the trees are abandoned to the vicissitudes of the weather.

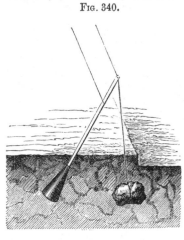

Fig. 340.

Mode of Fixing Iron Support, &c., shown in preceding Figures.

I describe this more for what it suggests than anything else. Some similar arrangement is badly wanted with us, and should not be difficult to contrive. By having a few lines of choice Apples trained on the low cordon system at each side, and two good rows of Pear trees, a great deal of valuable fruit could be protected at the same time by making some arrangement whereby the whole could be covered with cheap canvas.

SHADING FOR CONSERVATORIES.—A mode of shading conservatories and glasshouses by means of laths and slender rods of wood is common in France, and several inquiries about it induced me to obtain specimens of the various kinds used, and have them figured. The illustrations represent small portions of this shading of exactly the full size. The large one of laths united by

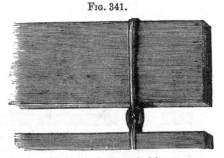

Fig. 341.

Portion of Lath Shade for roof of Conservatory: full size.

wire is frequently used for the outer side of the roof of conservatories, in which position it is supposed to save a great deal of trouble as compared with the common modes

of shading by canvas and like materials. It is sold in
lengths about a yard
wide, but may be
readily adapted to the
roof of a conservatory
of any shape, and fitted
into the smallest
nooks on curvilinear
roofs. The two smaller
sizes woven together
by twine would seem

FIG. 342.

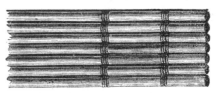

Shade of very slender rods of wood and twine:
full size.

better adapted for the inner sides of windows in corri-
dors and conservato-
ries. The three kinds
are made by M. Mus-
serano, Rue du Fau-
bourg St. Denis, Paris,
and the large one, M.
Lebeauf, 6, Rue Vesale,
Jardin des Plantes.

ATTACHING WIRE
TO GARDEN - WALLS,
TRELLISING, &c.—If

FIG. 343.

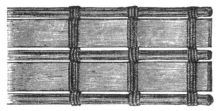

Shade of small Laths and slender Rods united
by twine : full size.

there be any one practice of French horticulturists more
worthy of special recommendation to the English fruit-
grower than another, it is their improved way of placing
wires on walls, or in any position in which it may be
desired to neatly train fruit trees. So many have been the
failures in British gardens as regards the placing of the
wire to which to affix the trees, that the system has been
given up as useless and too expensive, and many have said
that the old-fashioned shred and nail are yet the best.
But there is a very much better and sounder way, and I
am completely converted as to the value of the French
mode of wiring here illustrated. In the first instance,
several strong iron spikes are driven into the brickwork
at the ends—in the right angle formed by two walls—
nails with eyes in them being driven in in straight lines,
exactly in the line of direction in which the wire is

wanted to pass. The wires are placed at about ten inches apart on the walls, and the little hooks for their support, also galvanized, are fixed at about ten feet apart along each wire. The exact distance between the wires must, however, be determined by the kind of tree and the form to be given to it. If horizontal training of the branches be adopted, the wires had better be placed to form the lines which we wish the branches to follow; if the branches are vertical, as in Fig. 243, we need not be so exact. The wire—

FIG. 344.

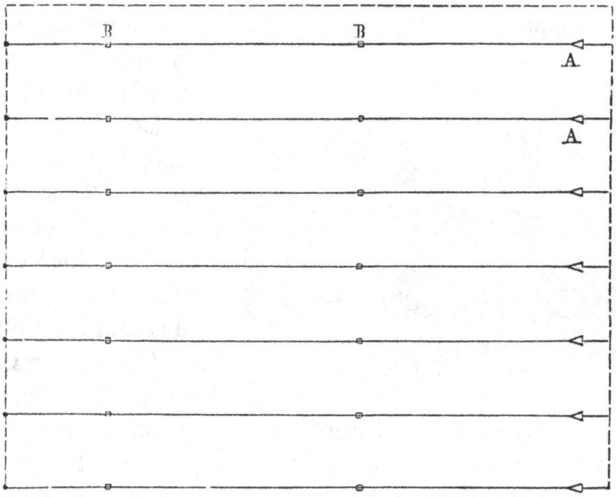

Mode of arranging wires on walls for training fruit trees with vertical or horizontal branches. A, Position of raidisseur; B, Nails with eyes, through which the wire is passed.

about as thick as strong twine—is passed through the little hooks, fastened at both ends of the wall into the strong iron nails, and then made as straight as a needle and as tight as a drum, by being strained with the raidisseur. The wires remain at about the distance of half an inch or three quarters from the wall.

If we consider the expense of the shreds and nails, the cutting of the former, the destroying of the surface of the walls by the nails, and the leaving of numerous holes for vermin to take refuge in; the great annual labour of nailing, and the miserable work it is for men in our cold winters and springs,—it will be freely admitted that a change is wanted badly. The system of wiring a wall above described is simple, cheap, almost everlasting, and excellent in every particular; and it must before many years elapse be nearly universally adopted in our fruit gardens. A man may do as much work in one day along a wall wired thus as he could in six with the old nail and shred. As to galvanized wire having an injurious effect on the fruit trees trained on it, it is simply nonsense; I will not therefore waste space and the intelligent reader's time by discussing it. Given a concrete wall, as described elsewhere in this book, smoothly

Fig. 345.

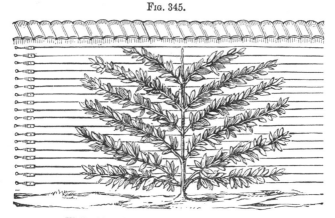

Wall with Galvanized Wires for training Trees.

plastered, and wired thus, what fruit trees could be in a more excellent position than those upon it? The temporary coping taken off after all danger from frost was past, every leaf would be under the refreshing influence of the summer rains, all the advantages of walls as regards heat would be obtained, the syringing engine would not be counteracted by countless dens offering dry beds and comfortable breeding-places to the enemies of the gardener and the fruit tree, while the appearance of the wall would be all that could be desired.

The wire and the raidisseur are also efficiently used so as to do away with any necessity for nailing in training the

Peach and other trees, when trained as cordons, as shown
in the accompanying figure. When the lines which the

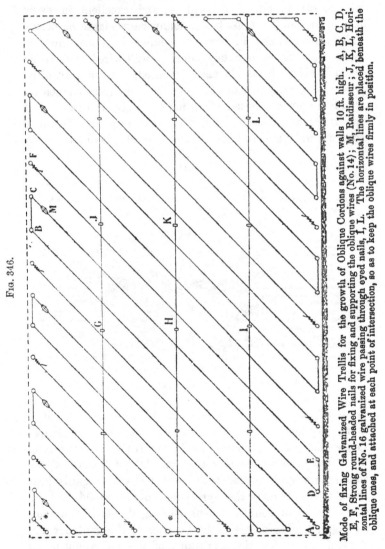

Fig. 346.

Mode of fixing Galvanized Wire Trellis for the growth of Oblique Cordons against walls 10 ft. high. A, B, C, D, E, F, Strong round-headed nails for fixing and supporting the oblique wires (No. 14); M, Raidisseur; J, K, L, Horizontal lines of No. 16 galvanized wire passing through eyed nails, I, L. The horizontal lines are placed beneath the oblique ones, and attached at each point of intersection, so as to keep the oblique wires firmly in position.

wires are to follow are fixed upon, bolts and eyes are driven
in, the wire is fixed to and passed through them, and then
made firm, as shown in the illustrations.

The French apply the term "espalier" to their wall trees, and in adopting the word from them we have transferred it to trees standing in the open, but trained in a similar manner. They term our espalier "contre-espalier," but the terms wall tree and espalier are distinctly and generally understood among us, and therefore it is better to employ them in their usual sense. The simplicity and excellence of their mode of making supports for espaliers will be better shown by the figures in the account of Versailles than by verbal description. The mode of making trellises for espalier trees now being extensively adopted in France is far superior to our own mode, and owes its excellence to the abundant use of slender galvanized wire and the little tightening implements, or raidisseurs.

Fig. 347.

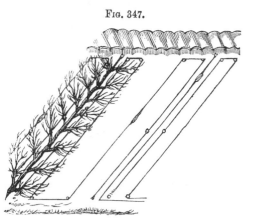

Wall Wired for Cordon Training.

The wire, which is so universally useful for the fruit garden, is sold in twenty-three different sizes. Of this an intermediate size, 12, 13, or 14, is that best suited and usually selected for strong and permanent garden work, albeit a mere thread to the costly bolt-like irons we use. The sort suited for walls is sold at about 3*l.* 6*s.* for 100 kilogrammes, equal to a little more than 220 lbs. English. Each kilogramme (a trifle more than 2 lbs. 3 oz.) of this affords more than 131 English feet of wire. The price given is that for the second quality of wire; the first quality of the same pattern costs about 6*s.* 6*d.* more for the 220 lbs. Thus, of this wire of the very best quality, and such as, if placed properly in its position, is as permanent as it is useful,

200 lbs. avoirdupois may be obtained for less than 4*l*., and 220 lbs. will extend a distance of 13,123 English feet. This size will also suit well for espaliers, No. 12 being strong enough for the coarsest trees on walls or trellises, and 13 and 14 quite large enough for general work. A size, or even several sizes smaller, will suffice for dwarf trellises with three rows of wire or so, to accommodate very dwarf trees of any shape the cultivator may desire. Since writing the above, I have been informed by Messrs. J. B. Brown and Co., of 90, Cannon-street, that they supply the chief Parisian houses who furnish French fruit-growers with this useful galvanized wire. The sizes appropriate for cordons, trellises, &c., are sold at from 30*s*.

FIG. 348.

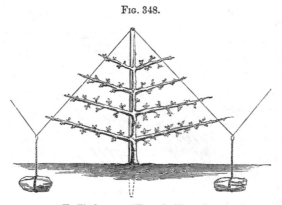

Trellis for young Trees in Nurseries.

to 44*s*. per mile, so that the material is cheap enough. It is sold by weight, No. 12 size being 31*s*. per cwt. of 1241 yards, and a smaller size 34*s*. per cwt. of 2031 yards. Those who merely want a little wire for experimenting with cordons can buy it by length, the smaller size at 2*s*. 3*d*., the larger at 3*s*. 6*d*. per 100 yards.

In some of the best fruit nurseries I noticed a simple and effective kind of trellising used for training young wall and espalier trees. It is useful in enabling the French to keep in stock trees for these purposes to a greater age than is the case in our own nurseries, and for various purposes should prove useful to the grower of young fruit trees. A larger and modified application of the same plan would do well for large espalier trees; indeed, I have seen it applied with good effect, and it perfectly suits a method which is not uncommon in France, of keeping the upper branches of trees, trained horizontally, shorter than the lower ones (Fig. 364), so as to secure perfect vigour in the lower branches. This trellis may be established at a trifling cost by using light posts of rough wood, or, if permanent, and greater strength be desired, of T-iron. In either case the posts must be firmly fixed. The wire should be passed through a hole or strong eye in the top of the pole, and fixed with stones or irons in the ground. In order to train the shoots straight, their rods may be extended from the post to the wires with but little trouble. Other illustrations of the neatest and best trellises in use in French gardens occur in several parts of this volume. Those in the Imperial gardens at Versailles cannot be surpassed for appearance and durability.

EDGINGS FOR PARKS, PUBLIC GARDENS, SQUARES, DRIVES, &c.—The edgings in gardens have a very important bearing on their general aspect, and often on their cleanliness. Hosts of people with gardens are continually looking out for a good edging, and many are taken in by the aspect of those made of tiles, material, &c. Any variety of brick, imitation stone, or terra-cotta edging, is the ugliest and most unsatisfactory thing that can be admitted into an ornamental garden. Massive edgings of stone around panels, in geometrical gardens, are of course not included in those alluded to. Pottery edgings are enough to spoil the prettiest garden ever made, and are as much at home round a country seat as a red Indian at a mild evening party.

Looking at them as they are carefully arranged by exhibitors in one or two of our public gardens, you may possibly think they are clean, symmetrical, and everything to be desired. But when brought home and arranged round the borders their true charms begin to display themselves. Being all of an exact pattern, they must be arranged so as to look quite straight in the line. If they wabble about, one this way and one that, the line is not agreeable,

QQ

even granting that the things themselves are tolerable.   It
is difficult to "set" them easily and cheaply, so that they
will remain erect.   To have them set by a mason is a plan
resorted to by some; but it is simply a way of wasting
money.   Of course, a good workman may arrange them
neatly enough by ramming down the soil firmly on each
side; but even then, they are, after all, the worst variety of
edging known.   They are also often of a texture that
cracks into small pieces with the first frost, though there
are some much more tenacious.   The expense in the first
instance is heavy, and one way or another they become unsatisfactory, till there is no tolerating them any longer, and

Fig. 349.

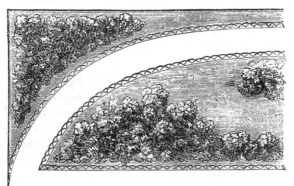

Showing the effect of Rustic Iron Edgings in the public gardens.

they are thrown by with the old iron or the oyster
shells.

The reason why people have resorted to them is, that the
edgings ordinarily used often prove disappointing and dirty,
and they long for something that will be neat and tidy at
all times.   To abuse a bad thing without offering a better,
or any at all, is often no better than to stand still and tolerate
a nuisance; but in this instance I am able to recommend a
capital permanent edging—everlasting, in fact, and with
nothing that could offend the most critical taste.   It is
simply made of rustic rods of cast iron, in imitation of the
little edgings of bent branchlets that everybody must have

seen. They are evidently cast from the model of a bent branchlet, generally about as thick as the thumb, but they are of various sizes. The marks where the twigs are supposed to have been cut off are visible, and altogether the thing looks as rustic as as could be desired, is firm as a rock when placed in position, and, in a word, perfect. These irons are of course stuck in the ground firmly, and as shown

FIG. 350.

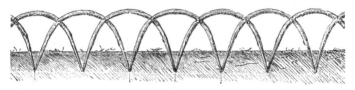

The large Iron Edgings (nineteen inches in span) used in the public parks.

in the figures. They may be set up by anybody. The fact that they are not stiff and ugly tile-like bodies prevents their offending the eye if one or two should fall a little out of the line here and there. But this is nearly impossible; for at the place where every two sticks cross each other they are tied by a scrap of common wire.

They should be so plunged in the walk, or by the side of

FIG. 351.

Cast Iron Edgings (twelve inches span) used in the public gardens.

the walk, that about six inches of the little fence appears above ground. This, however, may be varied with the size of the subjects which they are used to encompass; six or seven inches is the height given for edges for ordinary purposes. They are equally useful for the park, pleasure-ground, or even the kitchen-garden. In parks and pleasure-grounds, however, we usually have edgings of grass, and therefore it may occur to the reader that they are useless therein ; but the little fences of bent wood which furnished the idea for these iron edgings were generally used to prevent grass near drives and walks from being trodden upon ; and of course those now recommended will answer the purpose better. However, it is in much-frequented places along drives, and in public gardens and parks, that their chief merit will be found. They may be seen in every public garden in Paris, from the gardens round the Louvre, where you may notice them obscurely running along outside of the Ivy edgings, to the slopes of the Buttes Chaumont and the more frequented parts of the Bois de Boulogne ; and they must ere long be as widely adopted in England, for it is impossible to find a better or more presentable edging. In all squares or lawns where croquet is played they will, if set rather deeply, be found peculiarly useful in preventing the balls from running over the beds and breaking the plants. In some London squares I have recently noticed the beds raised bodily to a height of fifteen inches above the level, as a protection against croquet balls. All this trouble might be saved in a few minutes by placing these rustic iron edgings around the beds.

THE CLOCHE.—This is simply a large and cheap bell-glass, which is used in every French garden that I have seen. It is the cloche which enables the French market gardeners to excel all others in the production of winter and spring salads. Acres of them may be seen round Paris, and private places have them in proportion to their extent—from the small garden of the amateur with a few dozen or score, to the large one where they require several hundreds or thousands of them. They are about sixteen inches high, and the same in diameter at the base, and cost in

FIG. 352.

The Cloche as used in Winter-Lettuce culture.

France about a franc a piece, or a penny or two less if bought in quantity.

The advantages of the cloches are—they never require any repairs; they are easy of carriage when carefully packed; with ordinary care they are seldom broken ; they are easily cleaned—a swill in a water tank and a wipe with a brush every autumn clear and prepare them for their winter work. They are useful for many purposes besides salad growing ; for example, in advancing various crops in spring, raising seedlings, and striking cuttings ; and finally, they are very cheap when bought in quantity. But of course it is only in market gardens that they will be required in numbers ; in some small gardens not more than a few dozen will be wanted. Every garden should be furnished with them according to its size ; and when we get used to them and learn how very useful

FIG. 353.

The Cloche as used in the raising of seedling plants.

they are for many things, from the full developing of a Christmas Rose to the forwarding of early crops in spring, I have no doubt they will be much in demand. It is not only in winter that they are useful, but at all seasons, both in indoor and outdoor propagation and seed sowing. In France seedlings of garden crops likely to be destroyed by birds or insects are frequently raised under the cloche, and the same practice will occasionally be found advantageous in this country.

FIG. 354.

Cloche with knob.

Usually the cloche is made without a knob, as that appendage renders their package a much greater difficulty and increases the cost, so that practical men use only the one without the knob, like the specimens first figured. One with a knob may, however, be had, but it is not to be recommended.

Allusion has been made at p. 146 to the use of small bell-glasses with openings at the top. It would be a great improvement if some cloches were made in like manner, and this particularly for propagation in tan-beds and hot

pits. The opening would afford very slight though beneficial ventilation, and give a means of carrying or shifting
the cloche with one hand only. I am informed that there
will be no difficulty in making them thus without additional
trouble or expense, as soon as the firm who will undertake
their manufacture in England have full preparations made.
When not in use the careful cultivator puts his cloches in
some bye place, in little piles of half a dozen in each, a
piece of wood not more than half an inch thick and an inch
and a half square being placed between each, so as to prevent
them from settling down on each other. Workmen used
to them carry two or three in each hand in conveying

FIG. 355.

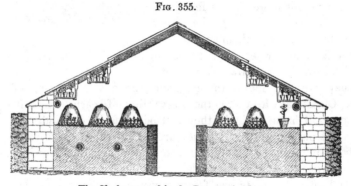

The Cloche as used in the Propagating-house.

them from place to place, by putting a finger between each.
In commencing to use them in our gardens it would be well
to see that they are placed in some spot where they will
not be in danger of breakage. The cloche must not be
confounded with the dark and very large bell glass that was
in common use many years ago in our market gardens, and
which may yet be seen here and there. These were even
dearer than the hand-glasses by which they were driven out
of use. The French cloche does not cost one-fourth so
much as a hand-light—and moreover does not, require both
painter, glazier, and plumber for keeping it in repair. It
will prove a distinct improvement in every class of garden.
How to procure these cloches has hitherto been the great

difficulty. Many have been deterred from employing them by the trouble, expense, and loss consequent on ordering them from France, and I have always despaired of their becoming useful to cultivators generally till they were produced in England at a cheap rate. Even if the carriage was not as heavy as it is, the risk of conveying such very fragile articles across the Channel is such as would prevent us from getting them in a satisfactory way.

I am pleased to announce that Messrs. E. Breffit and Co., proprietors of the Aire and Calder Glass Bottle Company's Works in Yorkshire, well known for its productions, are making preparations for their manufacture on an extensive scale. They will be able to supply them soon, and will have an abundant stock by the time it is necessary to employ them over next winter's crops of Salads and other vegetables requiring their protection. They propose to sell them at from 10*d.* to 1*s.* each, according to the quantities required, and a small addition for package and carriage will put them down in every part of the kingdom. Messrs. Breffit and Co. have offices at 83, Upper Thames-street, E.C.; stores at Free Trade Wharf, Broad-street, Ratcliff, E., 120, Duke-street, Liverpool—the seat of manufacture being at Castleford, near Normanton, Yorkshire. At any of the addresses orders will be taken and executed as soon as possible. It is fortunate that the manufacture of the cloche has been taken up by a firm with every means of carrying it on in the best manner, and with stores in the north and in Liverpool, as well as in London. They will be supplied to the nursery and seed as well as to the glass trades at wholesale prices.

In consequence of the fact that many of the articles mentioned in this chapter have not till recently been obtainable in this country, it may save trouble in inquiries to state that Messrs. J. B. Brown & Co., of 90, Cannon-street, E.C., have undertaken to keep a stock of them, including raidisseurs, chairs, shaded seats, iron implements, edgings in rustic iron, galvanized wire, and everything necessary for the making of improved fruit trellises, etc.

CHAPTER XXVII.

NOTES OF A HORTICULTURAL TOUR THROUGH PARTS OF
FRANCE IN 1868.

DURING the past summer I visited various parts of France, with the view, as usual, to observe interesting examples of fruit culture and horticulture generally. These notes, although dealing with several localities far removed from Paris, will yet be considered quite admissible here, from their relation to the fruit question in which we are all so much interested.

LYONS.—Great improvements in the way of creating promenades, parks, squares, and gardens are visible in this fine city as well as at Paris, and the noble public garden situated outside the city, near the banks of the Rhône, is one of the finest I have ever seen. A detailed description of it would require too much space, but a few notes of the chief points observed during a hurried visit may be of interest. Entering it on a bright midsummer morning at seven o'clock, the first thing that struck my eye was the feathery spray from the quantity of self-acting hose employed in keeping the turf green—no easy task in these parts. It is the same plan as that used in Paris which is elsewhere described; and it has the same agreeable result—fresh green grass at all times. The place is different to any we have in England, partaking of the characters of park, botanic garden, and zoological garden, all so combined that a high degree of beauty everywhere prevails. Here you get a glimpse of a railed field, with cattle grazing as in a well-kept pasture at home—a few minutes more, and an immense garden comes in view, presently to be succeeded by a group of conservatories, a fruit garden, a vineyard, a collection of

herbaceous and alpine plants, and so on, while frequently
magnificent masses of colour meet the view.

The purely scientific portion is not allowed to disfigure
the garden, while it is as useful as in public gardens where
it is allowed to destroy every trace of repose or naturalness.
It is arranged in circular beds, handy for reference and also
for keeping the plants distinct. The fruit garden is not
large, and aims more at showing the various forms of fruit-
trees than obtaining crops of fruit. Of course it is impossible to get a good result as regards produce, and grow in a
small space many varieties in all sorts of forms. The vase
form was very perfect here; it seems more ornamental than
useful. A novel way of growing Peaches was in operation.
It was a trellis backed with the neat straw mats so common
in French gardens. At the top a slight provision was made
on which to place a narrow straw mat, so as to protect
the trees in spring. Lathyrus ensifolius, a fine hardy
perennial kind, was observed here in flower. There is a
good deal of glass in the gardens, the larger conservatories
having Palms and fine-leaved subjects planted out and
arranged with admirable effect. Some of the smaller houses
are on a plan analogous to the very useful ones at La
Muette—a house placed at right angles to the others offers
free communication to all, and the workmen when removing
tender plants from one house to the other have not to
expose them to the open air. The houses are mostly
shaded with a very strong and thick but small-meshed netting, made out of the fibre of Lygeum Spartum. It seems
a decided improvement on the lath shadings so often used
on the Continent.

There seemed great activity and good management in the
hot-house department, and a capital feature was displayed
near it—a large trial ground. This disposed in parallel beds,
contained numbers of most things in the way of ornamental
plants for summer gardens, from Pelargoniums to Clematises. There is a memorial column in the garden,
and between the wide steps of this column and its actual
base, a rather narrow cavity is left, from which springs a
healthy ring of Laurel. It is a novelty, and in good taste.

In many of the well-planted and shady promenades of Lyons white stone seats are adopted instead of wooden ones, and very well they look—simple stone slabs, with stone supports of course.

Towards the close of the Paris Exhibition of 1867, a noble Peach tree was shown by M. Morel, trained as a Palmette Verrier, and so well trained as to excite the admiration of all who saw it. M. Morel lives at Vaise, a suburb of Lyons, and to visit him was one of the chief objects of my journey to this city. He is not a grower for the market, or a person who devotes himself exclusively to the culture of the Peach, but a general nurseryman. The wall on which his Peaches are grown is on an average thirteen feet high, and it is made of very cheap material— the common earth of the garden. First of all a foundation is made, and the wall raised a little above the surface of the ground with stone, so as to guard the chief material from injury by frost and wet. Then the earth is laid in and well battered down between boards, and on every layer of earth there is deposited about an inch and a half of mortar. The layers in one wall were about one foot deep; in another— the better wall of the two—they were about two feet deep; and between each layer the thin seam of mortar could be distinctly seen. The walls are about eighteen inches thick, and capped with a coping of tiles, under which are inserted iron supports for protection in spring. Wires are run along these, so that the mats may be conveniently supported.

It is worthy of particular notice that while the Peach does very well about here as a standard tree, good cultivators find distinct advantages not alone in growing it against a wall, but also in well protecting it when in flower. M. Morel considered that it is of decided advantage in three ways— firstly, in securing a crop by preserving the flowers from destruction by frost; secondly, by saving the trees from the malady caused by frosts and sleety rains falling on the young leaves and budding shoots; and thirdly, by the tendency which a wide temporary coping has in making the tree push more vigorously in its lower than in its upper parts. A wide mat at the top of the wall in spring obscures the light to a considerable extent from the points of the branches at the top, and this prevents the sap from running to the top as it generally does. However, a good trainer can always take care of that, and I merely mention these things to show that, even in a climate much better than that of Paris, protection to the wall is considered a necessity. The trees away from walls are often attacked by gum and the "maladies caused by the cold of spring," to use M. Morel's words. Does not this suggest the true cause of the miserable aspect of many Peach trees where careful protection in spring is not resorted to ? In numerous large British gardens, with plenty of means and time for less important objects than fruit culture, the walls are often left exposed or with the most meagre protection—a net, with a clear space at top, so that there is nothing whatever to prevent radiation or the cold sleety rains of spring from descending straight on the young leaves and flowers ; while in many continental gardens, with but a solitary man to attend to them, careful protection is regularly given.

Parallel with the Peach wall runs a trellis for training espalier trees, and this is also protected in spring, and in a very simple way. A cross bar is supported by the upright of the trellis, and lines of wire are run through it. Two double lines of wire are employed, so that the neat straw mats used for protection may be inserted between them, and be thus kept quite firm. The portions of the walls here occupied by the old and established trees were perfectly covered with the healthiest and the finest subjects I have ever seen ; even the bases of the stems and the branches had shoots trained over them, so that their surface was not awkwardly visible, as is too often the case. The forms most employed are the Palmette Verrier and the Candelabrum, which is simply made by training branches vertically from a horizontal shoot running near the ground. The pruning is done in winter, when time and weather permit, and not in spring, as is generally the case. There can be little gained by waiting till streams of sap are ascending through the branches, and a very little discernment suffices to distinguish the various kinds of buds in winter as well as in spring. The walls of earth cost about a franc fifty centimes a metre; of stone, five francs. The earth walls, when well made, last for a couple of centuries. I saw a house near at hand constructed in the same way, which had been erected one hundred years. An important point in M. Morel's culture, is that he does not stop the leading shoots, except when one happens to be weak. The main branches are left at a distance of about twenty-two inches apart. As to the state of fruit growing about here, it is still in want of much improvement. It is not easy, as M. Morel remarked, to get a new idea into the heads of the humbler classes of people; but a considerable advance has been made during the past generation. Thirty years ago, said M. Morel, a small Pear, of moderate quality, was sold for three sous—now it is worth about one. This good result has been brought about by the popularization of really good varieties.

I was much disappointed in the departmental school of fruit culture here. The aspect on entering was most doleful, and the walls in some places wretchedly covered. This, however, was explained by the fact that ver blanc had nearly destroyed the garden. This is a pest which we are spared in England, and of its destructive nature English cultivators can have no idea. Some Peaches against the walls looked very well, but many had perished from the ravages of the dreaded grub, as had numbers of Pears, Apples, and all sorts of fruit trees. I learnt nothing in this garden, but from its upper part the beautifully diversified nature of the environs of Lyons may be seen, everywhere dotted by well-built villas, and in the distance a ruined arch—one of the many traces of the great aqueduct constructed by the Romans to convey water to the city from the distant mountains. Hereabouts are also traces of the great Roman roads which radiated from hence to the Pyrenees, the Rhine, the Atlantic, and Marseilles. What a mighty grasp these old Romans had of the world! The efforts of modern conquerors seem puny compared to those of the civilizers of the olden time.

L'Ecole Régionale de la Saulsaie.—There are many

small departmental schools established with a view to spread a knowledge of rural pursuits in France—this is one of a few establishments with a more extended aim.   It is situated in the department of Ain, a couple of hours' journey by rail and carriage from Lyons, and is principally an extensive school of agriculture.   M. Verrier, after whom the form of tree known as the Palmette Verrier was named by Professor Du Breuil, was chief of the fruit-growing department; but at the time of my visit he had been dead for more than a year.   The first sight of this, and indeed most similar places in France, is not assuring ; there is a want of the finish which we Britons are in the habit of putting on country seats, farms, and gardens, and, if I may so speak, a hungry look about the place.   However, a garden is to me interesting and worthy of notice according to what it teaches or suggests, and, as great men have not been above accepting lessons from very humble ones, we must not conclude that because a place, from want of funds or other causes, is not so perfect as we had been led to expect, that it is therefore unworthy of inspection.

The first thing observable in the fruit garden was an almost total failure of the Pear crop—cold rains at the time of flowering had accomplished it.   What a " paradisiacal climate !"   The situation here is somewhat elevated, and thoroughly exposed, in consequence of being flat and not surrounded by sheltering woods or forests, and, although much further south than Paris, there can be little doubt that many parts of England are far more favourable to the production of fruit.   Apples were a better crop, but they too, strange to say, were a failure—in consequence of the very strong sun.   In any case, they had fallen off to a great extent.   The original specimens of the Palmette Verrier are to be seen here, and very fine some of them are.   A marked difference existed between the Easter Beurré Pear against walls and the same variety grown away from their shelter and protection.   Away from the wall and without protection the trees were a total failure, or in cases where they bore fruit it was diseased and useless.   Against walls, where the trees had been efficiently protected in spring, the trees and fruit were in perfect condition. This speaks for itself, and tells us, with many other things I have seen in France, that it is to well-managed walls we must look for the main improvements in the culture of our finer fruits. Here the French actually find that walls are not only a benefit but a necessity for some hardy fruits, yet we have been going on for years planting Pears in quantities away from walls, and paying little proportionate attention to the kinds that ought to be planted against them.

As the place is very much exposed to storms, peculiar expedients have been resorted to, so as to secure the trees against their influence. The practice of training trees with the branches crossed and intertwined by way of mutual support, was to be seen here in a large way, both in the case of Apples and Pears. Cheap laths and sticks are first used to train the trees into shape, and after they have attained their full size, crossing and supporting each other, the other supports are removed or allowed to rot. There were many Apple trees trained on this principle, and so well and firmly that there could be no doubt whatever that it is perfectly practicable and good, and that the objections which have been urged against it were entirely groundless. It was said by some that the branches would destroy each other by friction; there was ample evidence here that this was not the case, even with the strong winds that are nearly always blowing. When I mentioned the objection to the chief he was much amused, and simply pointed to a fine line of Apple trees, eight feet high, mutually supporting each other without the slightest injury. It is thus clear that we may not only much improve the appearance of our espaliers by adopting the system which I have figured and described elsewhere, but make the trees self-supporting. Better formed or more presentable espaliers than these could not be seen, and, as they were well interlaced one with the other, the strong wind blowing on one of the first days of July did not affect them in the least.

In the case of very large pyramids planted here, another expedient to protect them from the wind was adopted. As is not unfrequently the case in French fruit gardens, the branches of the pyramids are brought regularly in straight lines from the bole of the tree—that is, the branches form four, five, or six wings, as the case may be; five is perhaps the most usual number.   In the case I am describing there are four wings to each pyramid; but the branches, instead of being stopped, as is usually the case, are trained in straight lines from one pyramid to the other, so that they cross each other, forming a wall of trellis work, an opening being left at the bottom under which one may pass.   In one spot there were regular little squares formed thus between every four trees—in fact, a green wall of from twelve to fourteen feet high enclosed the visitor.   I never met with this elsewhere, and it was very well done.

An expedient to give additional support and strength to the espaliers was, when employing the double trellis, to let the two sides meet at top and lean against each other—thus, $\wedge$—instead of placing them vertically, as is the custom. A line of trees trained in the vase form were united one with the other by a strong arched branch, the branch springing from the top of the vase; and this simply because the place is open to fierce winds, which would render such exposed trees insecure without some support.   I question if any garden could afford a better test of the effect of wind on trained trees.   Some that were standing singly looked like very neat summer-houses.   They were pyramids, with the branches brought out from the main stem in six lines, the branches in each line being of course placed exactly one above the other.   Trees are trained thus so that the air and light may fully benefit all parts of them.   The character of a pretty bower was imparted to the space between every two wings of the tree by simply carrying an arched branch from wing to wing overhead.

There is here a very well-furnished Peach wall, made of common earth firmly pressed between boards in the making, and with a foundation of rough stones to prevent the humidity sapping the base.   This kind of wall is good enough for its object, will last for ages if well made, and may be coated and coped so as to look as ornamental as any other.

It is even cheaper than the system of concrete wall-building elsewhere alluded to. About Lyons I saw miles of wall built in this way, and numerous houses as well, and I am certain that it can be employed with advantage in growing fruit for the market, and for private use. The earth must be well battered down between boards, and it should not be either too sandy or clayey. The coping here is of tiles, not sloping down on both sides of the wall, but running clean from front to back, the higher side being reserved for the most important crop. Beneath this coping wooden supports, for accommodating a neat straw mat in spring, project about twenty inches from the wall. This may seem to the English cultivator an awkward and untidy mode of protection, but these mats are very neatly and cheaply made in France (as has been already mentioned), and they are of great use in many ways, from placing on the north side of a line of espalier trees to covering frames, and making a temporary coping for walls. For frames alone their introduction would be a benefit to us, as they afford a much better, neater, and cheaper protection than bass mats ; and as these have latterly become so dear, they should prove the more acceptable. Espaliers are here occasionally protected with the neat straw mats by simply projecting from the main support two little stays of iron or wood, which carry a rude and cheap span of framework, on which the mats are so placed in spring that the wind cannot blow them off. In looking at a fine specimen of Beurré d'Amanlis here, twenty-eight feet long and eight feet high, with three crowns wrought above the general level of the espalier, and loaded with fruit, M. Morel, the Peach grower, laughingly observed that that kind of tree was not at all a good one for the nurseryman ; the upright and oblique cordons planted against walls, and closer than people plant Cabbages, were far better.

A young plantation of Asparagus here looked somewhat like one of Celery. Trenches and plants were so distanced that each stool was a yard apart from its fellows in every way, and each plant was as carefully staked as if it were a

Dahlia in the garden of a careful amateur of that flower. There is a rather well-kept sort of botanical garden here, for the purpose of showing native, useful, culinary and other herbs, and when passing through this the superinten- dent said, pointing at a solitary plant of Rhubarb, " You eat that in England !" What a difference a few miles or mere accident sometimes makes ! Here is a vegetable second to no other, and which a race so distinguished in the kitchen should best know how to appreciate, and yet it is almost unknown to them ; and what a loss that is, we only can understand. There is a School of Dendrology here, with the trees planted in their natural orders, and, generally speaking, good facilities for teaching young men with a taste for rural pursuits.

DIJON.—The home nursery of Leconte was the only one I visited, and a very neat and well-kept one it is. It is an oblong piece of ground, about four acres in extent, and well walled in on every side, the walls being well coped with overlapping tiles. All the space on both sides of the walls was planted with oblique cordon Pear trees, trained on single galvanized wires, attached to two strong nails in the walls. They were young trees, but the walls were very nearly covered; the crop was nothing to speak of. The trees, however, are too young to judge much by at present. A wall about fifteen feet high was nearly covered with oblique cordon Pears, and as they had so much room to rise, the position seemed particularly suited to them. Near at hand they were grown to the same height by projecting a trellis above the garden wall, so as to form a very high screen of cordon Pears above it. This was done by erecting strong uprights of iron to the required height above the wall, and then running galvanized wires from the bottom of the wall to a strong horizontal wire or rod passing from upright to upright at the top. Looking along the long side or middle walks, cordon Apples could be seen stretching without in- terruption from one end of the garden to the other, the effect being very pretty indeed. They were planted a few inches inside the box edging, and between it and lines of handsome pyramidal Pears, conifers, &c., and, as usual, chiefly to fill up neatly and permanently a space that otherwise could not be usefully occupied. They are about three years old, supported on the usual slender galvanized wire, and in many parts bore a very fair crop, though in others they were nearly ruined by thrip. The little Lady Apple was particularly fine ; but generally the apples were like the pyramidal pears and apples, a failure as regards crop. However, of the two the cordons bore the best crop. The Pear, as a horizontal cordon, was not so good as the Apple; in fact, a failure, as is usually the case.

ANGERS.—This famous old town is known almost everywhere for its vast nursery gardens; there are, it is said, one hundred nurserymen, small and great, in its neighbourhood. It is a fine climate, this of Anjou—so genial as to develope the Tea plant in perfect health out of doors, and with sun enough to spice the air with the fragrance of that splendid evergreen, Magnolia grandiflora, which may be seen used as a promenade tree in the Place immediately outside the main entrance to the nurseries of M. A. Leroy. There are many noble specimens and lines of this plant in the nurseries, which are of vast extent—too much so, indeed, to permit of one visiting them, unless with plenty of time to spare. The Camellia does perfectly well in the open air, and is grown to an enormous extent, nearly two acres of ground being devoted to the production of young plants, 25,000 being grafted every year. At the time of my visit (July 18), nearly all the beds were shaded from strong sun by a thin spray of branches fastened between hurdles. Many other things are propagated in great quantity—Pears for example. Of one single variety, Easter Beurré, the enormous number of 40,000 plants are annually " worked," to use a propagating phrase. Of Duchesse d'Angoulême 25,000 are yearly required; of Williams's Bon Chrétien, 25,000; that excellent Pear, Louise Bonne d'Avranches, is also required to the extent of 25,000 annually ; and Doyenné d'Alençon to 20,000 plants—so that the number of one kind of Pear used is alone sufficient to form a nursery of itself. Observe the enormous number of Easter Beurré (Doyenné d'Hiver) required. This is the Pear which we import in such vast quantities from France in winter. In the region around Paris this kind must be grown against sunny walls. I need not add that it is folly to attempt its culture in any other fashion in England.

For three weeks before the date of my visit fifty workmen had been employed in budding here. The fruit trees are budded as we bud Roses, and those in which the buds fail are grafted in spring. In this way a year is gained. There is a splendid collection of pyramid trees grafted on the Quince stock, many of them of great size and perfect symmetry, the ground being rich and deep, and suiting the Quince to perfection. Every kind of fruit sold or recognised as a variety of any merit is grown here; Pears to the amazing number of one thousand and twenty-eight varieties; Vines, five hundred and fifty distinct varieties; Apples, eight hundred ditto; Peaches, two hundred and fifty, including forty-five of the best American kinds; and so on. The Apple is planted to a considerable extent as a horizontal cordon, and many varieties bear abundance of fruit, some of the finer Russian kinds being gathered before the date of my visit, 18th of July, 1868. The following varieties were bearing abundantly as cordons: Joanneting, Astrakan, Winter Pearmain, Archduchess Sophia, Court Pendu Plat, President Dufoy, several kinds of Reinette, several kinds of Calville, Transparent, and many others. It is scarcely necessary to enumerate kinds, as nearly every first-rate variety does well when trained in this way and grafted on the true French Paradise stock.

Of the 450 acres of nursery ground in M. A. Leroy's establishment much is devoted to the culture of Conifers and ornamental trees and shrubs, the peculiar feature of the culture being that the Conifers when young are grown in pots, for the sake of securing safe transport. The pots are all plunged in the earth, and when the plants become large they are placed in rough baskets and plunged in the same manner. This is a better preparation than that of the pot, secures safe carriage, and does not cause contortion of the roots as pots do. These baskets are made in quantity by the firm ; were it not so, the expense would prove too great to admit of this mode of culture.

There is a specimen of Cedrus atlantica, about thirty feet high, a graceful open tree, quite silvery, and dotted all over with little nipple-like male cones—one solitary and large female cone standing alone amongst them. It is found to grow very much faster than the Cedar of Lebanon in these grounds, and is so beautiful as to recommend itself to every lover of Conifers. Some plants grown indoors in England do very well here out of doors—Poinciana Gilliesi and Lagerstrœmia indica being particularly beautiful in July. Bambusa mitis grows very freely, and has here proved the best of the hardy Bamboos. The culture of fruit trees against walls is far from satisfactory. It would appear that the fine climate affords an excuse for careless-ness in this respect; but I made slight efforts to see the fruit culture of the district, as the climate is so unlike our own that observations made in it are not at all so applicable to our own culture as those gathered in the colder parts of France.

The public garden here is a purely scientific institution, directed by M. Boreau, a botanist well known for his knowledge of European plants. He is particularly fond of the wild French Roses, and described one of his own finding (R. conspicua) as the most beautiful and showy of all the wild Roses. Another interesting native plant was an unusually large and vigorous Solomon's Seal, named Polygonum intermedium—a subject that would repay culti-vation. St. Dabeoc's Heath, found in Connemara, in Ire-land, occurs in a wood about fifteen miles from this city.

The nurseries of M. Louis Leroy, in the Route de Paris, are also very extensive, though most of his ground is at a considerable distance from the town. The standard Mag-nolias in this nursery are excellent, and a singular graceful Conifer (Taxodium sinensis) claims attention from its novelty. A specimen of Wellingtonia here, only eleven years of age, is superb—twenty-five feet high, and a noble tree in every respect. There is a small public fruit garden in the town, quite recently planted, and promising to be useful in the future. Here, again, the horizontal cordons were in good bearing, though scarcely more than a year old.

I.was much indebted to M. Anatole Leroy, of the nurseries in the Route de Paris, for his valuable assistance in enabling me conveniently to see the horticulture of this neighbourhood, on the interest of which I have barely touched.

NANTES.—The Jardin des Plantes here is quite a change from what we are accustomed to see in French towns, and is well worthy of a visit and of imitation. It is a beautiful garden, in the highest sense, while it is instructive at the same time, and quite a credit to the town for the way in which it is kept. It is distinguished from the old style of French public garden by the almost total absence of straight lines, being varied in all its parts, and well and tastefully planted on that style for which the best name is the "natural." It is embellished by one of the finest groves of standard Magnolias (grandiflora) in Europe, if not the finest, and their noble flowers perfume the whole place. The planting is very tastefully done on the grouping system, while along some of the walks alternate beds of Camellias and Azaleas are placed, each bed being edged with Hepaticas. The grass was as green and as freely dotted with daisies as could be desired—the London parks at the same date being as brown and parched as the desert. This results from the excellent system of watering everywhere adopted in French gardens. Fine-leaved plants, Cannas, Ricinus, and the like, are abundantly used, and effect a noble improvement, as they everywhere do.

A particularly noticeable feature in the garden at Nantes is the way the rockwork is managed. It is not suitable for true rock plants, nor capable of being embellished by them, but its artistic effect is good, because natural, and there is a good deal of it. The principle adopted is that of letting the rock suggest itself, rather than piling it up in wall-like masses; solitary rocks peep out of the grass here and there by the margin of the water, and presently a group appears, the whole being intertwined by creeping and trailing shrubs in a way agreeable to the eye of taste. The water is not

so well managed, being rather formal and serpentine, not
following the true line of the brook or rivulet; but this
fault is not so perceptible here as in other French gardens.
The quality that we know of as breadth—that which we miss
so much at Kew and many other fine gardens, and see so
well exemplified in others of much smaller dimensions—is
finely shown in this garden, the eye resting on wide green
sweeps of grass, margined by varied and receding outlines
afforded by trees, shrubs, and flowers.   Cercis australis
forms a very ornamental tree in these grounds, the long
shoots drooping gracefully.   There is a small fruit garden and a small but useful botanic garden, both wisely
and effectively cut off from the general scene, and not
thrust under the eye of the public, to weary it with the
sight of a scientific arrangement of plants, which is as unnatural and ugly to the human eye as anything can well be.
Where you plant like subjects together, it is almost impossible
to have any of the freshness or variety of a true garden.

ROUEN.—This district is so near home, and its climate
so very much like our own, that even those possessed of the
erroneous idea that the climate of northern France is a paradisiacal one will admit the utility of studying the culture.
I first visited the nursery of Mr. J. Wood, an English
nurseryman, established here forty years.   Speaking of
fruit-growing in France and England, these were his
words: " For every single fruit tree sold in England there
are one thousand sold in France!   Every cottager with
ten square yards of ground buys and plants fruit trees.   If
it were not so you would not get so much French fruit in
England."   Generally, he said, the culture of wall fruit was
carelessly performed in that region, with the exception of
the Pear.   Fine old specimens of Pears against the walls
of chateaux afforded quantities of good fruit.   Some of the
walls here were covered with Pear cordons trained diagonally.   In reply to a query as to the merits of this
particular phase of the cordon system, Mr. Wood remarked,
" It has been a good thing for the nurserymen."   Precisely.
As a tree must be planted at every eighteen inches or so, it
is a very expensive proceeding; but a " good thing for nur-

serymen," of course. Alpine Strawberries are grown here
in considerable quantity, and preferred to the common kind.
By covering the ground with a little short manure to prevent evaporation, and giving abundance of water in dry
weather, they get them to fruit from early summer to late
autumn, gathering plenty of fruit all through this prolonged
season. When gathered fresh from the plants, and used
with the usual accompaniments, the best varieties of the
Quatre Saisons are certainly excellent. They are insufficiently known in England, where they ought to be very
extensively grown, as the climate is even more suitable to
them than that of France, and they would form a very
agreeable addition to the dessert at all times, especially
when the other strawberries are past.

In the market here specimens of the Reinette Grise, an
excellent apple, were selling in June at three and four sous
each; they were perfectly firm, and the flavour of the best.
Considering how valuable is an apple that keeps in the condition described to the month of June, we need scarcely say
that this deserves to be widely known in England. It
is well suited for culture on the cordon system. A
peculiarity in the mode of growing young Camellias and
Azaleas deserves notice. The spring after being grafted
they are planted out in pits or frames in light peaty earth,
and the result is that they form presentable specimens in
half the time that is usually required when they are grown
in pots. But they are all destined for pots, and this course
is pursued simply to get them ready in quick time. They
grow so fast that after six months they must be taken up
and replanted at greater distances apart. This phase of the
culture chiefly concerns the propagator and nurseryman;
but when large plants become shabby from having been in
dwelling rooms or any other cause, they are cut down and
planted out in the pits in the same way, the result being
that they soon return to a perfectly healthy and vigorous
state, and may, after a year or so, be again placed in pots
or tubs. This hint should be useful to amateurs whose
specimens are so frequently in that state when it becomes
doubtful whether they should be thrown away or kept.

As at Paris, great numbers of Dracænas and fine foliage plants generally are grown here. These also are at first planted out in frames and pits, in very light rich soil, and thus grown into healthy little specimens before being potted and used for furnishing. Weak and puny little plants put out in spring become nice stocky specimens in autumn, and are then taken up, potted, and placed on a gentle hotbed, where they soon root vigorously into the pots. It needs very little discernment to see the reason why things do so well planted out thus. They are not liable to the vicissitudes which things suffer in pots; they grow from a moist surface, which, as every plant-grower knows, is so congenial to the health of plants; and, to put it simply, they are under much more natural conditions than plants confined in pots.

Epiphyllum truncatum is grown here in quantity and variety, there being a dozen kinds. As many know, this is a first-rate plant for winter decoration. On its own roots it pushes but little; grafted on the Periskea it makes the free-flowering and vigorous plants which we sometimes see in England, and which are becoming every day more popular. Here, however, they employ, in preference to the spiny and slender woody stem of the Periskea, a species of Cereus, with a thick roundish stem, which forms a worthy pillar for the rich head of shoots this beautiful winter-flowering Cactus makes upon it. This stock should be used generally in England for this purpose. The Epiphyllums were tried upon Opuntias, but without success, the only kind that would grow upon them being E. Ruckerianum.

Of the public gardening at Rouen, that which pleased me best was a very small bit which formed a setting, so to speak, to the statue to Pierre Corneille. It is placed half way over the bridge, where that structure rests on the island in the Seine. A space on a level with the bridge, like a huge recess, is adorned with the statue. Around the stony base on which it rests runs a border of glistening Irish Ivy, elevated on a little plateau above the small lawn. The grass is bordered by a line of dwarf bushy Roses, springing from a band of the same "ould Irish" Ivy; behind all

there is a belt of shrubs, and around all a gravel walk, the
little space being kept private.   There were four large vases
in it, each containing a fine specimen of the New Zealand
Flax, but it is possible they had been placed here specially
for the Emperor's visit.   In any case, this mere spot, half-
way over a bridge, looked
one of the sweetest bits of
town gardening I have seen.

Fig. 356.

In the Botanic Gardens
at Rouen the first things
encountered were some of
those enormous and un-
meaning masses of Cannas,
Tagetes, &c., which the
French sometimes make
even larger and more hideous
than we do. There is, however, a neat fruit garden and
a good specimen of what
they call a school of botany
—*i.e.*, an arrangement of
hardy plants scientifically
named and arranged.   It is
only fair to say that the
plants are kept more dis-
tinct and well named in
those divisions of French
public gardens than is the
case in Britain, where they
are too often allowed to
grow wild through each
other.   Sabal Adansoni is
placed in the open air here

Pear Tree with the branches trained in
lines exactly above each other, and all
the points united by grafting.

for the summer, and seems a palm of such rigid and
leathery texture that it should do for general use in
that way.   The fruit garden contains a good many of what
are called model trees, and many cordons; but on the whole,
while there is much that is curious, and this division is well
kept, it really affords little instruction.   Specimens in the U

form abound.   The only noticeable feature was a trifling one —placing willow wands in the exact bend and direction in which it is desired to conduct the chief branches.   That once done, little remains but to tie the young shoots in the desired direction.   A great deal of the fruit had fallen, in consequence of the extreme heat, and of the soil being sandy.   There was one pretty good specimen of a winged pyramid—*i.e.*, a pyramid having the branches trained in five vertical lines, and with the points united by grafting, as in Fig. 356.

Wide edgings of Ivy are used as a margin to the Rose beds, and with a very good effect.   The rustic iron edgings, so much used about Paris, are also employed here.   Native Orchids are grown in the botanical division, among them the Lizard and others that are rare with us.   The common and the Irish forms of the Ivy are placed

FIG. 357.

Plan of preceding figure.

under exactly similar circumstances, a portion of each lying flat upon the ground, and another being conducted up a stake.   The decided superiority of the Irish kind can be seen at a glance.   It is not without reason they have selected it for the public gardens of Paris.   The better kinds of herbaceous Pæonies were planted in the grass at about a yard or so from the margin of the shrubberies—a good position for them, and when they decay no blank space is left.   The hardy Irises are grown in vases—a plan worth pursuing where early summer gardening is practised, but they should be in all cases associated with and springing from dwarfer plants.   The contents of the houses here were perfectly miserable.   The ground was almost covered in some parts with the dead bodies of cockchafers, and along one part of the railroad near Rouen I noticed a wood nearly a mile long quite stripped of leaves by this pest.   Not long before, I had thought, in passing through a rich, green, and well-wooded valley, what a transformation it would be if we could see the trees suddenly stripped of their summer robes and made to stand bare as in winter; and here it was with a vengeance.   It is no exaggeration to state that many great towering Lombardy Poplars, Oaks, and Birches, were stripped as bare as if it had been a December instead of a June morning.

The new garden in the centre of the town, named after Solferino, is very pretty, the trees being apparently well established, though it has been made only five or six years. This is in consequence of the excellent machinery for removing large trees which is now in use in most large French towns. The garden is embellished by a small piece of water and a really well-constructed seam of rockwork. A few boulders peep from the turf on one side of the water, and on the off-side the high rocky bank seems to have been worn away by time and water, every trace of art being concealed by trailing Ivy and Evergreens. The beds of Roses in this garden were covered with green moss gathered from neighbouring woods. It adds a good deal to the appearance of the beds, and by keeping them moist of course prolongs and improves the bloom.

TROYES.—This old and interesting town, from which Troy weight takes its name, is interesting to the horticulturist and fruit-grower from the Brothers Baltet having extensive nurseries in and near it. The home nursery, which is situated in the town, is extensive and rich in advanced specimens of pyramidal and other Pear trees. Soon after entering, horizontal cordons are seen in large numbers running along the back of the borders which margin the sides of the central walk, these borders being occupied by flowering plants. The cordons form a neat finish at the back, and bear a plentiful crop of fruit, though they are not established trees such as one sees in a private garden, but on the other hand subject to the transplantings, sales, &c., to which nursery stock is liable. In addition to those in this position, horizontal cordons were seen in many parts, and, where established, bearing wonderfully well. Thus the Lady Apple, well established on the Doucin stock, bore fruit almost as thick as the pretty little Apples could sit on a cordon not closely pinched in ; but on the Doucin the shoots grew too vigorously, and did not preserve that compact appearance and habit which is so desirable in these trees. If the soil were very poor and light and dry, the tendency to over vigour would be repressed, and the Doucin prove the most desirable stock.

What a happy thing it is that stocks which possess such admirable qualities are known and easily procured ! Your soil is rich deep loam, wet, cold, or stiff. Use the French Paradise, and you obtain large and beautiful fruit. But plant the same on a very poor, dry, hungry, or calcareous soil, and it is almost useless. But then we have the Doucin, which suits the poor soils to perfection, to fall back upon, and thus the best results may be produced on soils of very diverse and even very bad qualities. I measured some of the larger Apples here, and found that many were as much as ten, eleven, and eleven and a half inches in circumference on the morning of the 13th of July, though they were still green and swelling, and not to be gathered till October. Spring frosts occur here frequently, and my guide mentioned the absence of frost during the month of May of the past year as a very extraordinary occurrence. Here, as in every garden, the cultivator remarks, the cordons " are good and take up little space." Of course, in a large public nursery like this, little lines of trees under the eye of numerous daily visitors, who may at times buy such of them as they fancy, cannot be exhibited in the perfect state I have seen them in private gardens ; besides, a number of kinds are planted, and not those known to be best worth growing, and yet sufficient proof of the excellence of the system was here afforded.

The Pear was not growing as a cordon, although the Apple was so abundantly grown in that way, the Pear being considered unsuitable; and this I am strongly inclined to think is the case, from having observed the results of numerous plantations of horizontal cordon Pears. I have, however, known excellent crops to be gathered off Louise Bonne trained thus, and doubt not that a small and choice selection would be worthy of planting, especially where they could be safely protected in spring. One of the first things that meet the eye of the visitor is a nice crop of Beurré Clairgeau on a hedge formed of that variety. Several similar hedges are formed beside it, and arranged rather closely

together, so that plants may be placed between them for
the sake of shade. As clipped hedges of arbor vitæ are frequently employed in France for giving shade in summer, it
need scarcely be remarked that the substitution of hedges
of good varieties of Pears would be an improvement. Of
course the thing could be done in England as well as here.
At first stakes are used to support the trees, and indeed,
some must be employed till they have attained their perfect
development; but afterwards, if properly trained, they will
support each other perfectly, and they may be pruned and
kept to look as neatly as if supported by a costly trellis.
Alongside one of the main walks a young specimen of a

FIG. 358.

Name formed by Pear Trees.

very carefully and neatly made curtain of this kind may be
seen. These hedges bear as freely and well as any other
form of Pear tree whatever. The Beurré Clairgeau line was
worked on Quince and Pear stocks alternately. The trees
on the Quince were little better than dead; those on the
Pear were fine, full of fruit, showing in a marked manner
that the variety requires the Pear stock. There are several
curious attempts at forming the proprietors' names with
trees, and away from walls too, in this nursery, one
of which is here figured. There are, so far as I could
learn, no large gardens devoted to fruit culture in the
neighbourhood, but multitudes of small proprietors, with

specimens of good varieties in their gardens: many of these send their surplus fruit to market, and the same is the case over a large part of France.

If the Pear grafted on the Quince is planted in ground dry and not fertile, the tree pushes with little vigour and often makes hardly any progress. It is apt to carry when very young a superabundant quantity of fruit which soon exhausts it, and it will live but few years. This impoverishment may be prevented by " liberating " the tree and by causing it to grow on its own roots —a practice much recommended by M. Baltet. The process is as follows :—In spring from three to six vertical incisions are made a little above the junction of scion and stock, as herein represented, and about an inch or so long, and deep enough to slightly penetrate beyond the bark and slightly into the wood of the tree.

Fig. 359.

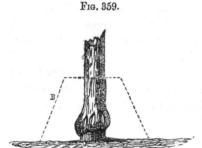

Mode of inducing a starved Pear on the Quince Stock to emit its own roots. B, Outline of mound of earth.

Afterwards a little heap of light and rich soil is raised around the stem sufficiently to cover up the incisions. It is then made pretty firm and covered with a couple of inches of old dung, so as to preserve it from getting dried up by very warm weather. Roots will quickly descend, and the tree will soon attain fresh vigour on its own roots, and become quite independent of the Quince. The old root and its influences will disappear in course of time. As the union of stock and scion is usually

Fig. 360.

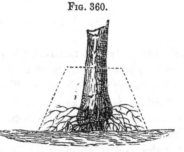

Result of the operation.

placed a little above the surface, and as the new crop of roots given off by the Pear will enter the ground from this position, it need hardly be said that the surface must be gradually raised towards the base of the tree by means of suitable turfy loam, so as to encourage the new roots. Considering the inconsiderate way that the Quince has been recommended for all soils, this mode should prove useful.

In another part of the town the Ecole Normale of the department has a garden behind it for the purpose of teaching the pupils fruit culture. Here double or super-imposed cordons of the Lady Apple bore fruit in great abundance. The walls were made of the dried stems of the common Reed, nailed between rough and cheap wooden framework, the mass of stems being about two inches thick. A flat board nailed along the top at about seven feet from the ground, afforded about eight inches of coping. The wall of the school for about four feet from the ground was very neatly covered with dwarf Peach trees which bore a fair crop, and neatly covered a space generally left naked.

BOURG-LA-REINE.—The very extensive nurseries of Jamin and Durand in this neighbourhood are full of interest to the fruit-grower. In addition to the nursery proper there are two fruit gardens—one belonging to M. Durand and the other to M. Jamin, both recently formed, and likely to prove of much interest to the fruit-grower by and by. Many French nurserymen have in addition to the ground devoted to the raising and training of young trees a private garden or " school " of fruit culture, in which the various kinds may be seen in a developed state. The garden recently established by M. F. Jamin has been well walled in, the walls of stone having a coping of overlapping tiles, which project about nine inches. This is, perhaps, as cheap and good a coping as any in use, and its effect is neat, much more so than that of other tile copings employed here. The walls are all wired closely and effectively with the galvanized wire and the raidisseur. The walls with the warmest and best aspects are planted with Peaches and winter Pears, and

herein is an instructive lesson. While continually talking of the fine climate of France, we have been going on for
years taking little interest in wall Pear culture, but planting
pyramids everywhere, and thus we have practically come
to be without a stock of the finer winter Pears, and compelled to import enormous quantities of them from France
at high prices.

Here I found a most experienced fruit-grower — one
who has also lived in good fruit-growing establishments in
England—who said emphatically that it is absolutely useless
to attempt the culture of the finer winter Pears, the most
valuable of all, away from walls, and that it is necessary to place such kinds as the Easter Beurré against
well-coped walls with a southern exposure, the soil being
of the finest description and the climate that of Paris. Of
course he could grow some of them in the open, but
then they would be uncertain and worthless; and he gives
an instance—Beurré Rance, which is first-rate against
walls. The collection of winter Pears had only been planted
a short time, and yet the crop was very good, every
young tree bearing as much as one could desire to see
upon it. It is finally intended that these Pear trees shall
assume the form known as the Palmette Verrier; but at
present the branches are trained diagonally—another instance of the excellent practice of allowing branches that
are finally to assume the horizontal position to grow first
in an ascending direction, so that they may be furnished
and formed with less trouble and in a shorter time, the sap
rising much more freely and naturally in young branches
that ascend obliquely than when they run in an exactly
horizontal direction. When the outer ones are long and
old enough to form the bend, and have their points directed
towards the top of the wall, then the current of sap is
drawn through as well as could be desired. Beurré Diel
is also planted against walls here—not that it may not be
grown in the open air, but its flowers are very liable to
be injured by frost in spring, and therefore it is placed on
a wall to secure a crop.

The mode of fixing the horizontal cordon here is the

most permanent if not the cheapest I have seen. Figure
155, p. 354, renders a description unnecessary. Several
hundred feet of wire may be placed between two of these
supports with the greatest advantage, and nothing can look
neater. The appearance of the horizontal cordon is very
much improved by this mode of arranging it, which is
to be preferred to that of using wooden posts. The
trainer remarked that any required length of wire
might be supported between two of these supports,
"even five hundred or six hundred feet." For short
distances it would not be necessary to have such strong
supports.

In this same garden the plan of adopting three rising
shoots from one base, instead of the cordon system, has
been carried out. It is applied to the Apricot and the
Peach, one kind being worked on each ascending branch,
and three kinds borne by one root. There was no indica-
tion of a disposition on the part of any of the trios thus
united not to grow agreeably together; indeed, they were
as equally balanced and healthy as could be wished.
Where it is desired, by nurserymen or private growers,
to have a goodly number of varieties in a restricted
space, this is proved here to be the best plan of all.
Western and southern walls require more protection
and wider copings than those with northern and eastern
aspects, the abundant rains being more dreaded than
the frost, and the western walls here have several inches
more coping than the eastern. It seems odd that cultivators living in such a perfect climate should take more
precautions against cold rains than we do in these watery
islands!

M. Durand's garden contains a collection of the choicest
grapes grown in France, and though quite recently formed,
already contains promising trees of many kinds. This
plan of devoting a special garden to fully formed fruit
trees is worthy of imitation by our nurserymen. I am
particularly indebted to M. Durand for valuable assistance
in seeing the gardens of the neighbourhood—those below
ground as well as on the surface—for we descended to-

FIG. 361.

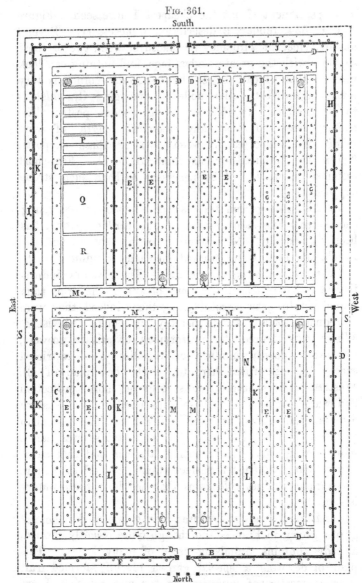

Plan of Fruit Garden for the North of France, designed by M. J. Durand,
of Bourg-la-Reine, 150 yards long and 80 broad.

REFERENCES TO PLAN.—A, Cisterns for water supply. B, South wall with
Peach trees trained in the fan, palmette, and candelabrum forms. C, Espaliers
of Pears in the palmette form, the trees at eighteen feet apart. D, Lines of
Apples trained as horizontal cordons, planted twelve feet asunder and a few inches

gether into coalpit-like caves to see the mushroom culture.

Fig. 362.

Pear Tree shown at the Paris Exposition of 1867, by M. Croux, of Sceaux.
All the points of the branches have been united by grafting.

On mentioning to M. Durand my wish to have a plan of what he would consider a good example of a fruit gar-

within the margin of the beds. E, Pear trees in the columnar form, planted at ten feet apart. F, North wall with Cherry trees, planted at about sixteen feet apart. G, Bush Apples on the Paradise stock, planted at six feet apart. H, East wall with winter Pears such as Easter Beurré, Crassane, and St. Germain. I, South wall of Peach trees with five erect branches, planted eight feet asunder. J, North wall with Cherries in the palmette form, planted about sixteen feet apart. K, West wall of summer and autumn Pears. L L, Interior walls of the garden. M, Pyramidal Pears planted at twenty feet apart. N, East wall with Apricots—horizontally trained trees planted twenty feet apart. O, West wall with Plums, planted at twenty feet apart. P, Gooseberries. Q, Currants. R, Raspberries. S, Boundary trellis, which may be covered with Vines, or Pears if in a cold climate.

den in Northern France, he was so good as to design one specially for me, and I have much pleasure in giving it here.

SCEAUX.—In the same neighbourhood are nurseries belonging to M. Croux, and a very good school of fruit culture apart from the large home nursery. It is nearly two acres and a half in extent, and established about six years. Many of the trees are trained into very curious forms. The cordons here have grown too strongly, and every second stem is severed. They had of course been previously firmly grafted one to the other. Cydonia sinensis against walls has fruit a foot long in favourable seasons, but is simply a curiosity. Several kinds of Ribes, including the gooseberry, are grafted on the red currant, and there are various other curiosities. The remarkable-looking specimen of training seen in the preceding illustration was shown by M. Croux at the Paris Exposition of 1867, and there much admired. The plant nurseries of MM. Thibaut and Keteleer in the same neighbourhood are well worth seeing.

CHATILLON, FONTENAY AUX ROSES.—Visitors to Bourg-la-Reine or Sceaux may on the same day conveniently visit the garden of M. Chardon in this village. The owner is an amateur, and has a most interesting little garden of fruit trees. In addition to the common and well-known forms, he has many specimens trained over walks and bowers, and altogether the garden is well worth a call from anybody visiting Paris who wishes to see what may be done with fruit trees by an amateur in his spare hours.

FIG. 363.

Trellis over walk covered with Pear Trees.

SUISNES (Brie-Comte-Robert).—The nursery of M. Cochet here is an interesting one for the fruit-grower, and the owner is a very popular horticulturist. Apples, on the horizontal cordon system, are planted here in large numbers,

in places where before they used to have high box edgings.
They were among the best cordons I have
seen in France, some bearing as much fruit
as they seemed able to properly develope; yet
M. Cochet considered it a very thin crop, and
said they frequently have them almost as thick
as they can stand along the line. This was

Fig. 365.

Fig. 364.

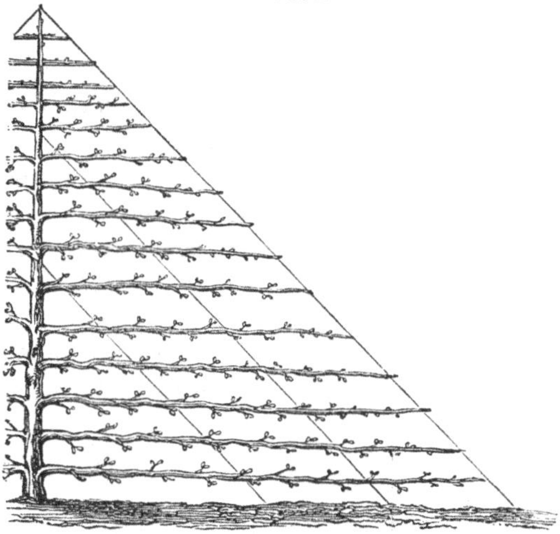

Pear Tree with horizontal branches, becoming shorter
towards the apex of the tree, and supported by slender
galvanized wires stretched from a stake at back of the
tree to pegs or stones in the ground.

Mode of support-
ing stake for trees
trained as shown
in the preceding
figure.

the case in 1867. Several walks are margined here with
two instead of one line of cordons, the inner line being
about three inches higher than the
outer one. Of course many variations may be made thus, but I have
as yet seen nothing to alter my
opinion that the single line, well
conducted and rather freely developed, is the best of all, though
there are many positions and cir-

Fig. 366.

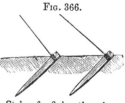

Stakes for fixing the wires
shown in Fig. 364.

cumstances in which two lines, superimposed cordons or other modifications, will prove desirable. M. Cochet has planted

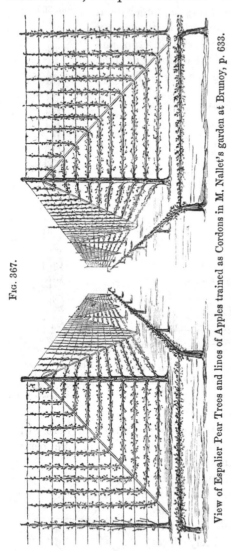

Fig. 367.

View of Espalier Pear Trees and lines of Apples trained as Cordons in M. Nallet's garden at Brunoy, p. 633.

almost every good variety of apple on this principle, and finds they all do well on it. All are grafted on the true or French Paradise stock. Some of his fruit from three year old plants was remarkably fine, and nothing could look prettier than the handsome apples along the side of the walks. Of course a much more regular and better

Fig. 368.

Plan of Espalier in pre-ceding figure at corner of line.

effect could be obtained by lines of one kind only, and the training and pruning of them also would be more likely to be performed in the best manner. In nearly all parts of the garden there was abundant evidence that the horizontal cordon for Apples is the best improvement effected in open-air fruit culture for years.

A line of Pears trained thus may also be seen, but it is a failure, although there was a fine crop hanging on one specimen of the Belle Angevine. When grown in this way the Pear usually manifests a disposition to shoot up "gourmands," or shoots very like those of Willows, from the bend. On walls where the sap has room to spread, this inconvenience is of course not present. The young Pear and other trees here in preparation for wall and espalier culture are beautifully trained in line by means of tightly strained galvanized wires. By this means trees fit to place against walls immediately, and without a leaf or shoot out of place, may be picked out at any time. A good many handsome

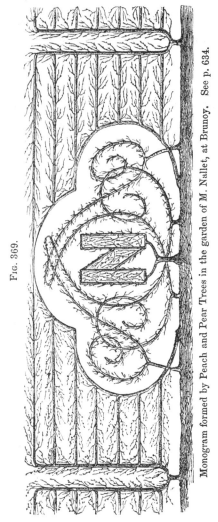

Fig. 369.

Monogram formed by Peach and Pear Trees in the garden of M. Nallet, at Brunoy. See p. 634.

Palmette and other trees are to be seen, but particularly remarkable are those trained " en fuseau," or in what is sometimes called the columnar form. This is simply a

tree trained to a single stem, or a vertical cordon, the top
being allowed to grow as high as it likes, and thus close
columns of leaves and fruits are formed as much as fifteen
feet high.   Nothing could exceed the fine condition of many
of these trees, perfectly laden from top to bottom in many
cases, and in many more bending arched to the ground with
the weight of their fruit.   They were not staked, but when
they are grown in a regular fruit garden it is the custom to
securely connect them near the top by a line of wire, so that
they cannot bend down with the weight of the fruit.   Their
advantages are that fruit and leaves enjoy abundance of sun
and air.   The fruit is said to be better flavoured than from

FIG. 370.

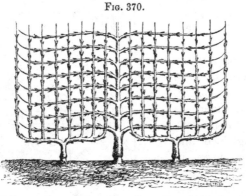

Portion of Self-supporting espalier of Pear Trees, formed of horizontal and vertically trained trees, the points of the horizontally trained tree grafted by
approach to the outer branches only of the vertical ones.

the Pyramid tree, in which there is usually a good deal of
shade, while they are perhaps the easiest of all forms to
conduct, and a great many kinds may be grown on a small
space.   Their drawback appears to be the great height to
which they attain; pruning, and the gathering of the fruit
are not so facile as is desirable.

In many French gardens a peculiarly simple and neat way of
training espalier Pear trees may be seen (see Figs. on p. 629),
and there were good examples both here and in the next place
described.   It consists of a stout stake for the main trunk of
the tree, and of wires running from this to stones or pegs

buried in the ground. That the roots of the tree may not
be hurt by a large stake, this is sometimes supported by
the stem, as shown by Fig. 365. Besides, the support for
the wires and younger branches is only required towards
the top of tree; hence another reason for not fixing the
stake in the ground. It is quite easy to project little
stakes from the stouter parts to the young growing branchlets of the tree, and thus keep the points perfectly trained
in the desired direction. On the first of November every
year, M. Forest, one of the many professors of fruit culture
in Paris, and a very popular and excellent one, gives and
illustrates here a lecture, which is attended by from three

FIG. 371.   FIG. 372.

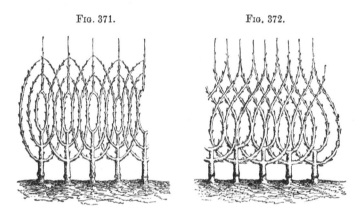

Espaliers of Pear Trees with the branches grafted by approach.

to four hundred gardeners from various parts in the neighbourhood. I may add that about here the best workmen
are paid 3f. 75c. a day; others less skilled or less able receive a franc or so less.

BRUNOY (SEINE ET MARNE).—There is here a very remarkable fruit garden belonging to an amateur, M. Nallet—
a garden which will repay a visit at any time of the year.
It is only a few minutes' walk from Brunoy station, passed
on the way to Fontainebleau, and within an easy distance
of Paris. It is an oblong piece of ground, walled in and
with a straight walk through the centre, bordered by two
lines of handsome pyramidal trees, cut off from the walk by

long horizontal cordons, lines of tall trellis-work running at right angles with the main walk, and accommodating an infinite variety of trees—many fanciful, and many of the best and most useful forms. Numbers of horizontal cordons were in fair bearing, but the proprietor complained that the crop was one of the worst he ever had. He considered that, taking bad years with good, an average of ten fruit per yard run of each line might be calculated upon. The cordons are never protected, and, here as elsewhere, furnish numerous places almost useless, and which would otherwise not be occupied at all. It is utterly impossible to give the reader an idea of the variety of form to be seen in the fruit trees, therefore we will confine ourselves to the most remarkable. The garden offers recreation to its amiable owner, and he, while not neglecting the very best forms, also amuses himself occasionally by transforming one or more trees into the monogram of his wife's name and his own. The columnar form elsewhere figured is very well developed here, some of the specimens approaching eighteen feet in height. They are regularly staked, and high lines of wire connect them by the tops, so that they are held firmly together.

Fig. 373.

A Pear Tree from handsome Espalier trained to form the name NALLET.

Numerous Palmette trees occur here, and it is noticeable that the lines which the trees are to follow are laid down at first with willow or other slender flexible rods. A Plum tree, trained as a Palmette Verrier, was very ornamental, the lines of fruit darkening the long, neatly guided branches. Nearly 1000 lbs. of galvanized wire have been used in this garden. Curtains of Pear and other trees, trained on slender trellises of this wire, are very well formed. The

Peach is grown to some extent against the walls, and successfully, some of the trees looking almost as well as those at
Montreuil, though the walls are not so high. A large
portion of the wall space is devoted to oblique cordons of the
Easter Beurré, and these were in excellent bearing ; they had
been planted six years, were about twelve feet long, and bore
from ten to fifteen fruit each. Planted at twenty inches
apart, and confined to one stem, which is never cut back
at the point if the wood be ripe, they soon cover the wall,
and, the good fruit of this variety fetching a high price, a
quick return is afforded by the
trees. There can be no doubt that
this is the best phase of the cordon
system against walls, and, as the
same plan has been carried out on
all the walls of the new fruit garden
of the municipality of Paris in the
Bois de Vincennes, there must be
some good reasons in its favour.
Several small walled gardens are
being made in connexion with the
chief one of M. Nallet, and here
again the greater portion of the
wall surface is devoted to Easter
Beurré, the plantations being one
and two years old. Six years ago,
the first trees in the garden were
planted, and I doubt much if any
fruit garden in existence better illustrates what may be done
with good management in a short time.

FIG. 374.

Pear Tree trained in the Crinoline form, ten feet high.

The practice of grafting by approach the branches of
the Pear trees is extensively employed here, as shown in
Figs. 370, 371, and 372. The figures will better explain
the mode of training and the aspect of the trees in the
garden than any description. I am much indebted to
M. Nallet for his kindness in sending me accurate sketches
of some of his most remarkable trees.

A distinct and apparently useful form of tree I met with
here for the first time. It is called the crinoline form, and

is made by taking eight branches from the base of the tree, and bringing them outside a circular hoop, allowing one main stem to ascend erect. The branches, after growing a little above the hoop, which gives a desirable uniformity to the base, ascend at regular intervals to the top, where they are neatly united to the erect shoot. The figures will explain this form, but the stake in Fig. 375 has been made much too large by the engraver. It should be of iron. This figure is in other respects a good representation of a handsome specimen in M. Nallet's garden. Each branch being kept distinct, and the tree being well opened up by this system, the effect was very good indeed, and the crops too, considering that they were a failure

Fig. 375.

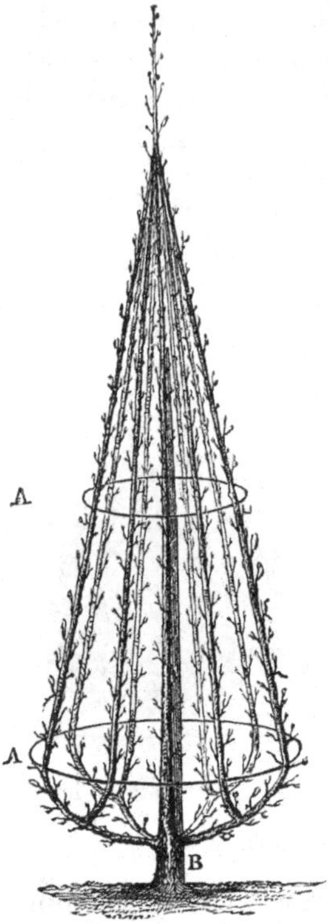

Pear Tree in Crinoline form, seventeen feet high and six feet in diameter.

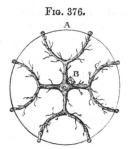

Fig. 376.

Plan of Pear Tree, shown in Fig. 375.

throughout France during the past year. A specimen of the Duchesse d'Angoulême trained thus was one of the most pleasing looking trees I have ever seen. The method has more advantages than would appear at first sight—the long fruiting branches being thoroughly exposed to the sun and light from bottom to top, the branches being held firm, and the tree being altogether a decided improvement upon the pyramid for important positions in gardens.

INDEX.

Cardoon, the, culture of, 522

Carrot, small, of the Paris market, culture of, 520

Catacombs, the, 111

Caucannier, M., his establishment for forcing asparagus, 514

Cauliflower, kinds grown round Paris, 519

Cauliflower, forcing of the, 524

Caves for storing plants under the Jardin Fleuriste, 151

Caves, mushroom, 472

Centaurea babylonica, 217

Cercis australis, 36

Chairs, garden, 563

Chamærops excelsa, 219

Champignonnistes, 471

Champs Elysées, 1

Change, necessity for a radical one in our city improvements, 116

Chasselas grape, long wall of, at Fontainebleau, 252

Chatillon, 628

Chevallier, M., his garden at Montreuil, 442

Chicory, improved variety of, 517

Chloride of calcium, use of, in fruit room, 566

Church gardens and cemeteries, 104

City graveyards, gardening in, 106

Clianthus Dampieri, 142

Climates of Paris and London compared, 548

Climbers trained up the stems of trees, 57

Clipping trees a barbarism, 260

Cloche, its use in lettuce culture, 490

,, the, 596

,, ,, where to obtain, 599

Cockchafer, ravages of, 618

Colonnade, the, at Versailles, 246

Columnar training, 391

Concert of the Champs Elysées, 5

Conservatory in Jardin d'Acclimatation, 34

Conservatories, wooden shading for, 586

Conservatories, frequently unsatisfactory, both as regards contents and design, 280

Contrast between the keeping of the grass in the parks of Paris and London, 37

Copper beech, the, unwisely recommended for town planting, 117

Cordon, horizontal, advantages of, 350

,, mode of supporting the horizontal, 336

Cordon, mode of supporting the horizontal, 354

Cordons, Mr. James Barnes, of Bicton, on, 351

Cordons, Mr. J. A. Watson on, 352

Cordon, pinching the shoots of the horizontal, 341

Cordon, management of the, 339

,, spiral, the, 366

,, system, how it may be advantageously adopted on fruit-wall borders, 352

Cordon system, objections to, answered, 348

Cordon system of fruit growing, 334

,, what is it? 334

Cordons, grafting to unite, 342, 343

,, more than five miles of, in Imperial Gardens at Versailles, 432

Corn salad, 498

Courtelière, 254

Courtois-Gerard and Pavard, MM., 516

Covent Garden, 540

Crambe cordifolia, 220

Crinoline form, a good one for the Pear, 635

Crowded streets and bad arrangements not confined to the central parts of London, 117

Cucumis perennis, 220

Cuscutas, the, in Jardin des Plantes, 75

Cyrtanthera carnea, 96

DATISCA cannabina, 220

Deciduous trees, reasons why they succeed in cities, 161

Deciduous trees, the best for London, 161

Dijon, 609

Dimorphanthus manchuricus, 142

Doucin stock, 355

,, ,, description of, 357

Dracænas for room decoration, 265, 275

,, the, 195

EASTER Beurré pear, double grafting of, 430

Easter Beurré pear grown on walls in France, 322

Echeveria metallica, 195

Edgings for parks, public gardens, &c., 593

Edgings of ivy, how to make, 310

Elm, large-leaved Weeping, a fine town tree, 172

Elm, the, as a city tree, 170

Elymus arenarius, 221

Endives, used in Paris, best kinds of, 517

English cemeteries, 107

THE END.

LONDON:
SAVILL, EDWARDS AND CO., PRINTERS, CHANDOS STREET,
COVENT GARDEN.

Printed in the United States
By Bookmasters